INTERNATIONAL UNION FOR ELECTRICITY APPLICATIONS

XIII INTERNATIONAL CONGRESS ON ELECTRICITY APPLICATIONS

INTERNATIONAL CONVENTION CENTRE
BIRMINGHAM UK

PROCEEDINGS VOLUME 1

INTERNATIONAL UNION FOR ELECTRICITY APPLICATIONS

XIII INTERNATIONAL CONGRESS ON ELECTRICITY APPLICATIONS

INTERNATIONAL CONVENTION CENTRE

BIRMINGHAM UK

16 - 20 JUNE 1996

OBJECTIVE OF UIE

To promote and develop applications of electricity in accordance with achieving energy efficiency, protection of the environment, economic viability and social acceptance.

Every four years the UIE organises an International Congress in collaboration with its members for Industrialists, Scientist and Engineers interested in the applications of electricity.

PREVIOUS CONGRESSES

Scheveningen	1936 - 1947	Liege	1976
Paris	1953	Cannes	1980
Stresa	1959	Stockholm	1984
Wiesbaden	1963	Malaga	1988
Brighton	1968	Montreal	1992
Warsaw	1972		

PROCEEDINGS PUBLISHED FOR THE BNCE & UIE BY
THE INSTITUTE OF MATERIALS

Book 642
Proceedings published for the
BNCE and UIE by
The Institute of Materials
1 Carlton House Terrace
London SW1Y 5DB

Disclaimer
The BNCE, UIE and The Institute of Materials do not
accept responsibility for the accuracy of the information
given in these Proceedings and state that the opinions
expressed are those of the authors.

ISBN 1 86125 007 X

British Library Cataloguing-in-Publication Data
A catalogue record of this book
is available from The British Library

Printed and bound at
The University Press
Cambridge, UK

President
W. Waring (GB)

General Delegate
R. Wolf (F)

Vice-Presidents
H. Albinsson (S)
C. Gaya Goya (E)
G. Michel (F)
Y. Shindo (J)

Past Presidents
R. Felix (F)
F. Lucke (D)
C. T. Melling (GB)
B. Sochor (PL)
E. Tiberghien (B)
P. Guesne (F)
M. Setterwall (S
C. Gaya Goya (E)
J. Finet (CDN)

General Secretary
G. Vanderschueren

UIE MEMBERS

Argentina	Finland	Japan	Romania
Austria	France	Netherlands	Spain
Belgium	Germany	New Zealand	Sweden
BrazilGreat Britain	Poland	Switzerland	
Canada	Italy	Portugal	United States
Czech Republic			

and including the following organisations:

CESI (Italy)
ECNZ (New Zealand)
EDENOR (Argentine)
Electricidade de Portugal
EPRI (United States)

Imatra Voima (Finland)
RWE & VDEW (Germany)
VEO (Austria)
UCAR (United States)

UIE STUDY COMMITTEES – PERMANENT

Applications   Information and communication   Education and Research

UIE WORKING GROUPS – 1992-1996

Energy saving in Industry by efficient use of Electricity   (WGIA)
Energy saving by efficient use of Electricity in Buildings   (WGIB)
Power Quality   (WG2)
Waste Treatment and Clean Processes   (WG3)
Electricity in the Food and Drinks Industry   (WG4)
Load Management in Industry   (WG5)

These groups will be presenting the results of their work as part of the Congress Programme.

UIE XIII INTERNATIONAL CONGRESS PAPERS COMMITTEE

Chairman
Prof L Hobson (GB)

Secretary
Dr R Johnson (GB)

Members
Mr W Boone (NL)
Mr M Maljean (B)
Dr A Heaton (GB)
Dr L Monier (CDN)

Mr C Moore (GB)
Mr H Ottosson (S)
Mr R Poiroux (F)
Prof D Spreng (CH)

Further information is available from:
The General Secretary, UIE, ESPACE ELEC CNIT, BP 10 - 2 Place de la Defense, 92053 Paris la Defense, FRANCE   Tel: (1) 41 26 56 48Fax: (1) 41 26 56 49

BNCE

BRITISH NATIONAL COMMITTEE FOR ELECTROHEAT (BNCE)

UIE XIII CONGRESS ORGANISING COMMITTEE

Chairman	– Mr Walter Waring
Executive	– Mr Michael Thelwell
Chairman International Papers Committee	– Prof. Les Hobson
Technical Sessions Manager	– Dr Roy Johnson
AV Manager	– Mr David Johns
Technical Visits Manager	– Mr Mike Dorrington
Registration, Tours and Social	– Conference Associate and Services International Limited (CASIL)
Exhibition	– Clayden Exhibitions / Joe Manby Ltd
Press and Publicity	– CMPR
Congress Proceedings	– Institute of Materials

The BNCE organises a range of activities and represents members through a number of committees and groups.

- Education and Training Committee
- Commercial Committee

- Dielectric Heating Group
- Induction Heating Group
- Infra-red Heating Group
- Industrial Heat Pump Group
- Plasma Arc Processes Group

Further information is available from:
BNCE
30 Millbank
London
SW1P 4RD
Tel: +44 171 344 5917
Fax: +44 171 344 5996

FOREWORD

The International Union for Electroheat (UIE) is a world wide organisation which is committed to the efficient and economically viable use of electrical energy in all branches of industry and commerce. It promotes the use of a wide range of diverse applications of electrical energy, many of which are at the forefront of technology. These advances are promoted with full recognition of the need to protect our environment and to achieve a social acceptance of the technologies used.

The UIE has developed over many decades and has representatives from almost all the industrialised countries of the world. It fosters the exchange of information between prospective users, equipment suppliers, electrical utilities and experts in both Universities and industrial research establishments by:-

- organising symposiums, seminars and conferences
- conducting technical and economic research
- publishing specialised texts
- co-operating with international bodies to establish industrially recognised standards
- facilitating the transfer of technologies between different countries.

Every four years the UIE organises an International Congress which acts as its focal point in the dissemination of information to all interested parties but especially to prospective users of the technologies. This year it is the responsibility of the British National Committee for Electroheat to organise this event. It is 28 years since a Congress of this magnitude was held in the UK and it is of paramount importance that this opportunity should not be missed. The Congress has been organised in such a way as to appeal to users within individual industry sectors whilst providing a forum for the exchange of ideas across the disciplines involved.

The proceedings of the Congress provide details of the state of the art in all the topics relevant in the field of electroheat and should be indispensable to the following:

- every type of prospective user of electrical energy as an industrial or commercial process
- manufacturers and installers of electrical equipment
- companies involved in the production and distribution of electricity
- consultants and researchers in relate fields involving the efficient, economic viable and environmentally friendly application of electrical energy.

I would like to take this opportunity to thank all those who contributed to the production of these proceedings and to the success of the Congress itself, especially the authors, speakers, session organisers and chairmen. The Congress could not have taken place without the assistance of the UIE and its National Committees, especially the BNCE, and I would also like to thank my colleagues on the International Papers Committee for all their efforts in producing what I hope will be an invaluable record of the state of the art electroheat technologies.

Prof. L. Hobson
March 1996
Chairman International Papers Committee

INTRODUCTION

These two volumes contain 150 oral and poster papers which form the basis of the presentations at the UIE XIII International Congress on Electricity Applications held at the International Convention Centre in Birmingham from 16-20 June 1996. The Congress is organised by the host country and hence its theme, size and activity relates to the location.

The XIIIth Congress is the first to be held since the UIE changed its objective from electroheat to electricity applications reflecting an increasing number of industrial processes using electricity but not requiring the production of heat.

The 14 Technical Sesions are arranged according to Industrial Sectors (Metals, Food, Chemicals, Waste Treatment) which enable delegates and readers of these Proceedings to consider the alternative techniques available for a single process. The programme also includes sessions on energy efficiency, load management and research and education which are important aspects of the UIE activities.

The papers were selected by an International Papers Committee following a Call for Papers to which over 200 proposals were received from over 26 countries worldwide. The information represents the current range of applications and the state of development in various countries. There are also presentations from UIE Working Groups which have in many cases produced detailed reports to assist UIE members in their work.

The papers are reproduced as submitted by authors in English or French with abstracts in both languages. Some authors produced papers in their own language with subsequent translation to English or French.

The Organisers and the International Papers Committee record their gratitude to all the authors and collaborators for their contributions and participation in the Congress.

Roy Johnson
Secretary to International Papers Committee

VOLUME ONE
CONTENTS

EE	Promoting Energy Efficiency and Load Management
EE	Promotion de la Performance Énergetique et de la Gestion de la Charge
RE	Research and Education
RE	Récherche et Enseignements
MI	Electric Arc Steel Making
MI	Production d'Acier par four à Arc Électrique
MII	Metal Processing and Heat Treatment
MII	Metallurgie et Traitement Thermique
MIII	Strip Heating
MIII	Chauffage de Bandes
MIV	Surface Treatment
MIV	Traitement Superficiel

VOLUME TWO
CONTENTS

FDI   Food and Drink I
FDI   Industrie Agro-Alimentaire I

FDII   Food and Drink II
FDII   Industrie Agro-Alimentaire II

FAB   Fabrication
FAB   Fabrication

FIN   Finishing
FIN   Finissage

BS   Building Services
BS   Bâtiments

CC   Ceramics and Chemicals
CC   Industrie Ceramique et Industrie Chimique

WTI   Waste Treatment I
WTI   Traitement des Déchets I

WTII   Waste Treatment II
WTII   Traitement des Déchets II

Promoting Energy Efficiency and Load Management

Contents

Load management in industry 1
Grattieri, W. UIE Working Group WG5

**Load management in industrial firms – a service from
RWE Energie AG for industrial customers** 7
Thomas, R., Klöckner, R. (RWE Energie AG) GERMANY

Demand side management and energy savings by electricity 15
Jansen Wm. J.L., De Wit, L.R., Boone, W. (KEMA) NETHERLANDS

Long term impact of efficient motors 23
Contant, J., Maherzi, S. (Hydro-Quebec), Genois, J., Lafrance, G.
(INRS) CANADA

Marketing electricity in the competitive New Zealand market 31
Drew, S. (ECNZ), Slack, G. (Mercury Energy Ltd) NEW ZEALAND

Energy services – a new concept for electric utilities 39
Wälchli, T. (E.B.M.) SWITZERLAND

**Power for efficiency and productivity databases – a powerful
tool for problem solving and analysis** 51
Hulls, P. (Electricity Association) UK

**Substitution of fuel by electricity using induction in Swedish
industry – presentation of some cases** 59
Ritums, A. (Vattenfall AB), Brandt, H. (HEAT TECH Induction AB)
SWEDEN

Energy auditing and its tools 67
Nobre, E.C. (CEMIG), Limaverde, L.C. (Electrobrás) BRAZIL

**Industrial electricity use characterised by unit processes
– a tool for analysis and forecasting** 77
Söderstrom, M. (Linköping Institute of Technology) SWEDEN

Overall energy management in small and medium scale industry . 85
Eerola, P., Lehikoinen, S., Sippola, J. (Imatran Voima OY) FINLAND

Load Management with neural networks 93
Klöckner, R., Thomas, R. (RWE Energie AG) GERMANY

Efficiency monitoring of induction motors without torque transducers 101
Lewis, C. (ECNZ), Walton, S.J., Penny, J. (Innovative Developments Ltd) NEW ZEALAND

Improving electrical efficiency in electric arc furnace steelmaking 109
McKellar, R.D., O'Rourke, B. (Ontario Hydro), Reesor, R.A. (Ontario Hydro Technologies) CANADA

Energy savings in industry by efficient use of electricity 117
Hoe, J.M. UIE Working Group WGIA

Electric Load Management in Industry

BY THE UIE/TTE WORKING GROUP ON LOAD MANAGEMENT IN INDUSTRY

ABSTRACT

Industrial Load Management consists of the implementation of technologies and techniques, such as process modification/rescheduling, capacity addition, thermal energy storage, automation, electrotechnologies, cogeneration and facility management, in order to improve the customer's load shape. The UIE Working Group has identified the present state-of-the-art of Load Management techniques and determined in which ways they can be implemented. The results are reported in a brochure titled "Electric Load Management in Industry" available at the stand of UIE. The aim of the brochure is to give readers an overview on how the above measures can be implemented and how they can make best use of the consultancy services and expertise provided by the utilities. The brochure includes 21 actual examples in different types of industries, explained in terms of implementation, cost, benefits and users' reactions.

RESUME

Afin d'améliorer la courbe de charge des sites industriels, la gestion de l'électricité met en oevre des méthodes et des techniques comme la restructuration et l'automatisation des procédés de fabrication, l'augmentation de leur capacité de traitment, le stockage d'énergie thermique, les techniques électriques décentralisées, la cogénération et la gestion informatisée. Un groupe de travail de l'UIE on a établi l'état de l'art et determiné ses possibilités d'implantation. Le résultat de ce travail est rassemblé dans une brochure intitulée "Electric Load Management in Industry", disponible au stand de l'UIE, et dont le but est de donner au lecteur une vue générale des méthodes et techniques évoquées ci-dessus. Les distributeurs d'électicité proposent leurs services de consultance et d'expertise pour en faciliter la mise en pratique. La brochure comprend 21 exemples concrets dans différents types d'industries, détaillés en termes d'installation, de coût, de bénéfices et de réactions d'utilisateurs.

EE 2

UIE XIII Congress on Electricity Applications 1996

Working Group Membership:

Chairman: Walter Grattieri (I)

Secretary: Jean-François Reynaud (F), Hans Ottosson (S), Alfred Demouselle (B)

Members: Terry Dunne (UK), Hans-Rudolf Hagmann (CH), John Kevers (NL), Stefan Lindskoug (S)

Correspondent: Koei Aoki (J)

INTRODUCTION

Electric Load Management (here simply called Load Management) can be defined as any action taken by the customer and/or the electricity supplier to alter the load profile in order to gain from reduced total system peak load, increased load factor and improved utilization of valuable resources such as fuels, or generation, transmission and distribution capacity. From an electricity user's point of view it is a question of taking advantage of incentives and favourable pricing regimes so as to conveniently modulate load curves and therefore gain significant savings, with no adverse effect on product quality or productivity.

Industrial Load Management consists of the implementation of technologies and techniques, such as process modification/rescheduling, capacity addition, thermal energy storage, automation, electrotechnologies, cogeneration and facility management, in order to improve the customer's load shape. The UIE Working Group on Load Management in Industry has been studying this subject for the last three years: literature has been reviewed, rate systems analyzed, data collected, electricity users and equipment manufacturers interviewed, so as to identify the present state-of-the-art of load management techniques and to determine in which ways the load control can be effectively implemented in the various Countries for the different types of industries. The results are reported in a brochure titled "Electric Load Management in Industry" available at the stand of UIE.

CONTENT OF THE BROCHURE

The aim of this brochure is to give the readers an overview on how Load Management can be best implemented by industrial electricity users. Large industries usually have sufficient in-house knowledge on load management, while small and medium size industries often lack the necessary information and technical skills. For this reason, the content of the brochure is specifically focused on typical applications for small/medium sized industries. However, utility marketing and distribution personnel, energy consultants, and process/equipment manufacturers will also find the content of the brochure of interest.

The brochure is divided into several parts.

The first chapter deals with the special features of electricity and explains some basic concepts of Electric Load Management.

In chapter two, the main factors influencing the cost of the electrical energy are explained.

Chapter three points out the importance of load management for the efficient use of the production, transmission and distribution system. The impact of load management on the electrical system and the advantages resulting for the customer are discussed.

The interaction between the customer and the utility is stressed in chapter four, from a marketing approach.

Chapter five explains how the different industrial processes affect the load in the electrical network. It also explains what organizational and technical measures can be undertaken to control the load. Examples of load management in selected industries are also presented.

Chapter six deals with the case studies, a collection of actual examples of implementation of load management techniques at several industrial facilities from the Countries represented in the working group. Each example is explained in terms of implementation, cost, benefits, and the reaction of those concerned. Table 1 classifies the examples analyzed.

Table 1: List of the case studies reported in the brochure

Type of Industry	Company Name	Country	Main Products
Agriculture	Greenhouses, North Brabant	The Netherlands	Vegetables
Agrifood	Abattoir de Verdun	France	Beef and sheep meat
Agrifood	Sieras, Yonne	France	Ready-to-eat dishes
Agrifood	Allied Mills Ltd. Liverpool	Great Britain	Flour and by-products
Agrifood	Fukumusume Sake Brewery	Japan	Sake products
Building materials	Ciments Lafarge	France	Cements
Chemicals		Italy	Synthetic resins and insulating enamels
Chemicals	Takeda Chemical Industries	Japan	Industrial chemical products
Electrical and electronics	Takaoka Electric Manufact.	Japan	Electric machines and appliances
Electrical and electronics	ABB Turgi	Switzerland	High power converters, electronic equipment
Metal works	Fonderie de Saint Dizier	France	Cast iron products
Metal works	Fonderie ROZ S.p.A.	Italy	Cast iron products
Metal works	Tokyo Kinzoku Co. Ltd.	Japan	Non-ferrous metal products
Metal works	Gunnebo fastening AB,	Sweden	Nails and other iron products
Metal works	Lundgrens gjuteri AB,	Sweden	Cast iron products
Paper and wood	Ansgarius Svensson	Sweden	High quality timber
Paper and wood	Widmer-Walty AG,	Switzerland	Recycled paper and corrugated paper
Plastics	General Electric Plastics	France	ABS plastic granules
Quarrying	Cava Francesca S.r.l.	Italy	Inert meterials
Quarrying	Steinbruch Mellikon AG	Switzerland	Limestone
Waste treatment	CO.BE.A. S.p.A.	Italy	Recycling ferrous and non-ferrous metal

A final appendix introduces the rate structures across some European Countries and other points related to Load Management, including bibliography.

In writing this brochure we have intentionally focused on the general conditions that make load management a viable practice in industrial facilities. Less is said about specific equipment, devices, and installations, as these may become obsolete in the short term and may not be readily available everywhere. Detailed and up-to-date information on the subject can easily be obtained from local manufacturers, producers, or distributors. It is our hope that readers will find this brochure a useful tool for the design and implementation of load management projects. Utility staff can play an important role in guiding customers towards making correct choices. In this perspective energy users are invited to apply to the local Utility with any technical or commercial problems that may arise.

LOAD MANAGEMENT IN INDUSTRIAL FIRMS - A SERVICE FROM RWE ENERGIE AG FOR INDUSTRIAL CUSTOMERS

DR. REINER THOMAS, DR. RALF KLÖCKNER
RWE Energie AG

ABSTRACT

Today, electrical energy is an important production factor in many sectors of industry. Among other things, the costs of power consumption have to be minimized, too, in order to improve the cost structure. Since the demand tariff is usually used in industry, the total electricity costs are made up of a demand component (proportional to the annual peak demand) and an energy component (proportional to the number of kilowatthours consumed). The payment for demand and especially for energy can vary at different times of day and in different seasons depending on the generation structure of the electricity utility and its use.

Whereas the share of the electricity costs, which depends on the amount of energy used, is mainly determined by the production process and can therefore be changed by shifting production to different tariff periods (the same production output requires the same amount of energy assuming that all energy-saving potentials have been exploited to the full), the demand rate can be influenced by the structure of electric power consumption.

As a rule, care should be taken that the consumption of electrical energy is as even as possible and that demand peaks are avoided in particular. RWE Energie offers advice free of charge on the subject of industrial load management. The individual advice to an industrial customer is described.

RÉSUMÉ

Dans de nombreux secteurs de l'industrie, l'énergie électrique constitue un facteur de production important. Afin d'améliorer la structure des coûts, les frais d'achat d'énergie électrique doivent également être réduits. La réglementation puissance-prix employée la plupart du temps dans l'industrie, établit les coûts globaux de consommation d'électricité comme suit: une part de puissance, proportionnelle à la puissance annuelle maximale, et une part de travail, proportionnelle au nombre de kilowatts-heure consommés. Le prix de la puissance, et surtout du travail, peut varier selon l'heure du jour ou la saison, conformément à la structure de production de la compagnie d'électricité et à la façon dont elle est sollicitée. Alors que la part des coûts de consommation d'électricité dépendante de la quantité d'énergie fournie est déterminée dans une large mesure par le processus de production et peut donc être modifiée par un décalage de la production à des horaires tarifaires différents (même quantité de production requise - à condition que tous les potentiels d'économie d'énergie soient exploités - même énergie), le prix de la part de puissance peut varier selon la structure du prélèvement d'énergie électrique. Il faut fondamentalement veiller à ce que l'énergie électrique soit prélevée de manière aussi constante que possible, les pointes de puissance devant en particulier être évitées. Dans le cadre de son offre en prestations de services pour ses clients industriels, la RWE Energie AG fournit à titre gracieux des conseils sur le thème de la gestion de la charge. La façon dont se déroulent ces projets de conseil sera expliquée ultérieurement.

1. LOAD MANAGEMENT

Any modern load management aims to make optimum use of demand-saving potentials so as to minimize electricity costs without adversely affecting the production process in any way. Depending on the respective outline conditions, various approaches are possible. The measures that can be taken to cut electricity costs by load management can be essentially broken down as follows:

- Planning of plant operations.
- Controlling the production process.
- Automatic load management.

These methods are covered in greater detail in the following. The understanding of the various methods of load management requires detailed knowledge of the billing methods of the respective electricity utilities. The calculation of the energy rate is exclusively determined by the amount of kilowatthours consumed and their breakdown into normal-tariff and low-load-tariff periods. The demand rate is directly proportional to the peak demand of a given billing period, usually one year. The annual peak demand is obtained depending on the respective electricity utility as mean value of the two (or three) highest monthly peaks. These are in turn recorded by a maximum-demand meter usually as quarter-hour, half-hour or hourly demand mean values and are stored across the billing period.

1.1 PLANNING OF PLANT OPERATIONS

The planning of plant operations is of course mainly guided by production engineering and energy economics. Reduction of the maximum demand may be achieved by shifting operation from a given month with a high demand set up to months with low maximum demands and by shifting production sequences from daytime, when the highest demand normally occurs, to the night, which may require the introduction of a second or possibly third shift. An additional savings effect would result here from the reduction of the normal-tariff share in favour of a higher low-load-tariff share in the overall consumption of electricity.

1.2 CONTROLLING THE PRODUCTION PROCESS

In order to spread the load as evenly as possible during the shift, the control of the production process should not only be focused on the production-process-related requirements but should also give due consideration to the minimization of energy costs.

Unless inevitably required by the production, care should be taken not to operate unnecessarily many plants with a high connected load at the same time. This can be ensured, for instance, by following fully automatically a production schedule drawn up according to the aforementioned criteria. Should this be impossible, plants with a high connected load, for instance, the simultaneous operation of which is not absolutely required, can be interlocked against each other. A detailed analysis of the actual conditions is required as a matter of principle before any interference with the production process. The pros and cons of changing the prevailing conditions can only be weighed afterwards.

1.3 AUTOMATIC LOAD MANAGEMENT

If the optimization potential under 1.1 and 1.2 is exhausted, a maximum-demand monitoring system or, as a more comfortable solution, a peak-load optimization system guarantees, if properly applied, compliance with a fixed or variable maximum demand and, as a result, ensures relatively fixed and calculable electricity costs.

Since not all operational sequences can be planned to the last detail, a certain incalculable demand potential will remain left at any rate even if the in-plant flow of operations and the production with all the directly or indirectly connected consumers are planned and controlled in the best possible way.

The use of power relay, maximum-demand monitor, optimization computer or process computer depends on the respective conditions and should only be planned and implemented after detailed analysis in close contact with respective energy utility. RWE Energie offers advice free of charge in such cases.

2. PLANNING AND USE OF A LOAD-MANAGEMENT SYSTEM

2.1 REVIEWING THE ELECTRICITY BILL

In any industrial enterprise, the annual electricity bill should give rise to a review of the electricity demand situation. The data listed on the bill may already point to potential deleterious developments in the consumption of electrical energy.

The following questions can be answered by means of the bill:

- Do the monthly peaks scatter very much?
  The bill shows the respective highest demand set up in the individual months. If these values scatter very much, this may be indicative of a very uneven and hence inefficient consumption of electrical energy.

- Is the utilization time very low?
  Utilization time is the quotient of the electrical energy in the billing period and the chargeable demand (mean of the two highest monthly peaks). A constant consumption over a billing year would result in a utilization time of 8,760 h. With three-shift operation, realistic values are about 5,000 h, for single-shift operation the value is often below 2,000 h. Of course, these values do not only depend on the working time, but also on the respective industry, the type of production equipment used and other outline conditions and can therefore only be regarded as rough reference values.

- Is the low-tariff share very small?
  For three-shift operation, the low-load-tariff share should typically not be much smaller than the normal-tariff share. Even if the plant is operated with a single shift, it is often possible to move part of the electricity consumption to nighttime hours without impairing the production sequence.

- Are there any plans for changes such as investments, shutdowns, new buildings or extensions, which will have an impact on the electricity consumption?

If the answer is yes even to only one of these questions, a more detailed analysis of the electricity consumption situation should be carried out by the respective energy utility.

2.2 ANALYSIS OF THE ELECTRICITY CONSUMPTION SITUATION

If a detailed description of the in-plant flow of operations and the production sequence does not or not clearly reveal any savings potentials, the electricity consumption should be recorded, possibly over a longer period of time and separately for several consumers. Afterwards, the recorded data will be analyzed. In so doing, assumptions are made, in a departure from the existing consumption conditions, on the future demand when various savings possibilities are used. Different scenarios are simulated and their impact on the electricity costs is determined.

All reasonable saving measures will subsequently be discussed with the staff in charge of production and the in-plant flow of operations in order to work out a solution which is satisfactory both in production-engineering and energy-economics terms. If a load management system is to be purchased as a result of the discussion about potential savings, the energy utility can give support in drawing up the performance specifications.

The planning and design of the load management system will then be carried out in close cooperation between the customer and the suppliers.

Typically, trial operation will be necessary, especially when it comes to installing large-scale systems. All the operating parameters can be optimized during this period.

The operating parameters, which may have to be readjusted, should changes be made again to the operating/production sequence, are finally set after completion of the trial operation.

3. CONCLUSIONS

This paper is meant to give a rough overview of load management. It covers both organizational measures and technical systems for load management.

The technical systems presented fall into several categories in terms of their applications, performance and prices. A distinction was made here between power relay, maximum-demand monitor, peak-load optimization computer and process computer with integrated load management. For each type of system there is an optimum application so that none of these systems can be favoured right away. An optimization computer will certainly in general be more efficient than a power relay. From a purely economic point of view, the installation of a power relay may often be the economically more viable alternative, especially in the commercial sector, whereas it may be worthwhile in larger enterprises to use a more expensive system as the energy consumption grows. Following a detailed analysis, however, the use of a load management system may possible also turn out to be inefficient.

In reality, analysis and planning will of course have to be much more detailed than in this description. Since the present documentation is just meant to give an overview, the problems at hand have not been described in depth. Additional relevant data will have to be taken into account for the economic efficiency appraisal when it comes to concrete applications.

In sum, it is fair to say that the use of a load management system may well be advisable in many enterprises and municipal facilities. Since two cases are never completely identical, however, not pat solution can be offered here to reduce the demand peak. This is why an experienced adviser of the cognizant electricity utility should be consulted as a matter of principle so as to avoid potentially misdirected capital spending.

Demand side management and energy savings by electricity

W.J.L JANSEN, L.R. DE WIT, W. BOONE
KEMA, Arnhem, the Netherlands

Abstract

Demand-Side Management (DSM) activities are actions of utilities and non-utilities to influence the amount and timing of energy use. DSM is now also useful for energy conservation programmes. Gas-fired CHP (Combined Heat and Power) was recognized as a good opportunity to achieve a decrease in primary energy consumption and at the same time extend the services of the utilities not only to gas and electricity, but to heat as well. If enough CHP was built to cover the heat demand, more electricity would be generated than is needed. It is stated that more use of electricity offers on one hand the possibility of an increased use of CHP plants with the accompanied energy savings and on the other hand results in increased energy savings because of the introduction of energy efficient electrotechnology.

Résumé

Les activités Demand Side Management consistent en des actions de services publics et privés dans le but d'influencer le volume et le minutage de l'utilisation d'énergie. A présent, le DSM est également utile pour les programmes de conservation d'énergie. La CHP (combinaison de chaleur et de puissance) à combustion de gaz a été reconnue comme étant une bonne occasion d'obtenir une réduction de la consommation d'énergie primaire. Si un volume suffisant de CHP était produit pour répondre à la demande de chaleur, on se retrouverait avec un excédent d'électricité. Il y est affirmé qu'une utilisation plus importante d'électricité fournira, d'un côté, la possibilité d'une utilisation accrue d'usines CHP, avec les économies d'énergie qui en découlent, et que, de l'autre côté, cette utilisation aura pour résultat des économies d'énergie plus importantes grâce à l'introduction d'électrotechnologies efficaces sur le plan de l'énergie

INTRODUCTION

The definition of Demand-Side Management (DSM) can be given by the following statement: DSM activities are actions of utilities and non-utilities to influence the amount and timing of energy use. Tariffs have always played an important role in DSM. The stimulating effect of tariffs is enormous and this part of DSM was already introduced in the Netherlands at the beginning of the seventies by introducing differentiated tariffs to promote electric boilers. In the Netherlands where distribution utilities have to purchase their electricity from the producers, the intermediate rate between distributors and producers is very important. To cover the capital costs of generation the wholesale prices were set by dividing the total costs of production by the maximum load demand, which resulted in a price of 230 NLG/kW. This price for distribution utilities was, to some extent, copied in the retail prices as well and related to peak times in electricity demand. This resulted in shifting load of industries from peak time to off-peak time. This stimulation of load shift was very effective and resulted in smoothing the load during the year, as can be seen in figure 1.

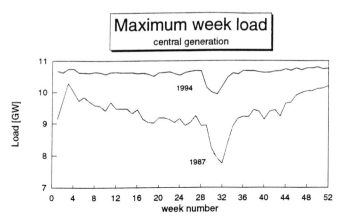

Figure 1   Maximum week load of central generation for 1987 and 1994

At this time it also became very profitable for industries to build combined heat and electricity production plants (CHP). The distribution utilities could then buy electricity from CHP plants during peak hours to avoid the capital costs of central production.

MARKETING ASPECTS

To make full use of the potential of DSM options it is very important to have knowledge about the customer. Questions have to be answered such as: how much energy is the customer using and at what time, how much is a customer prepared to pay for a DSM option, which techniques is a customer using? If we cannot answer these questions, the outcome of the calculations of the DSM impact is not reliable. A good example of unexpected outcome is the CFL (Compact Fluorescent Lighting) programme in the Netherlands. Because of the very high efficiency of the CFL and the lower energy costs of using them, people installed these lamps to lighten their gardens during the night. In this way, to some extent, the load is increased instead of reduced. This principle is known as the rebound effect. When services become cheaper people start to make more use of them. To set up better DSM programmes in the future, electricity companies need to know more about their customers use of energy. Introducing DSM to industrial customers requires knowledge of industrial processes. The present estimated potential for energy savings based upon economically attractive options is 30% of the current energy use [De Beer et al, 1994], but this potential will not be reached. This means that more factors than economic reasons are influencing the customer to participate in DSM programmes. Especially for industrial customers factors like reliability and a proven process are often more crucial than economics. Residential DSM programmes for energy conservation aim mainly for a few large energy consumer services like heating, cooling and lighting. By stimulating the awareness of customers in combination with subsidies, most targets can be reached. In the industrial sectors the problem is more complicated. The number of energy services is much larger, which means that no general approach can be given. An IRP (Integrated Resource Planning) study showed that with DSM options only the CO_2 emissions can be reduced by 11 Gton, with a more integral approach, taking more energy carriers into account, CO_2 emissions can be reduced by 16 Gton [Sep, 1994]. In this study DSM options in the form of electrotechnologies were not yet considered.

NEW CONCEPT OF DSM

DSM actions in the past have proven their success in electricity supply as can be seen from figure 1. Because of the tariff structure, the efforts in load management had accomplished that the load factor had increased to 80% in 1994; DSM is now more useful for energy conservation programmes. DSM was originally developed to optimize electricity supply and, therefore, only took into account the electricity use. If DSM is only applied to electricity, effects to be gained are relatively small. At this moment awareness is growing that DSM should be extended to all energy carriers. In the Netherlands in most cases only electricity and gas have to be taken into account. If energy consumption has to be minimized, the idea of separate targets for all energy carriers has to be abandoned. Instead of that, integrated strategies to minimize the consumption of primary energy fuels have to be developed. In the beginning of the nineties CHP (Combined Heat and Power) was recognized as a good opportunity to achieve a decrease in primary energy consumption and at the same time extend the services of the utilities to include not only gas and electricity, but heat as well. However, at the moment there are still separate targets for electricity reduction and gas reduction. As we will discuss in this paper, this approach will lead to sub-optimalization of energy supply.

NEW THOUGHTS

If enough CHP were to be built to cover the heat demand, more electricity would be generated than is needed. To overcome this problem, heating by the heat pumps in residential areas while in industry efficient electrically driven processes in the form of electrotechnologies like heat pumps, dielectric, infra red, induction heating, etc. can be introduced. A problem in promoting energy efficiency by electrotechnology is the price per energy content of natural gas and electricity. The average electricity price for an industrial customer in the Netherlands is 12 ct/kWh, which means 3.4 ct/MJ. The price of natural gas for the same customer is 22 ct/m^3, which means 0.7 ct/MJ. The Primary Energy Ratio (PER) of an industrial process is the ratio of the primary energy consumption W (MJ$_{thermal}$) of a process driven by primary fuels to that of the same process driven by electricity [Jansen et al, 1995]. The primary fuel needed to generate electricity consumption E (kWh$_{electric}$) is taken into account in the primary energy ratio PER, i.e.

$$PER = W / (3.6 \, E / \eta_{power\ station})$$ (1)

So with respect to primary energy consumption, to be energy efficient with electricity in processes we need a PER larger than one. However, if we want an electrotechnology to be economical only on the basis of energy saving, then a PER of 4.9 is needed. This means that a large potential of energy saving will not be used. The PER value is dependent on the efficiency of the electric power generating stations ($\eta_{power\ station}$); if this efficiency is improved, for instance by using combined heat and power, even more processes could be driven by electricity from an energy-efficiency point of view. It should be anticipated, however, that electrotechnologies like mechanical vapour compression, heat pump and air knife are only used because they are energy-efficient. Other technologies like induction, radio frequency, microwave and infra red are used in industry because they are cost-effective. Part of the cost reduction is gained by energy efficiency, but a large part is gained because processes are faster and better controlled.

As mentioned before, separate energy conservation measures for each energy carrier will lead to sub-optimization of energy use. In figure 2 a scenario is given in which heat pumps play an important role in combination with CHP [Kleinbloesem et al, 1992].

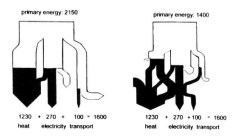

Figure 2   Residential heating with and without heat pumps

As is obvious in figure 2 only CHP and use of waste heat from the electric power stations will lead to a primary energy consumption of 2150 energy units. By generating more electricity and using this electricity to drive heat pumps to produce heat in residential areas, the primary energy consumption is reduced to 1400 energy units. If an even wider perspective is taken, by considering energy chains, more energy can be saved by integration of energy flows and by using electricity. It was demonstrated that introducing CHP plants and electric transport, electric heat pumps for residential heating and hot water supply, results in primary energy saving of 18%, reduction of CO_2 emission by 22%, an 8% reduction in costs and an increase in electricity consumption of 100% [Van Liere, 1995].

If we do not create new "rational" electric load, the amount of CHP which can be built is limited. Heat pumps for residential areas and electric transport are good ways of building electric load, but there are more opportunities in the industry to use electricity, by implementing the use of different electrotechnologies in industrial processes. Electrotechnologies on the one hand offer cost effective techniques for industry and on the other hand are promising with a view to energy saving. It is estimated that the application of electrotechnology in Dutch industries can lead to 6 PJ primary energy being saved [Jansen et al, 1995].

ENERGY SAVING BY USING ELECTROTECHNOLOGY

ENERGY SAVING

Through the use of electrotechnology, primary energy can be conserved in two ways:
- direct energy savings arising from the use of efficient technologies, such as the heat pump, mechanical vapour compression and the air knife.
- indirect energy savings arising from better controllability of the electrotechnology (PID control, fuzzy logic, artificial intelligence, etc) resulting in less waste, lower reject rates, lower standstill losses and hardly any start-up losses.

In the following a number of examples of energy efficient use of electrotechnology is given.

ENERGY SAVING THROUGH EFFICIENT PASTEURIZATION

In many dairy products, such as yoghurts or ice creams, mixtures of fruit and sugar syrups are incorporated. After mixing, the product has to be pasteurized, before being packed, at a temperature varying between 85 °C and 125 °C, depending on the mixture. In the conventional process this was carried out in a batch process in tanks heated with steam or hot water. To reach the required speed of the process, this type of heating caused local overheating near the tank walls, which resulted in an overcooked taste and a degradation of the aromatic components of the mixture. The new process is a continuous process based on the high energy densities possible with microwave technology. As a result of the volumetric heating (36 kW microwave energy, 2450 MHz), because of the more homogeneous process temperature and shorter process time, the quality of the product is improved. At the same time the average energy consumption of the whole process has been reduced by 75%. [UIE, 1992].

ENERGY SAVING IN DRYING BY RECYCLING THE LATENT HEAT OF EVAPORATION

The heat pump and mechanical vapour compression are well known for their ability to conserve energy in drying. With these electrotechnologies the latent heat of evaporation is recovered and the recovered heat is supplied to the drying process.

In conventional tunnel dryers vegetables and fruits are dried with a specific energy consumption of around 0.99 kWh per kg water evaporated; with heat recovery from the exhaust air, energy consumption can be decreased to 0.84 kWh/kg. With a heat pump the energy consumption can be drastically decreased to 0.29 kWh per kg water evaporated. This means a reduction in the specific energy consumption of 70%! Furthermore, due to the lower working temperatures the quality of the product (taste, flavour, vitamins, etc) is improved [UIE, 1988].

Mechanical vapour compression is used in the dairy, agriculture, food and petrochemical industry for example for concentrating milk and wey, concentrating animal waste, drying sugar beet pulp and destillation of propene. Energy savings between 50 and 67% are possible by drying with mechanical vapour compression. [UIE, 1988].

ENERGY SAVING THROUGH DRYING WITHOUT EVAPORATION OF WATER

The air knife is a very efficient electrotechnology for drying. With a high speed air jet the bulk of surface water on a product is removed mechanically as a liquid rather than as a vapour. Specific electrical energy consumption is low and varies between 0.05 and 0.1 kWh per kg removed water. An example is drying of trays used for the delivery of bread and confectionery from the bakery plant to the retail shops. After cleaning they have to be dried. In the old situation the trays were transported through a 70 °C hot spray detergent wash and rinse system and after that dried in a gas-fired dryer. On leaving the dryer, the trays were still wet. In the new system the trays are totally immersed in a water/detergent solution at 45 °C and subsequently treated by a high pressure water rinse, again at 45 °C. Drying is now carried out by air knifes. Energy costs for drying were cut by 85% because of the lower energy consumption of the air knife and because of the lower cleaning temperatures possible with the air knife. With the air knife 32% of primary energy is saved [UIE, 1988].

CONCLUSIONS

Electricity has a large potential for energy saving through the use of gas-fired combined heat and power plants in combination with electrotechnology for heating in residential areas and industrial processes. In residential areas the heat pump is used for heating homes and for the supply of hot water. In industry different electrotechnologies such as induction, radio frequency, microwave, infra red, mechanical vapour compression, the heat pump and the air knife are used in an energy efficient way. Several examples are given with energy savings up to 75%. More use of electricity on the one hand offers the possibility of increased use of CHP plants with the corresponding energy savings, and on the other hand results in increased energy savings because of the introduction of energy-efficient electrotechnology.

REFERENCES

- BEER, J.G. de, WEES, M.T. van, WORELL, E., BLOK, K., 1994. Potential of energy efficiency improvement in the Netherlands up to 2000 and 2015. Department of Physics and Society, Utrecht University.
- JANSEN, W.J.L., DE WIT L.R., BOONE, W., 1995. Improved efficiency by application of electricity. Unipede, 2nd conference on the application of electricity, Barcelona, September, 1995.
- KLEINBLOESEM, B.A. & DIEPSTRATEN F.M.J.A. Sep wil "kwaliteit" in het energiebeleid. Energie & Milieuspectrum nr. 10, 1992, p16-20.
- LIERE, J. van, 1995. From Source to Service, a strategy for optimal use of energy. UNIPEDE, 2nd conference on the application of electricity, Barcelona, September 1995.
- SEP and IJsselmij, 1994. Integrated Research Planning, Executive Summary, August 1994.
- UIE, 1988. Industrial drying by electricity. UIE, Paris, 1988.
- UIE, 1992. Dielectric heating for industrial processes. UIE report, 1992, Paris.

Long Term Impact Of Efficient Motors

Jacques Contant and Sadok Maherzi
Marché industriel, Hydro-Québec

Julien Genois and Gaëtan Lafrance
INRS-Énergie et Matériaux[1]

Montréal, Québec, CANADA

Abstract

From 1989 to 1994, many public electric utilities in North-America, like Hydro-Québec, proposed demand side management (DSM) programs for the industrial market. Those programs have been involved mainly for motor's utilization because of their impact on the total energy consumption. Since then, electricity surplus, dereglementation and low energy prices certainly contributed to reduce budgets for DSM. In that new context, this paper focuses on two aspects of the impact evaluation of efficient motors: a) based on a huge survey achieved by Hydro-Québec, we first draw a detail picture of the motors utilization in the industrial market; b) according to various strategies, we discuss the possible impacts of efficient mechanical systems for the period 1995-2010.

Résumé

De 1989 à 1994, plusieurs services publics d'électricité d'Amérique du Nord, dont Hydro-Québec, ont lancé des programmes d'efficacité énergétique dans le secteur industriel. Ces programmes ont surtout concerné les moteurs à cause de leur importance dans la consommation totale d'électricité. Depuis, le surplus généralisé d'électricité, les bas prix de l'énergie et le contexte de déréglementation ont contribué à une réduction des budgets d'efficacité et à une modification des stratégies. Compte tenu de ce nouveau contexte, cet article présente d'abord un portrait complet de la force mécanique dans le secteur industriel en se basant sur une très importante enquête. Puis divers scénarios de prévision pour la période 1995-2010 sont présentés afin de donner un aperçu des potentiels d'économie dans le domaine de la force mécanique.

---

[1] Research supported by the Hydro-Quebec(load forecasting department) and the CNRC

2- Energy-consuming equipment in the Quebec industrial market

Hydro-Quebec performed an important survey in order to describe all equipments by process, by technologies and end-uses. This survey, also, tries to identify building-related energy uses (thermal envelope, HVAC, lighting), and to evaluate the electric consumption for the industrial market in Quebec.

Figure 1. shows a schematic view of the industrial electricity demand by end-uses in Quebec. Among the 11 707 existing industries in Quebec, a number of 1 247 industries were surveyed. Among them, a great number of large industries were visited making the results highly realistic. As a result, in 1991, large industries consumed 88% of the total annual electricity consumption. The major consuming sectors were the primary metal industry (45%); the pulp and paper industry (27%), and the chemical industry (8%).

In 1991, the total energy consumption was 642 320 Terajoules. Electricity was the most widely used energy source (38%); followed by natural gas (20%); oil (15%), and other energy sources (28%).This industrial profile suggests a clear idea on the importance of motor which represents 48% of the total electricity consumption in this industry. This example, on the industrial market in Quebec, gives an accurate picture of the situation in all of North America for two main raisons:

1. The motor production standards are the same.
2. The Quebec industrial sector pattern is quite representative of the industrial production process in North America.

3- How to optimize mechanical power systems?

There are at least three ways to optimize the mechanical power systems in industries:
   Case A) Replacing all standard motors with high efficiency motors.
   Case B) Using variable speed control on mechanical power systems.
   Case C) Performing regular and preventive maintenance on equipments.

Case A and B have been implemented by Hydro-Quebec as energy efficiency programs for mechanical power systems. We could mention here, the HEM program (High Efficiency Motor) and the EOS-PFCS program (Energy Optimization System - Pump, Fan and Compression System).

Introduced in 1991, the **HEM program** promoted the replacement of standard motors with high efficiency motor. Financial incentives of $450 per kW saved were provided to purchaser ($175/kW in 1994) with extra bonus of $75 for the distributor. The objective was to allow customers to recover their additional cost of purchasing the HEM within six months to two years. Incentives for distributors were provided to encourage them to maintain more HEM's in stock. This program will be dropped in 1996.

Figure 1

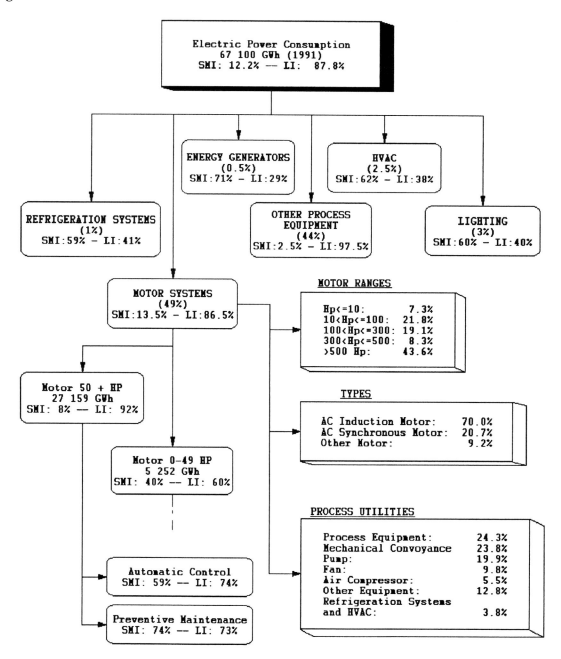

As stated by this program, a high efficiency motor is an induction three phases motor with an efficiency higher to the average efficiency of the same type in the market (Figure 2). In others words, in order to grant the subvention, the customer must purchase a motor with an efficiency greater than the market standard. In consequence, the supplier may deal with technical and financial limitations concerning the improvement on efficiency. He can offer different efficiency motors for the same power in order to satisfy the several customers' needs. This will have an impact on how to establish policies related to a specific efficiency program. For our study purposes, we have traced the "weighted mean" for both the standard and the high efficiency motor. As foreseen, the economic energy potential is better for low power than for high power motors. For instance, in the case where motor range is between 10 and 20 HP, the actual gain is around 5% whereas the gain is only 1.3% for ranges over 450 HP motor.

The second program launched in 1992 and called **EOS-PFCS** (Energy Optimization System - for Pumps, Fans and Compression Systems) is founded mainly by the fact the systems are not used to be operated at full load or semi-full load in continuous mode. In most cases, the load changes constantly during the process. This is mainly true for pumps, fans and compression systems. The major point on this program is to give technical support for those companies who need help when they want to optimize their mechanical systems. This program will end by the year 2000.

Figure 2

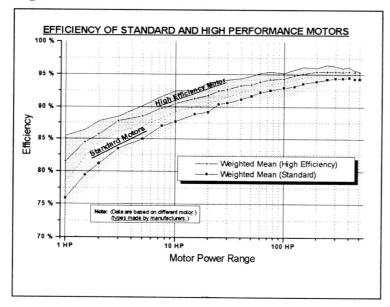

Depending on system types, Hydro-Quebec estimates the energy saving potentials between 3% and 40% which gives an actual economy of around 710 GWh. We have to point out here that, the program is only for customers having an electric consumption over 30,000 kWh/year. The pump systems represents only 20% of energy for the total motor's energy, the fan and compression systems are about the same. Generally speaking, the overall investment is greater than in the HEM program.

2- 4.- Evaluation approach

On a long term basis, the evaluation of high efficiency motor impact poses a number of challenges since the data are recent and incomplete. Hydro-Québec has completed, however, a successful evaluation of their industrial high efficiency motor programs[4] which gives an idea of the market evolution on a short term. In the future, we certainly have to mention that the Canadian government is considering some minimum levels for regulation of electric motors in 1996. These levels coincide with those contained in the U.S. legislation (1997) and are currently to come into effect in Ontario and British Columbia in 1996. But these laws do not give any guarantee for further mechanical system improvements.

In that context we tried to establish the possible range impact of efficient mechanical systems. The industrial demand for the period 1991-2010 has been achieved by an innovative equipment model[3] developed by INRS-Énergie / Hydro-Québec in 1995 and called MEDE-IND. This energy demand model can simulate individual industrial sectors, processes, equipment utilization and uses of technologies.

We used data, gathered by Hydro-Quebec for the industrial sector, into our model to simulate the **economic reference scenario** with the following assumptions:

a) based on latest Hydro-Québec forecast[4], we assume that the annual GNP growth rate for the industrial sector is 2,7% for the period 1992-2010

b) the current goals for the HEM (1996) and EOS -PFCS (2000) programs are realized, but no or little additional efficiency gains are predicted after the end of those programs.

This economic reference scenario is subdivided into two distinct groups:

- **Group 1.** Scenarios for motors, divided into scenarios 1a and 1b.
- **Group 2.** Scenarios for system equipments, divide into scenario 2a and 2b.

Scenario 1a: all standard motors are replaced by high efficiency motors in the period 1996-2010; this includes both Canadian and American legislation's impacts. This scenario is quite possible since the motor's life is about 15 years; it is still very optimistic, since no one can predict the same success of these laws for all categories of motors and processes.

Scenario 1b simply extends the success of the HEM program for the entire 1996-2010 period.

Scenario 2a is a theoretical case were all mechanical systems are optimized trough three measures:

i.) standards motors are replaced by high efficiency motors,

ii.) sharp system maintenance are achieved,

iii.) variable speed controls are used wherever needed. This scenario gives the maximum energy impact if newest mechanical technologies are used in all sectors and processes which seems quite impossible to achieved in the next 15 years;

Scenario 2b is built on the condition that the EOS-PFCS program is kept in place after 2000.

2- 5.- Possible impact of efficient motors

Table 1 presents different scenarios of efficiency impacts on mechanical system improvements for the 1991-2010 period in the industrial sector in Quebec. As expected, the possible impact of high efficiency standards is quite small compared to system optimization measures. According to **scenario 1a** results, the average efficiency maximum gain for the period 1991-2010 period in replacing all standard motors by high efficiency motors is only **1,6%**, while ,in **scenario 2a,** the mechanical system optimization gain reaches **10,1%** for the same period.

Table 1: Long term impact of efficient motors

Scenarios	1995 GWh	2000 GWh	2010 GWh	Average efficiency gain[3] (1991-2010)
Current goal				
HEM	100	100	100	0,2%
OES-PFCS	227	717	710	1,7%
Maximum potential[1]				
Scenario *1a*	100	240	710	1,6%
Scenario *2a*	100	1320	4100	10,1%
Trend scenario[2]				
Scenario *1b*	100	200	400	0,9%
Scenario *2b*	227	717	1400	3,4%

1. - *Standard motors are replaced by efficient motors.*
2. - *Hydro-Québec programs are extrapolated until 2010.*
3. - *Compared to total energy consumption without energy saving.*

However, these results do not give a realistic picture of possible impact of efficient motors for this period, neither the level of effort to achieve these goals; this is particularly true for the scenario 2a. In other words, even if it seems worthwhile to built programs to promote mechanical system improvement measures, the cost benefit analysis shows less gap between both program type.

For instance, the average efficiency gain for a trend scenario in replacing standard motors by high efficiency motors (*scenario 1b*) is **0,9% in 2010**, while the mechanical system optimization gain reaches **3,4%** (*scenario 2b*). Furthermore, if the Hydro-Québec EOS-PFCS ended in 2000, the possible impact of penetration of efficient motors may be as important as the HEM overall impact for the period.

Likely **the scenarios 1b and 2b give more realistic estimates than scenario 1a and 2a**. But even though, they do not have the same long term probability of success. Better standards and better quality controls of market products supported by legislation (scenario 1b) are quite easy to put in place. However, incentive programs based on technical analysis (scenario 2b), are less confident on a long term period for two main reasons:

i.) Due to electricity surplus and dereglementing industry, no electric utility will guarantee to support such a program for a long period; DSM fashion may well be abandoned forever.

ii.) Assuming a continuous maintenance and a mechanical system optimization supposes a continuous education on least cost benefit for all plant sizes and all sectors; furthermore, it also supposes that engineers install metering systems to follow the performance of each mechanical systems; present data on the subject do not allow us to estimate the real long term confidence of such measures.

It should be noted that our model version under-estimates the motor electricity consumptions in the future for two raisons:

i.) New environmental laws; for example, in the pulp and paper sector a recent law, which forces the industries to treat their water and chemical wastes, causes a significant increase of electricity consumption (around 10% increase for some plant).

ii.) Tendency to use more electricity in processes need more and more ventilators, pumps and motors.

In conclusion, this work has certainly permitted us to draw a very detail picture of the possible energy impact of efficient motors on a long term period. According to the announced laws to control severely the market, replacing standard motors by high efficiency motors seems probabilistic on a long run. However, even if we have one of the more complete data base, many questions are raising up.

a) How quick will the market react?

b) How can we strive consistency between "normal conservation trend" and program impact?

c) What is a realistic impact of efficient mechanical systems in the changing context of DSM (Demand Side Management) in North-America?

Solutions are known and advance technologies exist. But since the problem seems to be no more politic, the issue is now to know how these technologies will be adopted on a least cost basis. At the light of new data, answering these questions is at the top of our agenda.

References

1.- HYDRO-QUEBEC, *Profil sectoriel de la consommation électrique: marché industriel*, nov. 1993.

2.- HYDRO-QUÉBEC *Relevé des équipements et comportements énergétiques du marché industriel au Québec*, Montréal, oct. 1993.

3.- J. GENOIS, G. LAFRANCE, L. VIGEANT-LANGLOIS, *Modélisation du secteur industriel, rapport d'étape présenté à Planification Stratégique*, Hydro-Québec, déc. 1995.

4.- A. PARECE, W. KOTIUGA, S. HASELHORST, *Improving Estimates of motor program savings: applying two-stage sampling with in-field measurement of motor performance*, Energy Services Journal, I(1), 7-19, 1995.

Marketing Electricity In The Competitive New Zealand Market

DREW, Stephen - *Electricity Corporation of New Zealand Limited*, and
SLACK, Graham - *Mercury Energy Limited*
NEW ZEALAND

ABSTRACT

The introduction of competition at every level of New Zealand's electricity industry has forced generators and retailers to differentiate themselves from their competitors by supplying products and services that attract and retain customers.

The two largest players at the generation and retail end, the Electricity Corporation of New Zealand Limited (ECNZ) and Mercury Energy Limited, have offered energy services to industrial customers for several years. They have been delivered across the industrial spectrum, with particular attention to the meat, dairy and forestry sectors which give New Zealand its unique export profile.

ECNZ and Mercury Energy have successfully used energy services to differentiate themselves from the competition and create strong relationships with customers. The combined impact of both organisations working together has greatly exceeded what they could have achieved individually. However, both organisations now face the challenge of putting their energy services groups onto a more commercial footing.

RESUME

L'introduction de la concurrence à tous les niveaux de l'industrie de l'électricité néo-zélandaise a obligé les sociétés productrices d'énergie et les revendeurs détaillants à se différencier de leurs concurrents en offrant des produits et des services susceptibles d'attirer et de fidéliser les clients.

Cela fait déjà plusieurs années que les deux plus importants agents du secteur commerce de détail et production, Electricity Corporation of New Zealand Limited (ECNZ) et Mercury Energy Limited, offrent des services d'énergie aux clients industriels. Ces services sont offerts à travers toute l'industrie avec une attention toute particulière aux secteurs forestiers, laitiers et de la viande réputés pour avoir permis à la Nouvelle-Zélande d'atteindre cette unique position d'exportateur.

ECNZ et Mercury Energy utilisent avec succès les services d'énergie pour se différencier de la concurrence et établir de solides relations avec leurs clients. L'impact de la coopération de ces deux sociétés a permis d'accroître considérablement les résultats qu'elles auraient atteint en travaillant séparément. Ces deux sociétés doivent désormais répondre au défi d'orienter leurs groupes de services d'énergie vers un développement plus commercial.

THE ROAD TO DEREGULATION

New Zealand has a population of three and a half million people spread over an area the size of the United Kingdom. Its economy was in the past highly regulated with the government determining the operating context of most of the country's major service providers including the electricity supply industry.

For the last ten years, the economy has been going through a period of intense reform driven by deregulation, privatisation and the creation of a rapidly expanding open market economy.

International trade has undergone a shift with the development of new export markets in Asia and the trend towards adding value to primary products before they leave the country. Exporters are now proactively meeting international requirements for products tailored to specific market needs. Export earnings are expected to continue to increase, with the forestry sector contributing greatly to export growth.

Although manufacturing operations in general are still small by world standards and largely centered in and around Auckland city, New Zealand has some of the largest dairy and meat processing plants in the world.

ELECTRICITY SECTOR DEREGULATION

The electricity sector too has undergone a process of deregulation, albeit following in the wake of the telecommunications and transport industries. The first stage of the reform process involved changes to the ownership structures of power companies. Whereas they were previously in the hands of the community, largely through local body ownership, they now operate as commercial bodies. Many are publicly listed companies and some are now largely owned by off-shore utilities.

Secondly, the eradication of franchise boundaries on 1 April 1994 has resulted in a competitive environment for energy sales to all customers. Power companies are now competing to retain existing customers and gain new ones, particularly large industrial users and organisations with multiple sites.

The Government's withdrawal has allowed market forces to change the shape of the electricity supply industry. Whereas there were previously 52 retail power companies, there are now less than 40, with further contraction to about 10 suppliers expected.

The final stage of the deregulation process was the division of the country's national generation company, the Electricity Corporation of New Zealand, into two competing generators early in 1996. ECNZ, which previously generated 96 percent of New Zealand's electricity needs, now has 60 per cent market share. Its competitor, Contact Energy, while smaller, owns key power stations and resources, including important gas contracts. New generators, many with international backing, are also entering the market.

The gas supply industry is also going through a process of deregulation although the level of competition is comparatively low due to market dominance by a small number of suppliers.

Just as ECNZ is New Zealand's largest generator, Mercury Energy is New Zealand's largest energy retailer. It is based in the country's population and manufacturing centre, Auckland city, and supplies approximately one sixth of New Zealand's electricity. It is vertically integrating upwards by developing generation based on landfill gas, combined cycle gas-fired cogeneration and geothermal plant.

Mercury Energy was the first retail company to gain significant extra share of the retail market through the signing in 1994 of a supply contract with one of the country's major industrial groups operating in the forestry sector. It also now supplies the world's largest cheese factory, built in 1995.

Mercury Energy identified customer service as one of the key elements of retail competition and was the first retail supply company to introduce guaranteed performance standards backed by compensation.

ENERGY SERVICES

Early in the 1990s, ECNZ and Mercury Energy started working along the same path towards developing enhanced customer services, including a large element of technology assistance. ECNZ initially supported projects which delivered benefits directly to its major industrial customers and achieved significant success with the pulp and paper industry where the projects resulted in greatly enhanced customer relationships.

In 1992, an Energy Services Group was created within ECNZ to develop industrial products and services for all power companies to deliver to end users, particularly those supplying New Zealand's main export areas; meat, dairy and forestry. Key engineering staff with extensive industrial experience in these sectors were recruited to provide an interface with industry.

A range of programmes were developed to meet the primary concerns of New Zealand industry, particularly in achieving total energy savings, along with reducing the environmental impact of industry and improving productivity and product quality. Programmes were developed for evaporation and drying, motors and drives, water and waste water treatment, and refrigeration and heat pumps.

Mercury Energy started its programme with the opening of an electricity utilisation centre in 1990. This includes a showroom for demonstrating electrical technologies, a seminar room and a library.

Amongst the services provided by the centre are advice to customers on managing load to control maximum demand, understanding and applying tariffs, energy management and applying energy efficient technologies. Services being developed include advanced metering and monitoring technologies, power quality and reliability and advanced electrical technologies for improving productivity and product quality. Many of the customers helped by Mercury Energy now have a key advantage over their competitors.

Like many other power companies, Mercury Energy is being assisted in meeting its marketing objectives by delivering some of the products and services developed by ECNZ. Four of the programmes the two organisations are involved in are described below.

MOTORS

In order to offer assistance across the spectrum of industrial users, ECNZ has developed a package of products and services to aid the selection and operation of electric motors.

In 1994, it launched a Windows based software product, called Motor Selector, to help users select cost effective electric motors. The response to Motor Selector has been encouraging. More than 20 power companies, including Mercury Energy, now supply the software to their customers. It has also been sold in North America.

ECNZ's experience with Motor Selector, aided by customer feedback via power companies such as Mercury Energy, led to the launch of a second version early in 1996. This is the first in a new family of technology evaluation software which will aid in the selection of technologies like aerators, boilers and compressors. They will all be based on the same core software so people who are familiar with one of the products can easily apply the others. The software compares not only the total monetary cost but also noise, emissions and other subjective factors.

ECNZ has also developed a portable device, called Motor Monitor, to measure the performance of electric motors in situ. Mercury Energy has successfully trialled a prototype of the device with industrial customers to confirm the operating value of the unit.[1]

DEHUMIDIFIER DRYING

ECNZ is also committed to improving the energy efficiency of drying processes. Additional value can be added to many natural products like fruit, timber, deer velvet, sphagnum moss and fish by using a dehumidifier dryer to dry them in a low temperature, controlled atmosphere.

Although research has been done on designs for advanced dryers in New Zealand, a technology transfer partner was needed to develop the technology for commercial applications. ECNZ is fulfilling this role by bringing together key partners to deliver the technology to industry through power companies.

For example, ECNZ has worked closely with Mercury Energy on an advanced dehumidifier dryer to be used in developing drying schedules for a range of products including vegetables, fruit, seeds and timber. A portable dehumidifier dryer test rig was built and initially made available for Mercury Energy's use before being offered to all power companies.[2]

In this way, ECNZ is making advanced technology available to retail companies so they can assist their customers in applying the technologies.

INFRARED HEATING

Several years ago, Mercury Energy saw the energy efficiency benefits infra-red heating offered industry. It developed an infra-red test unit so its customers could determine whether the technology was applicable to their product. As a result of the unit, a number of Mercury Energy's customers have replaced gas-fired heating with infra-red technology, making savings both in installation and running costs and reducing their drying times.

Through its links with the Infra-Red Research Club at EA Technology and the Centre For Materials Fabrication at EPRI, ECNZ has provided technical expertise to Mercury Energy's projects and has now made portable test units available to other power company customers.

WOOD PROCESSING

A key contributor to the New Zealand economy is the forestry industry consisting of a very successful pulp and paper industry and a rapidly expanding solid wood sector comprising many hundreds of small sawmills, remanufacturers and furniture makers. The continued growth of the solid wood sector is seen as being pivotal to the long term expansion of the pulp and paper industry; a major load for ECNZ.

By late 1995, 11 power companies, including Mercury Energy, were partners in a programme developed by ECNZ for delivery to wood processing customers.

Projects have been established to solve a range of wood processing problems, not all of which are related to electricity supply. For example, ECNZ has supported the forestry industry in improving its sawing technology. It has been involved in a two year trial of a Mattison straight-line ripsaw at the Forestry Research Institute. The saw, which is very accurate and thus reduces sawmill waste, is now being evaluated by a commercial sawmill.

Other projects aim to determine the critical factors involved in sawing accuracy and optimising motor sizing in sawmills. A pilot superheated steam kiln has been built to prove the value of MVR technology in drying high quality products and ECNZ's Motor Monitor is being used in some sawmills.

Mercury Energy joined this programme to strengthen its ability to service its new forestry industry customer.

FUTURE CHALLENGES

For ECNZ, added value services like those supplied by the Energy Services Group, are a key component of its marketing strategy which gives it a competitive advantage over other generators. By continuing to provide power companies with the tools to help their industrial customers, ECNZ will offer a valuable service and create customer loyalty.

Mercury Energy has discovered that, as energy trading margins diminish, it is important to commercialise the application of technology solutions in addition to building relationships with customers. Mercury Energy is working to build the technology services division into a profit centre, taking advantage of its existing relationships with its large customer base.

ECNZ and Mercury Energy are both facing challenges from their competitors. They are working to identify and meet customer needs while putting their energy services groups on a commercial footing. Both organisations believe that by working together they can effectively develop and deliver services that create cleaner, faster, more efficient industrial operations. The relationships that result will be the foundation of electricity supply contracts in the competitive market.

REFERENCES:

1. C. LEWIS et al: 'Efficiency Monitoring of Induction Motors without Torque Transducers', UIE Congress 1996.

2. N. J. BARNEVELD et al: 'The Development of an Advanced Dehumidifier Dryer', UIE Congress 1996.

UIE XIII Congress on Electricity Applications 1996

EE 39

MR WÄLCHLI THOMAS
Vice-Director, Elektra Birseck Münchenstein (EBM), 4142 Münchenstein (Switzerland)

Energy Services - A New Concept for Electric Utilities

Les Prestations de Services dans le Domaine de l'Energie. Un Nouveau Concept pour les Distributeurs d'Energie Electrique.

....

ABSTRACT

Since 1979 the promotion of efficient use and application of energy is a further stratetic target of Elektra Birseck, Münchenstein (EBM), which is a medium sized swiss electric utility. To reach this goal, EBM has diversified in district heating. In regions with a high density of heat consumption, EBM supplies the heat with district heating. The base consumption is covered with cogeneration plants, the peak consumption with conventional boilers. The electricity produced by the cogeneration plants is transported by the existing electric grid into regions with a low density of heat consumption in order to use it for running electric heat pumps. In comparision to other systems heat pumps allow to use most renewable energy at lowest cost.

However, the realisation of our concept is handicaped by several economical and psychological obstacles. First of all there is to solve the conflict between the economical and environmental interests. With our advice-service and our concepts of heat supply specifically planned for the individual customer we are able to diminish these difficulties. Heat supply and operating management of heat systems by EBM mean many financial, administrative, technical and ecological advantages for the customers.

RESUME

Elektra Birseck Münchenstein (EBM) avec son siège en Suisse, est une entreprise électrique de taille moyenne. Depuis 1979, EBM poursuit, parallèlement à sa mission première de distribution, un objectif stratégique promouvant les opérations d'économie d'énergie et celles relevant d'une utilisation plus rationnelle de l'énergie. Afin d'atteindre ce nouvel objectif, EBM a entre autre diversifiée ses activités dans le chauffage urbain. Dans les quartiers présentant un besoin important en chauffage,

EBM fournit la chaleur par des centrales alimentant les réseaux urbains. Pour desservir les besoins de base des consommateurs de chaleur raccordés au réseau de chauffage, il est fait appel à des installations de co-génération; les consommations de pointe étant couvertes par des chaudières traditionnelles. L'électricité produite dans ces installations de co-génération est transitée à travers le réseau du distribution existant dans des zones de moindres besoins en chauffage afin d'alimenter des pompes à chaleur. En comparaison avec d'autres systèmes, ce sont les pompes à chaleur qui permettent actuellement une utilisation maximale à moindres coûts de l'énergie renouvelable.

Cependant, la mise en oeuvre du concept "co-génération et pompe à chaleur", développé par EBM et qui fait figure de pionnier au niveau de la politique ènèrgétique Suisse, est entravée par divers obstacles de nature économique et psychologique. Un service conseil compétent et le développement de concepts orientés vers les besoins d'approvisionnement en chauffage des clients, permettent dans de nombreux cas de trouver des solutions paliant aux difficultées liées aux intérêts d'ordre économique et à l'aspect de l'environnement. La desserte en chauffage ainsi que l'exploitation de l'installation de la production de chaleur par EBM offre aux clients un certain nombre d'avantages d'ordre financier, administratif, technique et écologique.

ENERGY AND ENVIRONMENT PROBLEMS

Nature needs millions of years to form fossil sources of energy like oil and gas. Today man is using up those limited resources in a very short time by a simultaneous extreme polution of the environment and a poor exploitation of the capacity to perform mechanical work which is stored in fossil sources (Fig. 1). As a result of these universal energy problems, the energy crisis of the seventies and the rising antagonism to nuclear power EBM engaged to promote an economical and rational application of energy already in 1979.

HEAT SUPPLY CONCEPT "COGENERATION AND ELECTRICAL HEAT PUMP"

The concept of cogeneration and electrical heat pump was consequently propagated and realised by EBM. Since 1982 the first block heating station was put into operation. Today this concept is pathbreaking for energy politics and the program "Energy 2000". In areas with a high heat supply demand the constant load of the heat consumption is covered by block heating stations and the peak demand by conventional oil or gas boilers. The heat is distributed by mains from the heat generation plant to the different buildings. With a specific average of at least 5 MWh per meter and year the installation and running of mains are more efficient than conventional and individual heat supply systems. The current which is produced by a block heating station is carried by the already existing distribution grid to areas with low heat consumption and is there used for the running of electrical heat pumps. Only the combination of block heating stations and electrical heat pumps, which use environment energy, performs a more efficient revenue of the energy raw material (Fig. 2) and besides the environment is less poluted (Fig. 3).

BASIC TERMS FOR THE COGENERATION AND ELECTRICAL HEAT PUMP CONCEPT

Because electric utilities are obligated by a law of the canton Basel-Landschaft to compensate exceeding electricity from private block heating stations well planned block heating stations can be runned economically. On the other hand seasonally different tariffs, which were claimed by the ecologically orientated groups and then forced upon by the politicians prevent a widespread use of electrical heat pumps.

Actually with the present exceeding current production in Europe additional expensive current is produced by block heating stations with fossil sources of energy without using at least 50 % of it in electrical heat pumps according to the target of "Energy 2000". This is an evident ecological retrogression.

CUSTOMERS' REQUIREMENTS

Customers do not require the supply of primary or secondary energy in form of oil, gas or electricity; they want a warm apartment and warm water. There are various requirements concerning the heat supply:

- economical energy supply
- high energy supply safety
- no or little investment expenses for the heat supply
- low administrative and low operating costs.

SERVICES OF EBM

As a modern and consumer-friendly service industry EBM offers an advisory service which checks the customers' needs. With that many differing impediments to the realisation of the cogeneration and electrical heat pump concept can be cut back. Because of the different evaluations of the sometimes opposite customers' objectives EBM works out specific heat service concepts which are distinguished by different elements (Fig. 4).

HEAT GENERATION PLANT, MAINS AND INVOICE FOR THE HEAT PURCHASE

The efficient use of energy, conservation and especially the use of renewable energy have their price. Depending on what the customer is ready to pay for the heat supply, a special ecological concept is chosen or else a more conventional one. The heat can be produced from case to case by boilers, block heating stations, electrical heat pumps, wood chips heating systems and solar panel systems. The heating-installation is set up in a suitable room which is put at disposal free of charge by the owner of the building on the basis of a servitude contract.

Are several buildings supplied with heat by the same heat generation plant center then the heat is carried by mains to a heat exchanger where the customers can take over the heat. Depending on the customers' requirements the entire heat supply for a building is delivered by one single heat delivery point or by several flat specific heat delivery points. The owners of buildings or flat tenants are charged individually for the supplied heat. The tariff is the result of a basic price and expenditure of work.

FINANCING AND OPERATION

The renovation of existing or the construction of new heat generation plants and mains can exceed the customers' financial capacities or intentions. Basically there are three possibilities for the financing of the heat supply:

- In the case of only one ore few heat purchasers the necessary investments for the heat generation plant and mains are effected by EBM. The costs are included in the heat tariff and proportionally paid by the customers. If desired, the customer can finance a part of the investment costs by paying the costs for the heat supply point. The consequence of this is a reduction of the heat tariff. The mains are runned by EBM.

- In most cases a joint-stock company is set up between EBM, the municipality and the bigger customers. Now the joint-stock company is responsable for the financing analogous to the financing by EBM. The technical and administrative management is runned by EBM on the basis of a management contract between EBM and the joint-stock company.

- If there is only one building which has to be supplied with heat, the investments can be effected by the owner. If the owner is not able to run for instance a block heating station, because he lacks a technical qualified staff or generally he does not want to be involved with the heat generation, so EBM takes over the management on the basis of a management contract.

ADVANTAGES FOR THE CUSTOMER

The heat supply and management of the heat production stations by EBM have different financial, administrative, technical and ecological advantages:
If EBM runs the heat production stations the owners or the managements of properties are relieved of responsability and trouble. The customer has not to bother anymore about the purchasing of fuel, burner service, cleaning of the boiler, cleaning of tank and chimney, checking of the exhaust results, resolving problems, repairs and maintenance, invoice, insurances and so on.

EBM is organized as a co-operative without pursuit of profit. EBM has not to make a profit; the service has only to cover all costs. Heat supply projects are planned for the long term. With EBM as heat supplier the customer can be sure to have a partner for many years. The fulfillment of the heat delivery contract is assured. The supply of electricity and heat is dependent on mains. Out of the traditional business of EBM many synergies on the technical as well as on the administrative side can be used if EBM supplies the heat. The consequence is an economical heat supply with a high supply safety.

The know-how of the EBM experts makes it possible that the heat supply is economical, safe as well as optimal in energetic and ecological respect. Compared with other operating authorities of block heating stations, the block heating stations of EBM realize the planned operating time and are in every respect above-average. Further the exhaust results of the conventional boilers and of block heating stations are periodically checked. Any divergences from the standard are quickly discovered. This makes a rapid correction possible, which is necessary for an ecological heat supply.

A well trained operating team garanties a supply right around the clock. Many operation problems can be solved by the EBM staff. Therefore the safe heat supply is dependent only for the least part on the suppliers of the installations. Any trouble of the heating installation is detected by a monitoring system and automatically reported to the headquarters of EBM. For the most part the problem can be settled by an engineer on stand-by with the help of a long-distance system (Fig. 5). If this should not solve the problem, the decentralized error analysis makes it possible that the right specialist is called in for the elimination of the fault. This increases the supply safety and reduces the running costs.

The heat supply projects, which are realized by EBM, are economically independent because of the existing know-how and therefore have not to be subsidized by the government or the canton respectively by the public at large or our customers of electricity as tax payers.

THE OPPORTUNITIES OF THE ELECTRIC POWER COMPANIES

The electricitiy and heat supply is dependent on mains. If the heat supply is runned by an electric utility so many synergies in the technical and administrative fields of the traditional business can be used.

Thanks to this competitive advantage the electric utilities can offer a more economical heat supply and assure a higher supply safety than other companies, which are not in the mains dependent energy supply business. This competitive advantage is a real enterpreneurial opportunity for every electric utility to diversify successfully into new fields of promising future. The electric utility should not miss those opportunities untested.

Conventional Oil or Gas Boilers

Oil/Gas 1000° C

Loss of the usable temperature gradient

Heating temperature 80°C

EE 46

Fig. 2

Efficient Use of Energy leads from the Conventional Boiler to the System of Cogeneration and Electrical Heat Pump

1 Conventional Boiler

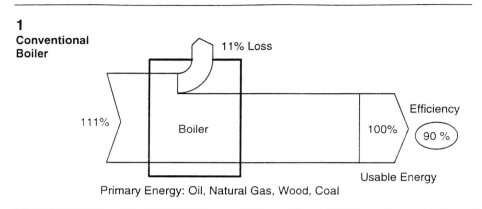

2 Block Heating Station

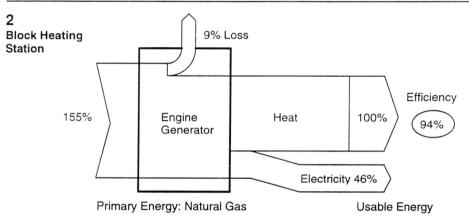

3 Combinet Plant: Block Heating Station and Heat Pump

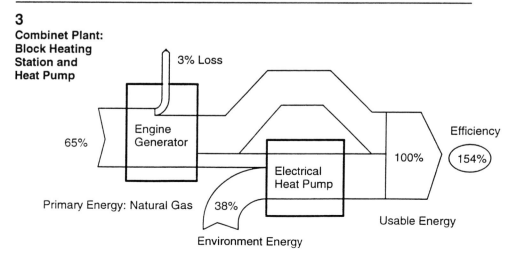

Fig. 3

Comparison of Emission between Conventional Boiler and Cogeneration plus Electrical Heat Pump

Conventional Boiler

Annual Emissions and Oil Consumption of 760 Apartments heated with Conventional Boilers: 400 t Oil

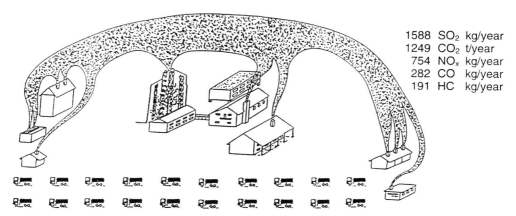

1588 SO_2 kg/year
1249 CO_2 t/year
754 NO_x kg/year
282 CO kg/year
191 HC kg/year

Cogeneration plus Electrical Heat Pump

Annual Emissions and Oil Consumption of 760 Apartments heated with Cogeneration, Electrical Heat Pump and Conventional Boilers for the Peak Demand: 10 t Oil

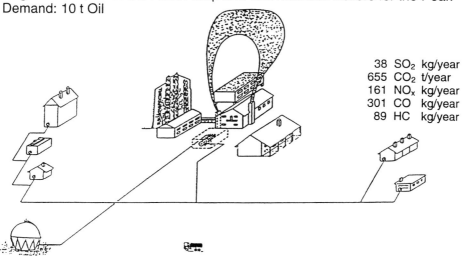

38 SO_2 kg/year
655 CO_2 t/year
161 NO_x kg/year
301 CO kg/year
89 HC kg/year

Possibilities to Realise Heating Projects

Criteria		Shape	A	B	C
1	Heat Production	Peak Demand	Conventional Boiler	Electrical Heat Pump	Wood Chips Heating
		Constant Load	Block Heat Station		
2	Number of Buildings to be Heated		One	Few	Many
3	Pricing System		Tariff	Actual Cost	
4	Operator of Heat Production Plant		EBM	Third Party	
5	Investor and Owner of the Heat Production Plant		EBM	Limited Company	Third Party

Fig. 4

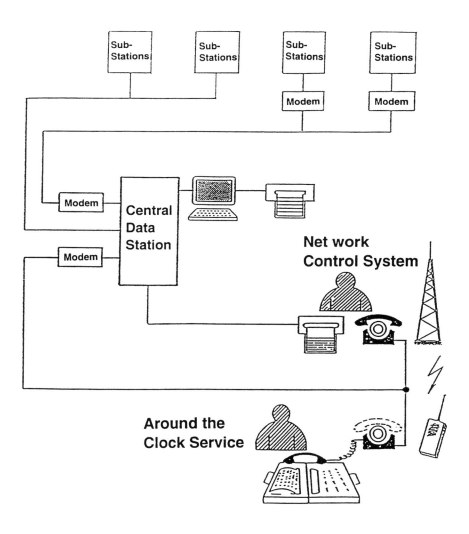
Fig. 5

Power for Efficiency and Productivity Database
- a powerful tool for problem solving and analysis

PAT HULLS
Electricity Association

ABSTRACT

As a result of running the Power for Efficiency and Productivity Awards, the UK electricity suppliers have for the last twelve years been gathering data on a wide range of applications of electricity in industry. The result - every Electricity marketeer's dream - a source of information revealing the motivation for and benefits of adopting their product, electricity.

Recording the information on a database with suitable means for easy interrogation has produced a powerful tool with a multiplicity of applications.

Whether the need is for:

        an analysis of a particular market sector,
        finding the answer to a customer's problem
or   proving the case for the positive environmental aspects electricity has to offer in a particular situation,

the solution can probably be found by the use of the PEP database.

RÉSUMÉ

Faisant suite au déroulement du concours PEP, les companies d'électricité en Grande Bretagne ont, depuis 12 ans, compilées des données sur les applications électriques les plus diverses dans l'industrie et dans le commerce . Le résultat - le rêve de tout vendeur d'électricité - est une source d'information qui révèle les motivations des consommateurs et les avantages qui sont liés à l'utilisation de leur produit: l'électricité.

L'enregistrement de cette information sur une banque de données qui posséde des moyens d'interrogation facile à utiliser, a produit un outil d'analyse puissant avec une multiplicité d'applications.

Quant il est nécessaire:

        d'analyser un secteur de marché
        de trouver la solution à un problême client
ou   de démontrer les avantages environnementaux de l'électricité pour une application donnée.

la solution peut probablement être trouvée dans la banque de données PEP

INTRODUCTION

Many companies around the world are experiencing increasing pressures to reduce staff numbers and yet at the same time provide the same or improved level of service to their customers. This process inevitably results in some loss of the older members of staff who possessed a wealth of knowledge in their areas of expertise. "Expert systems" were developed to record and make use of the knowledge held by skilled personnel at all stages of design, manufacture and maintenance. The advent of readily available computer technology made these systems ideal for passing on specialist knowledge to the less experienced. The PEP database is not an "Expert System" in the true sense of the word but does enable competent staff, without an in-depth technical background, to answer queries on a wide range of subjects. It may be that a food processor is experiencing problems with a baking process. With just a few questions to establish the product or material and the type of process it is possible to search to see if anyone in the past has had similar problems. The records may show a number of different solutions for the same problem. The next stage may be to uncover more detailed information. If an "in house" research or development centre is available then it can carry out further assessment of the problem. On the other hand the customer could be given details of suitable equipment suppliers as indicated on the appropriate record/records.

ACQUIRING THE INFORMATION

The Power for Efficiency and Productivity (PEP) Awards were initiated by one of the UK's electricity suppliers, Norweb, in 1983. In the following year this became a National event and has been running annually ever since. Building on this initiative, UNIPEDE ran an International "Eta" award in 1991 and 1994.

Apart from rewarding the efforts of the many entrants who have taken part over the years, the PEP awards have also resulted in the build up of a valuable data bank of industrial processing intelligence. Every year in the UK, on average, about 300 companies enter for the PEP awards. This means that the UK electricity industry has around 3 300 detailed records of manufacturing processes, across all sectors, that have been improved by the application of electrical techniques.

This "in-depth" data represents a valuable resource, not only in marketing terms but also as a means of analysing the contribution electricity has made over the years to improving production efficiency, reducing energy consumption and CO_2 emissions.

Each entrant to the competition completes an entry form which records the basic information about the company, address, nature of business, number of employees, etc. Of primary importance is the "before and after" situation ie the nature of the production problem and the technology used to effect a solution. Details of the operating costs before and after the change together with the capital cost of making the change provide an indication of the simple payback time. This is the "meat" of the entry form. For the purposes of the competition what is needed is quantifiable information rather than just a "good story". What is good for the judges of the competition is also good for the database.

There are usually a number of benefits obtained by the switch to electricity not all of which can be quantified. From a customer standpoint, energy saving may not have been the major objective but the change to electricity almost always produces one.

RECORDING THE INFORMATION

It should be stressed that the information on the forms is given in confidence. The information is therefore categorised in a standard manner. The name and full address of the customer is not included. This ensures anonymity although by the allocation of an individual record number it is possible for the administrators of the database to identify the customer. This makes it possible to seek permission for release of the customer's name or possibly to arrange a visit to the customer to see the process.

Subscribers to the PEP database receive a "runtime" disk that enables them to search the database but there are no facilities for them to add additional records of their own or to alter existing data.

As part of a plan to speed up the transfer of information to the database, this year's entry forms have been produced produced in both printed and electronic versions. Next year it is proposed to take the next step of modifying the entry form so that completed electronic entries can be added directly to the PEP Database.

SOFTWARE SYSTEM

The current version of the database uses the DOS version of "Foxpro". The database can be interrogated using a mouse to make selections and to operate controls on the screen. Most options can be operated from the keyboard but it is slower. However the next stage of development is to convert to Windows and Windows 95 versions of "Access". The objective of the change is to make the display more attractive and the operation more "user friendly".

USING THE DATABASE

The opening menu offers a choice of four ways of looking at the database:

1. Pre-defined searches
2. You choose any one field to search
3. Build your own multi-field search
4. Browse whole database.

1 Pre-defined searches

The first option "Pre-defined searches" is the most user friendly and is the preferred method for accessing data.

Having selected "Pre-defined searches" the screen as shown in figure 1 will appear.

This enables a search of the database by:

- **industry**,
- **process**/application,
- electrical **technology** employed
- **benefit**.

either singly or in combination.

```
------------------------------ Enter Search Criteria -------------------------
 SELECT BUTTONS FOR
  EXPANDED LISTS                                    SELECTED CRITERIA
                    *
  [ Industry ]                                      Industry   [        ]

  [ Process ]                                       Process    [        ]

  [Technology]                                      Technology [        ]

  [ Benefit ]                                       Benefit    [    ]

 LEAVE BLANK TO                  *
  DISREGARD                       Leave blank and enter SIC-Code -> [    ]
                                  for more exact choice of industry
 <Exit>  <Search>  <Reset>
 Use <Search> when selection is complete & <Reset> to clear selection
```

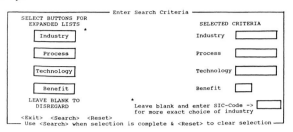

Figure 1

Highlighting any of these categories produces a "pop up" menu. Figure 2 shows the result of highlighting the "**Industry**" box..

```
------------------------------ Enter Search Criteria -------------------------
 SELECT BUTTONS FOR
  EXPANDED LISTS                                    SELECTED CRITERIA
                    *   |Aerospace        |
  [ Industry ]          |Agricul/Hortic   |         Industry   [Food    ]
                        |Boat Building    |
                        |Ceramics         |
  [ Process ]           |Chemicals/Oils   |         Process    [        ]
                        |Construction     |
                        |Electrical       |
  [Technology]          |Electronics      |         Technology [        ]
                        |Food             |
                        |Glass            |
  [ Benefit ]           |Leather          |         Benefit    [    ]
                        |Medical/Pharma   |
 LEAVE BLANK TO         |Metals:Engin'ng  |
  DISREGARD             |Metals:Foundry   | and enter SIC-Code -> [    ]
                        |Non-Manufacture  |act choice of industry
 <Exit>  <Search>  <Reset>
 Use <Search> when selection is complete & <Reset> to clear selection
```

Figure 2

Having selected say "Food" this will then be indicated on the right-hand side of the screen under the heading of selected criteria. A more precise selection of industry can be made by ignoring the "Industry" box and highlighting the SIC box at the bottom right hand corner and typing in the relevant SIC code, e.g. 15.84 for confectionery.

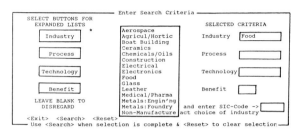

```
---------------- PEP DATABASE: DEFINED SEARCH - Screen 1 of 3 ----------------
Record No 092207
Size of site (employee band) [1-100]  Actual No     12        [  PGDOWN   ]
                                                              [Next Screen]
Industry  Food                  SIC Code (1992) 01.25

Products  Course fishing bait
Material  High protein mix balls 'boilies'
Process   Moisture removal

Original Technology
          Air drying initially followed by electric oven
Electrical Technology
     1st  Dehumidifier
     2nd

Equipment Rating 1st       9 kW   2nd         kW

  <  Top   >< Prior  ><  Next  >< Bottom ><  Exit  ><Reports ><Reselect>
Use <Reselect> to Change Search Criteria or <Reports> to View/Print Reports
```

Figure 3

A similar approach is used on the other boxes. For example if there is a requirement to search for applications involving the *drying of confectionery products* then it would be necessary to highlight "Drying/Curing/Firing" in the "**Process**" box. If there is uncertainty as to the technology that might be used or the benefits resulting then this is as far as setting the search criteria need go. Highlighting "**Search**" will initiate a search through the database. If

```
---------------- PEP DATABASE: DEFINED SEARCH - Screen 2 of 3 ----------------
Record No 092207    Plant Commissioned    01/10/91
                                                              [  PGDOWN   ]
Fuel Change  N    Original Fuel  1st     Elec                 [Next Screen]
                                 2nd
Supplier 1:    Drycool Environmental Systems Ltd
         2:

Capital Cost     Capital Saving      All Other        Simple Payback
   ()                 ()             Savings()        (months)
   16000                0                44538              4.31

Improvements in:
              Business Performance Y        Product Quality        Y
              Production Costs     Y        Energy Use/Output      Y
              Working Conditions   Y        External Environment   N

              <  Top   > < Prior > < Next   > <Bottom > < Exit   >
            Return to Screen 1 to <Reselect> or to choose <Reports>
```

Figure 4

successful then a screen as shown on figure 3 will appear. At the same time the number of records found which meet the search criteria will be shown briefly in the top right-hand corner of the screen.

In the example chosen there are six records. Each record has three screens of information as shown on figures 3, 4 and 5.

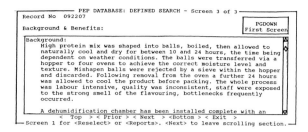

Figure 5

Screen three includes a "background" commentary. This feature really brings the information to life and gives a real feel for the company featured. Moving between records can be carried out from any of the three screens by highlighting "Prior" or "Next".

Other examples of constructing searches are:

Induction melting of iron
    INDUSTRY: select "Metals/Engin'ng"
    PROCESS: select "Melting/Holding"
    TECHNOLOGY: select "Induction"

Infra-red curing of paint on wooden furniture
    INDUSTRY: select "Timber"
    PROCESS: select "Heat setting of coatings"
    TECHNOLOGY: select "Infra-red"

What electrical techniques are used by the Glass Industry?
    INDUSTRY: select "Glass"
    PROCESS: leave blank
    TECHNOLOGY: leave blank

2 You choose any one field to search

This is useful for locating records in a particular field where a specific description is known to exist. For example an enquiry could be made for all those companies working with aluminium. The field in which to search would be "**Materials**". Typing in the first few letters of "aluminium" would then produce the required list. (figure 6).

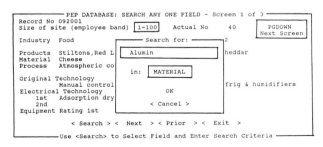

Figure 6

3 Build your own multi-field search

Equipment Suppliers

It may be that a list is required of all the suppliers of heat pump dehumidification equipment. In this case "Build your own multi-field search" should be selected. Scrolling through the selected list will provide details of all the equipment suppliers together with their addresses. A further refinement is to call for the suppliers to be listed in ascending order. This has the effect of arranging the records in equipment supplier alphabetical order. As all the records for the technique are displayed the most popular suppliers may be seen at a glance. This may be useful when deciding on a suitable company or companies to contact for further information. (figure 7).

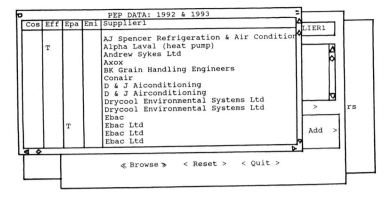

Figure 7

Statistical analysis

Although there are many well established markets for various types of electrical equipment it can be surprisingly difficult to determine details of equipment numbers, sizes, applications etc. The PEP database does provide a useful guide. It may be of interest to know, for example, the number of

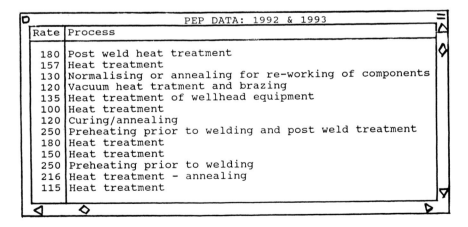

Figure 8

convection ovens with ratings ranging from 20 to 50 kW inclusive. (figure 8). Having found this basic information it is possible to add further filtering to determine for example only those ovens or furnaces used for heat treatment of steel.

4 Browse whole database

The last option on the menu is self explanatory. Bearing in mind the number of records likely to build up as the database develops, the browsing option is likely to be used mainly by the administrators of the database. Figure 9 shows just a small section of the database. In all there are about 90 fields plus the number of entries through which to scroll.

```
                    PEP DATA: 1992 & 1993
Dataset Year  Record no  Elec dist  Site size  Uk empl  Business
P       1992  092001     EMID       1              40  Traditional English C
P       1992  092002     EMID       2            5000  Concrete Structures
P       1992  092003     EEB        1             100  Industrial Building B
P       1992  092004     EEB        0                  Car Manufacturers
P       1992  092005     LE         2            2500  Rope Manufacture
P       1992  092006     LE         1              42  Processing high value
P       1992  092007     MANW       2             750  Sparking Plugs for Mo
P       1992  092008     MANW       1              10  Medical Devices
P       1992  092009     MID        2          130000  Automotives
N       1992  092010     MID        1                  Vegetable Processing
P       1992  092011     NEL        1             297  Aluminium extrusions
P       1992  092012     NEL        1             450  Ferrous and Non Ferro
P       1992  092013     NORW       2             900  Automobile Component
P       1992  092014     NORW       1              85  Sub-Contract Heat Tre
P       1992  092015     HE         1                  Powder bonded non-wov
P       1992  092016     HE         2           25000  Specialised Engineeri
P       1992  092017     SP         1                  Architectural Coating
P       1992  092018     SP         2             240  Furniture manufacture
```

Figure 9

Facilities are provided to print out hardcopy reports.

VALIDATION OF DATA

There are two schools of thought about verification of records. One is that a database should include all records of past and present applications, the other is that only those records of applications known to be in operation should be included. The latter approach can be very expensive to maintain and would undoubtedly see the loss of older records. This would then make the use of the database as a tool for statistical analysis less attractive.

In the PEP database all records of past and present applications are included. What really matters is that at the time of installation there was a valid reason for the change to electricity and that there was a sound commercial reason for its adoption.

Where a past application is of particular interest because it may well prove to be the solution to a customer's current problem then there is always the option for the database administrators to verify that the past application is still in use.

ENVIRONMENTAL IMPLICATIONS

In recent years the PEP entry form has required quantified information on CO_2 emissions. This information can be analysed using the PEP database. Indeed PEP entrants to date are collectively saving around 1MtC per year of carbon dioxide.

STEPPING STONE TO AN INTERNATIONAL DATABASE

UIE is in the process of producing an international data base. This will use a more limited selection of fields but the database will still be sufficient to identify solutions to problems and more importantly the names and addresses of the relevant equipment suppliers. Data from the PEP database and a similar input of records from France's Centre Français de L'Electricite (CFE) database will result in an initial database of over 5000 records. When the system is fully developed all UIE member organisations will be invited to contribute to the database resulting in a very powerful tool giving instant access to equipment supplier names and addresses world-wide.

Substitution of Fuel by Electricity Using Induction in Swedish Industry - Presentation of Some Cases.

ANDREJS RITUMS *Vattenfall AB, Sweden*
HÅKAN BRANDT *HEAT TECH Induction AB, Sweden*

ABSTRACT

SANDVIK BAHCO, ENKÖPING

The company is world famous for its production of Bahco wrenches since 1892, and installed its first induction hardening equipment in 1979 for large wrenches followed by a second unit in 1981 for smaller wrenches. After obtaining unsatisfactory results for the large wrenches the company decided to abandon induction hardening of large wrenches. The induction equipment for the smaller wrenches is still in operation after 15 years. In late 1995 the company replaced the fossil-fuelled hardening of large wrenches by new, more powerful induction hardening equipment.

ACCRA TEKNIK, ÖJEBYN

The company, based in the north of Sweden, was founded in 1993 and produces roll-formed steel sheet beams with complicated cross sections and continuous lengths up to 24 m. Induction heating is used in the hardening process, which is arranged in-line with the production process. The induction hardening provides the possibility of producing profiles of extraordinary high strength and small margins of tolerance which would be impossible otherwise.

LKAB, KIRUNA

The iron ore fines of the LKAB mine at Kiruna are transported 200 km by rail to the ice-free harbour at Narvik in northern Norway for shipment to Europe and the rest of the world. In winter, however, the material in the rail wagons freezes, and this causes major problems when emptying the wagons at Narvik. A feasibility study into replacing the current steam heating by induction heating has been carried out.

Remplacement des combustibles par l'électricité par induction dans l'industrie suédoise. Présentation de quelques exemples.

ANDREJS RITUMS   *Vattenfall AB, Sweden*
HÅKAN BRANDT   *HEAT TECH Induction AB, Sweden*

RÉSUMÉS

SANDVIK BAHCO, ENKÖPING

L'entreprise est connue mondialement pour sa production de clés plates Bahco depuis 1892. Elle a installé sa première machine de trempe pour grosses clés en 1979, suivie d'une deuxième en 1981 pour des clés plus petites. Après avoir obtenu des résultats de la trempe des grosses clés peu satisfaisants, l'entreprise décida d'en abandonner la trempe par induction. En revanche, la trempe par induction des clés plus petites fonctionne depuis 15 ans. Fin 1995, Sandvik Bahco a remplacé la trempe à l'énergie fossile des grosses clés par un nouveau et plus efficace équipement de trempe par induction.

ACCRA TEKNIK, ÖJEBYN

L'entreprise, installée dans le nord de la Suède, a été fondée en 1993. Elle produit des poutres d'acier à feuille enroullée aux sections complexes de longeur allant jusqu'à 24 m. Le chauffage par induction est utilisé dans la procédure de trempe, qui est une étape de la fabrication. La trempe par induction procure la possibilité de fabriquer des profilés d'une extrême résistance tout en ayant de faibles marges de tolérance.

LKAB, KIRUNA

Le fin minerai de fer de la mine LKAB de Kiruna est transporté par train sur 200 km jusqu'à Narvik, port libre de glace dans le nord de la Norvège, d'où le minerai est expédié par bateau vers l'Europe et le reste du monde. En hiver, pourtant, le minerai gèle dans les wagons, ce qui pose des difficultés pour le décharger à Narvik. Une étude de faisabilité en vue de remplacer l'actuel chauffage à la vapeur par le chauffage par induction a été faite.

SANDVIK BAHCO, ENKÖPING

INDUCTION HEATING FOR HARDENING OF WRENCHES

Background

The Sandvik Bahco company produces 1.8 milion wrenches per year. Since Bahco began to produce J P Johansson´s world famous invention in 1892, the company has produced about 95 million wrenches. To speed up the production process the company decided in 1979 to install induction heating equipment for hardening the wrenches.

Equipment installed in 1979 and 1981

The earliest installed equipment was designed for hardening large wrenches. It was a vacuum tube generator with the capacity of 160 kVA and generating a frequency of approximately 300 kHz. The equipment installed in 1981 was identical, but intended for small wrenches.

The inductor consists of 3 mm copper sheet, forming a single round, wide coil for the simultaneous heating of 10-12 wrenches. The coil induces an axial magnetic flux. The wrenches are heated to approximately 900°C and then quenched to 150°C in 4 seconds. During the first half second of quenching, the temperature of the steel must fall to below 550°C to achieve a completely martensitic structure at the surface of the wrenches. After quenching the wrenches are tempered at 350°C in a second inductor identical to the first one. After tempering the hardness amounts to 45-48 HRC (Rockwell).

The tempering generator has a capacity of 80 kVA generating a frequency of 350 kHz.

The vacuum tube generator intended for the large wrenches did not have enough heating capacity and was taken out of use within a year of installation. Instead, the large wrenches were hardened in fossil-fuelled furnaces until late 1995, when the Sandvik Bahco company reintroduced induction hardening because of the development of new, more powerful medium frequency generators.

The equipment for the smaller wrenches is the same and is still operating after 15 years.

Equipment installed in 1995

The medium frequency generator installed in 1995 is a solid state transistor-type generator with the capacity 210 kVA generating a frequency of 50 kHz. The transistor-based technology improves the overall efficiency by approximately 20% in comparison with the vacuum tube generators.

The inductor coil was originally designed to create a transverse magnetic flux in the wrench in order to achieve as high efficiency as possible. However, heating trials resulted in excessive heating of the thinnest parts (the waist). For this reason the design of the inductor coil was changed to create axial magnetic flux, resulting in lower efficiency, but with a frequency allowing heat distribution to be as even as possible in all parts of the wrench.

The generator for tempering has a capacity of 70 kVA generating a frequency of 25 kHz.

Advantages of induction heating for hardening
- Higher metallurgical quality
- The hardening is in-line with the production process
- Improved working environment; minor build-up of smoke
- More energy efficient; no heating of furnaces, heating only when needed
- Shorter throughput time
- Easier to control quenching
- Reduced treatment after hardening because of higher dimensional stability
- Easier to maintain the hardening equipment; low maintenance costs

Disadvantages of induction heating for hardening
- Tailor-made technology, which makes it inflexible
- Strong magnetic fields near the induction coils

ACCRA TEKNIK, ÖJEBYN

INDUCTION HEATING FOR HARDENING OF STEEL SHEET BEAMS

Background

The increasing customer demand for higher strength and less weight has formed the business idea of this company. To be able to offer the market steel components of high strength at a reasonable cost, these can be made of stainless or low alloyed steel sheet which is hardened to achieve the right strength. By using a new efficient production process, where the products are roll-formed and the heat treatment takes place in-line, the production process can be highly automated.

Production

The ACCRA TEKNIK company produces roll-formed steel sheet beams with complicated cross sections and in continuous lengths up to 24 m. Three TIG welding sets, 500 A each, in-line with the production process, allow for constructing tubes with non-circular cross sections.

The breadth of the steel sheet used is 70-500 mm, with a thickness of 1.5-5 mm. The steel grades normally are stainless chromium steels, SS 2304, characterized by high strength and good forming properties, even when cold, albeit difficult to weld, and low carbon boron steel (Boloc 02) with excellent welding properties, even when hardened. Boron steels can often be used without separate tempering because self-tempering occurs during the quenching when hardening.

The field of application is found for example in container and rail vehicle construction.

Induction heating

Induction heating is used in the hardening process, which is arranged in-line with the production process and after the beams have been formed. Induction hardening provides the possibility of producing profiles of extraordinary high strength and small margins of tolerance which would be impossible otherwise.

The stainless chromium steel is heated to 1070°C, holding time 3 minutes, and the boron steel to 900°C, holding time 5 minutes.

The maximum heating speed is estimated at 15 m/min depending on the shape of the cross section and the thickness of the steel sheet.

Quenching

The quenching zone is designed as a spray box. The steel profile is continuously quenched as it passes through. The quenchants are oil for the chromium steel and water or oil for the boron steel. The good hardening properties of the chosen steel grades allow for long quenching times.

Tempering

The chromium steel is tempered for 30 minutes at 180°C. The boron steel is self-tempered during the quenching process.

Induction equipment

In late 1995 a solid state transistor-type generator with the capacity of 560 kVA generating a frequency of 25 kHz and a vacuum tube generator with the capacity of 160 kVA and generating a frequency of approximately 300-500 kHz were installed. A third vacuum tube generator with the capacity of 120 kVA for increased production capacity is planned.

The induction coil induces an axial magnetic flux in the passing beam.

The 560 kVA generator is intended for heating up to the Curie temperature 769°C. After reaching this temperature, the heating efficiency at the present frequency becomes too low, because of excessive penetration depth to permit a further temperature rise. The 160 kVA generator comes into operation at a higher frequency and raises the temperature level to the hardening temperature.

Advantages of the induction hardening process
- Higher metallurgical quality
- In-line process leading to high production speed
- Continuous process which allows extreme lengths yet retaining dimensional stability
- Energy efficient process
- High strength/weight ratio because of evenly distributed tensions along the whole component
- Absence of lengthening joints improves architectural features

Disadvantages of the induction hardening process
- Strong magnetic fields near the induction coils

The quenching speed is dependent on the speed of the induction heating process

UIE XIII Congress on Electricity Applications 1996　　　　　　　　　　　　　　　　　　　　EE 65
LKAB, KIRUNA

REPLACING STEAM HEATING BY INDUCTION HEATING.
FEASIBILITY STUDY.

Background

The LKAB mine at Kiruna produces about 20 million tonnes of iron ore per year in the form of ore fines and pellets. About 5 million tonnes of fines are produced per year. The iron ore products are transported 200 km by rail to the ice-free harbour at Narvik in northern Norway for shipment to Europe and the rest of the world. In winter, however, the material in the rail wagons freezes, and this causes major problems when emptying the wagons at Narvik.

In order to reduce material freezing a heating system using steam has been employed in recent years. However, the capacity of this system is not sufficient to cope with the planned increase in production. One major disadvantage of the steam system is that it incurs high maintenance costs for electrical and mechanical equipment due to corrosion in the neighbouring plant.

Performance requirements

It was stipulated by the mining company that the heating system that would be able to replace today's steam system must be able to handle the planned average production level of 600 tonnes/hour. The temperature of the input material is stated to be +5 °C. The current steam system manages to raise the temperature of the material to +15 °C, and the temperature required to prevent freezing of material during the actual railway journey is +25 °C. The width of the belt conveyors is 1 metre, and their peripheral speed is 1.0 m/s. Since the installation is housed inside the plant, the physical dimensions of the heating equipment are an important factor.

Results

Two different efficient electrical methods were examined in the feasibility study; induction heating and IR heating (Infra Red).

Induction heating

Advantages:

- Very high power density, which means that the space required for the installation would be substantially smaller than that required for other heating methods.
- Relatively high efficiency (60-65%); electricity power requirement 5.1 MW.
- Easily adjustable, with the ability to adjust the power quickly.
- Low maintenance costs.

Disadvantages:

- High investment cost (about SEK 24 million, USD 3.6 million).
- Powerful magnetic field at the induction coils.

IR heating

Advantages:

- Lower investment cost (about SEK 12 million, USD 1.8 million) than for induction heating.
- IR equipment can be assembled directly above existing transport tracks.
- Adjustments may be carried out quickly.

Disadvantages:

- Higher operating and maintenance costs than for induction heating
- Low total efficiency (about 40%); electrical power requirement 7.8 MW.
- Requires the installation of a ventilation system.

Results

Heating fines using an efficient electrical method is technically fully feasible. Electrical methods are today rather expensive with regard to investment costs, but these methods are being developed continuously, which in the longer term will lead to lower investment costs. The results of the study have initiated a thorough and comprehensive compilation of offers from manufacturers and analyses of the possibilities of various efficient electrical methods

Energy Auditing and Its Tools

EDUARDO CARVALHAES NOBRE, *Electrical Engineer for CEMIG (Power Utility of the State of Minas Gerais) - Brazil*

LUIZ CLOVIS LIMAVERDE, *Electrical Enginner for ELETROBRÁS - Brazil*

This work presents a summary of the methodology and the process of development of **MARK IV**, a software which is part of the Energy Use Control for Cost Reduction project. This system quantifies several kinds of end uses of energy, both technically and economically, and furnishes the entrepreneur with results that will enable a thorough evaluation of the company's equipment from the point of view of energy conservation. From such outcome, every investment to be made, as well as the final results and capital earnings to be achieved, become fully known by the user.

MARK IV analyzes the following items: motors; lighting systems and installations; circuit distribution boards; transformers; air-compressors; air conditioners and refrigeration systems; furnaces; boilers; energy losses in pipelines and steam line valves. It also examines the consumer's energy bill.

On synthétise, dans ce travail, la méthodologie et le développement du software **MARK IV** (logiciel) du projet de contrôle énergétique pour réduction des coût. Ce système permet de calculer les chiffres téchniques et économiques de différents usages finales et fournir des resultats pour que l'entrepreneur puisse évaluer ses équipements quand à l'aspect de conservation de l'énergie. A partir de ces chiffres, l'entrepreneur connaîtra tous les résultats de ses investissements ainsi que son retour financier.

Le **MARK IV** analyse les aspects suivants: moteurs, éclairage, tableaux de distribution des circuits, transformateurs, air conditionné, air comprimé, réfrigération, fours, chaudières et pertes d'énergie en tubulations et valves dans les lignes de vapeur , calculent la facture d'énergie du consommateur.

1 - PROCESS OF DEVELOPMENT OF THE ENERGY CONTROL

FOR COST REDUCTION PROJECT

Some studies, which were carried out within CEMIG's realm, reveal the energy conservation as an income producing alternative of investment for the utility, over both medium and long terms.

By analyzing the impact of energy conservation in CEMIG's electric energy market, one finds that the cost / benefit ratio is of 1.55, that is, for every US$ 1.00 invested there will be a return of US$ 1.55.

An effective action in energy control for cost reduction is based on the following goals:

a) An accurate identification of the energy consumption in order to have an evolution follow-up and a selection of priorities among the possible energy conservation measures.

b) The management of the use of electricity as for time of day and season of year, seeking an adequacy to the tariff system in force;

c) Measures that should lead to an immediate economy, by means of actions that do not require remarkable investments, through the identification of opportunities for energy conservation;

d) Earnings from the income produced by the capital investments, in terms and conditions that should be compatible with those found in the money market.

The methodology adopted in the energy control project comprises the following stages:

a) Selection of the consuming units as for their energy consumption and other factors;

b) Preliminary contacts to be made with the consuming units in order to acknowledge them on the benefits and obtain their agreement to have the energy use control program performed in their installations;

c) Field survey, measurements, data gathering, etc. that will enable the energy use control;

d) The creation of a data bank, with the information collected from surveys;

e) The determination of energy saving possibilities with the help of computers;

f) Reporting to the consuming unit: letters and written reports which shall present figures relative to the savings to be earned, the steps to be taken and the capital investments to be made (if any);

g) A continued work before the Consuming Unit in order to have it develop permanent programs for energy conservation, and also to seek the maintenance and enlargement of the initial project.

The methodology employed to formulate the energy use control system was based on two segments: the first one comprises the technical analysis of the end use of energy in focus and suggests the appropriate conservation measures. The second one includes the economic evaluation of those suggestions and determines if they bear a competitive position in terms of those other kinds of investments available in the market.

2 - OPERATING APPLICATION MARK IV

The application MARK IV was developed in Clipper 5.1 and it comprises 6 (six) modes that, when integrated, can perform the functions: INPUT, ALTER and DELETE data in and from the questionnaires, check CONSISTENCY and PRINT lists, reports and letters to the surveyed consumers.

a) SMAD - DATA BANK MAINTENANCE
This service Inputs, Alters Or Deletes data gathered through the questionnaires performed in field. It also adds simulations generated outside the system (e.g.: Energy Bill Analysis, Tariff Simulation, etc.).

b) SCAD - CONSISTENCY CHECKING - DATA BANK
This function must be performed to check consistency of data entry. For doing so, a questionnaire choice must be made. A list with all End Use and any possible mistakes will be automatically produced.

c) SLDA - DATA LISTER
Make a questionnaire choice and a report with all data on the selected End Use will be automatically produced.

d) SAUF - END USE ANALYSIS
This analysis takes into account, according to the questionnaire and End Use that have been selected by the user, the energy conservation options relative to the equipment that has been surveyed. After the analysis, a technical report is produced for every selected End Use. The report brings the specific recommendations for each use. In this section the following End Uses are considered: Transformers, Distribution Boards, Electric Equipment, Lighting Systems, Air-Conditioners, Cooling Systems, Boilers, Furnaces, Thermal Losses and Air-Compressors.

e) SERF - FINAL REPORT

Before producing a final report, the user must have performed the End Use Analysis (item d). Starting from the questionnaire and End Use selection, a financial and economic assessment of the selected End Use will be performed. The report is composed of the following:

- A letter, addressed to the consumer, explaining the guidelines in the reports.
- A summarized report with the technical recommendations as per each selected End Use.
- Financial-economic analysis of the investments required for the implementation of the suggested measures. It informs the return rate.

We recommend that a list with the data collect in field be sent to the consumer along with the report,. It may be produced in the Data Lister Module (SLDA) and the technical reports are issued by End Use Analysis Module (SAUF).

f) SUTI - UTILITY SERVICE

This module is composed of six options and is intended to facilitate the operator's work:

- IMPORT - It requires the user to insert a floppy disk in drives. When ready, the program verifies if the questionnaire already exists in MARK IV's environment. If not, it will be imported;
- EXPORT - Such option will make it possible to copy a questionnaire into a floppy disk;
- CLEAR - It deletes all questionnaires from MARK IV's environment. There are several levels of safety. It is a very useful option for files maintenance, after a back-up has been performed;
- SUPPORT FILES - It prints support files for routine conference;
- DISPLAY - It permits a verification of the several reports generated by the modules that have been performed in the environment;
- PRINT - It prints the files generated by the system, that is, Consistency Checking, Data Lister and both the Technical and Economical Evaluations.

3 - SIMULATIONS AND ANALYSIS MADE BY MARK IV

The program performs the following analysis:

- TRANSFORMERS - The software analyzes the voltage unbalance, the current unbalance, overloads and taps adjustment. It also evaluates both under-loaded and over-dimensioned transformers (transformers of up to 1,500 kVA).

- CIRCUIT DISTRIBUTION BOARD - It examines both measured and nominal voltages, unbalanced voltage and unbalanced current, bus bars, insulators or connections found in bad conditions, cables that are in precarious conditions, misplaced circuit distribution boards, any inadequate installations, vibrations, any material found in a precarious state, the lack of grounding, open circuit boards, circuit boards subject to an aggressive environment. The program only performs a quantitative analysis..
- ELECTRIC EQUIPMENT - It examines overloaded motors, the motor is dimensioning in terms of the load, if feed voltage is lower than plate voltage, the motor-machinery transmission system, any electric connections in bad conditions, the actual insulation of energy cables and the general conditions of safety.The motors' dimensioning in terms of the load is limited to equipment of up to 250 hp.
- LIGHTING
  - ◊ Analysis of the environment and room cavity.
    - ∗ It analyzes the environment color, the height of fixtures, the natural light input (zenithal and lateral), the cleanliness of natural light input (zenithal and lateral), the kind of atmosphere in environment, the frequency at which environment is cleaned and cleanness of environment.
  - ◊ Lighting Installations
    - ∗ Analysis of fixtures.optical features, the need for a diffuser, kind of reflector and distance between fixtures;
    - ∗ Analysis of lamps - power, color reproduction rate and space between fixtures;
    - ∗ Analysis of reactors - type, power and power factor.
  - ◊ Further information
    - ∗ Lighting Input - Levels recommended by NBR 5413 (Brazilian Regulation Norm), natural light, artificial light and natural and artificial light input;
    - ∗ Utilization of Environments - Times when utilization starts and ends and the utilization during peak-hours;
    - ∗ Disconnection at Lunch-Break - Conditions and duration of disconnection.
- AIR- CONDITIONERS
  - ◊ Self Contained Window Sets
    - ∗ It assesses the average room temperature, the level of impurities in evaporator, condenser and filter, the clogging in the air outlet of the evaporator, any clogging in condenser's cooling down air flow, the adjustment of the thermostat and adjustment for

cold days, the outer air inlet for the cooling down of condenser, the direct insulation over condenser, incidence of direct sunlight into room and the sealing conditions of room.

- ◊ Central Air-Conditioner / Direct Expansion
  - * Generation / Distribution - It analyzes the room temperature as for the temperature adjusted in the machine, the equipment performance in terms of the temperature differential in the evaporator, the excess of outer air, level of insulation, the heat source in back-flow of air, any leakage of air, the cleanliness of both condenser and evaporator, icing conditions of frosting in the refrigeration circuits, the thermostat adjustment, the adjustment of outer air entry during cold days, the coupling conditions of motor and fan, any clogging in insufflation, temperature of air at the last insufflator's outlet, the protection against solar rays incidence, sealing conditions of the environment, the room temperature as per sector and the temperature differential in the evaporator per each equipment.
  - * Water Cooling Systems / Condensers - It estimates the Condensing Systems on what refers to the electric binding and water leakage, the cooling tower on what refers to any clogging in the air flow and the chemical water treatment as well as the overall performance of the condensation system. Only the direct expansion systems will be analyzed.

- REFRIGERATION SYSTEMS
  - ◊ Generation / Utilization
    - * It analyzes the frosting on evaporators and pipes, the storage system, the relation between the measured and the recommended temperatures, the proximity of heat sources, the incidence of solar rays, the type of lamp used in cold storage room, the lack of a thermostat, the lack of an air forcer fan, the automation of inner lighting system, clogging and clearance in condensers, the lack of a safety collar in air forcer fan, the decentralization of the helix in the air forcer fan, the alignment of motor / compressor set and the existence of any leakage in compressor. It also checks if the compressor has been installed above evaporator's level, the lack of an oil separator in compressor's outlet, the lack of closure of cold rooms or isles at the end of working shift;
  - ◊ Water Cooling Systems / Condensers

* It evaluates the temperature differential in condensers and towers, the loss of load in condensers, the presence of a manometer in condensers, malfunctions in the manometer located in condensers, the electric binding among towers, pump and compressors, any temperature higher than the nominal temperature in condensers and towers, any temperature lower than the nominal temperature in condensers and towers, the installation of a thermostat at the towers, any clogging to the air flow in the towers, the chemical treatment in towers and any leakage in the hydraulic circuit. Only the direct expansion systems shall be assessed.

- AIR COMPRESSORS
  * It checks the position in which the suction pipe in the compressor has been installed, the existence of a suction filter, the cleanliness conditions of filters in compressor, the cleaning frequency of air filters, the auto turn-off pressure adjustment, the suction air temperature, the general conditions of motor / compressor set, the general conditions of transmission in motor / compressor set, gradient of the compressed air networks, the existence of a purger, tube drainage, the general conditions and leakage in the connections, junctions, quick connections, the general plan of installation, the coupling of the secondary branches and, finally, the general conditions of the installation.

- FURNACES
  * It estimates the thermal efficiency based on the relation between the heat absorbed by the load and the supplied heat, the specific consumption - information to be stored in data bank for further consultation and use.

- BOILERS
  * It evaluates the burning that presents an elevated amount of air, the thermal efficiency, the combustion exhaust gases, any losses through the boiler's sidewall, lower steam production levels - idle capacity, nonexistent or insufficient insulation, recuperation of condensate, thermal insulation in condensate return, the utilization of the heat from condensate, the maintenance routine of boiler and accessories, feed water treatment, training of the operators and also the bed-load discharge outlet.

- ENERGY LOSSES IN STEAM NETWORKS
  * It calculates the energy conservation to be achieved from the correction and realization of an efficient thermal insulation.

- ENERGY BILL ANALYSIS
    * It analyses the specific consumption of energy, the load factor / average cost of kWh, the optimization of the power demand, the tariff adjustment and optimization, any corrections in the power factor and a reduction on the compulsory loan.
- ECONOMIC EVALUATION
    ◊ Financial Flow - It evaluates the financial flow as per end-use, the global investment flow and also the income and expenses simulation in terms of the global use, and it analyzes as the inclusion of measures processed out of MARK IV as well;
    ◊ Economic Analysis - It estimates the Annual Uniform Value simulation as per global use, calculates the income producing rate, analyzes the Pay-back simulation as well as the investment sensitivity and risks;
    ◊ Parameters Used in Analysis: - Exchange Rate, Minimum Appeal Rate, Price of Electric Energy and Useful Life;
    ◊ Consumer's Report - A letter which brings a summary of the financial evaluation of the energy conservation potential and the managerial report, along with a list of the conservation measures (for each use) and the corresponding financial evaluation. It also contains the technical report of the end use analysis and a list of the data gathered in field.

5 - OTHER TOOLS:

- **ACE** -.Besides the tools destined to perform the energy auditing, we can also use a program of tariff analysis called **ACE**. It fits the electric energy supply contract realized with the local power Utility to the consumer's demand. The Brazilian tariff system has its regular modalities as well as its seasonal / time-of-day tariff. That means a higher price for both energy and demand during a period of three hours inserted between 5:00 p.m. and 10:00 p.m..A CONVENTIONAL modality is presented to the consumers that are supplied in high voltage and use a power of up to 500 kW. It has only a single value of tariff for the demand in the whole period and another unique value for the consumption in the entire period. The GREEN tariff modality has a single tariff for the demand in the whole period and differentiated values for peak and off-peak consumption. The peak consumption tariff is around ten times greater than the other one. The BLUE tariff modality is the most convenient to the consumer that is able to modulate the demand at the peak time, because such modality bears values that are differentiated for both consumption and demand at peak and off-peak hours. The computer program **ACE** makes it possible to perform the simulation of the bill in three different modalities. It also informs the user the most suitable tariff modality for every month and the value of the bill in each specific case. It is a very useful tool for consultants and industrial managers, when decisions are to be made.

- **POWER FACTOR** - The tool for the correction of POWER FACTOR is another one that is very helpful to the user. With the monthly input of the active and reactive energy values, the program determines the values for the CAPACITOR BANK that are necessary to increase the amount of PF to the minimum level of 0.92, according to the regulation in force. For the year of 1996, when the power factor shall be determined by the hourly average, the program will have an upgrade in order to perform the determination of automated capacitor banks.

- **MAECE** - This program evaluates, in a simple and fast way, the economic gains realized from the use of energy efficient equipment and systems. The program allows:

  ◊ The calculation of the economic viability of the purchase of more efficient equipment and systems;

  ◊ The assessment of the return produced by the replacement of an installation or existing equipment;

  ◊ The calculation of the annual reduction of expenditures on the electric energy consumption;

  ◊ The consultation of economic evaluations performed in practical experiences, out of the software menu.

  Based on some information to be gathered such as the electric energy tariff, the utilization factor of equipment, prices, the power capacity and the useful life of the equipment, it is possible to determine several economic indexes as, for instance, the annual gains, the reduction of annual expenditures on electric energy, the pay-back time, profitability and other parameters. It also helps to evaluate the advantages of purchasing of an equipment or system that consumes less electric energy.

UIE XIII Congress on Electricity Applications 1996   EE 77

Industrial Electricity Use
Characterized by Unit Processes
A Tool for Analysis and Forecasting

M. SÖDERSTRÖM

Energy Systems Division, Department of Mechanical Engineering,
Linköping Institute of Technology, Sweden

ABSTRACT

Unit Processes could be characterized as the building bricks of industrial energy use. Unit Processes are defined with regard to their aim e.g. forming, drying, joining and pumping.

In analysing industrial plants, Unit Processes could be used to facilitate comparisons between plants in the same branch or comparisons between branches at the Unit Process level.

Using Unit Processes the forecasting of industrial plant energy use can be made in a structured way. This also means that statistical methods can be used to get forecasts at aggregated levels such as branch and industry level. Information on industrial processes and their development could preferably be collected and put in a database where the information is characterized by Unit Processes.

RÉSUMÉ

Les procédés unitaires sont les briques de construction de l´utilisation industriel de l´électricité. Les procédés unitaires sont definie par leur but par example faconnage, sèchage, assemblage et pompage.

Les comparaisons entre des établissements industriels dans la même branche où entre les branches industrielles sont simplifiés par l´utilisation des procédés unitaires.

La structure donne par les procédés unitaires peut amèliorer les pronostics de de l´utilisation industriel de l´électricité. Les résultats au niveau d´établissement peuvent donc être agregée au niveau de branche industrielle. On peut stocker l´information concernant les procédés industriels et leur développement selon les procédés unitaires. Les avantages comprisent comparaison simplifiè et retrouvé simple d´information être originaire des sources varièes.

THE UNIT PROCESS AS AN ANALYSIS TOOL

The Unit Process concept is one way of dividing industrial energy use in smaller parts. This division makes it possible to reach at a well defined structure and thus a uniform treatment when analysing the energy use of the industrial plant. The Unit Processes are defined starting out from the aim of the industrial process; to mix raw materials, to cool or dry a product, to make pressurized air or to transport goods. Concludingly, the Unit Processes are the components which together constitute the industrial energy use.

We distinguish between eleven production processes and eight supporting processes.
The eleven production processes are: Disintegration, Mixing, Cutting, Joining, Surface Spreading, Forming, Heating, Melting, Drying & Concentration, Cooling & Freezing and Packaging. The eight supporting processes are : Lighting, Compressor (Air), Ventilation, Pumping, Premises Heating, Premises Cooling, Tap Hot Water and Internal Transports.

AUDITING

The basic idea of organizing auditing work is that the set of processes used by a customer is typical for the economic activity of the customer which means that one may concentrate on a few large energy users; the type equipment of that economic activity.

For every equipment the following data is needed:
- Installed Power
- Power Utilization Factor
- Annual running Hours

Where the Power Utilization factor is defined as the average power demand when the equipment is run divided by the installed power demand for the equipment in question.

The type equipment at individual customers are sorted according to Unit Processes. This means that it will be possible to aggregate not only branches but also across branches for the same kind of type equipment.

We use a case study from the plastics industry. The schematic production flow of this plant is presented in Fig.1. In this figure is also shown the Unit Processes (*in italics*) corresponding to the equipment of this plant. The annual energy use and its distribution on the Unit Processes is shown in Table 1.

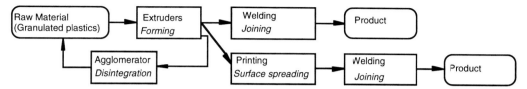

Figure 1. Production flow, equipment and corresponding Unit Processes of a plastics industry.

Equipment	Agglomerator	Extruders	Welding Machines	Printing Machines
Unit Process	*Disintegration*	*Forming*	*Joining*	*Surface spreading*
Installed Power [kW]	125	640	100	10
Power Utilization Factor	0,48	0,44	0,6	0,6
Annual Running Hours	1150	4970	3700	4970
Material Flow [ton per year]		1740		
Annual Energy Use [MWh]	69	1400	220	30

Table 1. Annual energy use for the production processes of the plastics plant in Figure 1.

When having performed a number of audits in the plastics industry the conlusion is that the following equipment is typical and is thus the equipment on which to concentrate the auditing work. The type equipment of a plastic products industry is: Injection moulding, air compressor, transport system, ventilation and lighting. These correspond to the following Unit Processes: Forming, Compressor (Air), Internal Transports, Ventilation and Lighting.

COMPARISONS

The Unit Process is a general concept for all branches of industry. This enables comparisons to be made of a given Unit Process e.g. forming between two or more branches. Unit Processes can also be used to compare between industrial plants within the same branch. A Unit Process - Plant matrix is created. An example of such a Unit Process - Plant matrix is shown in Table 2. In this case the matrix is filled with data on the specific use of energy. Alternatively such a matrix could be filled with data on the kind of energy used, the process technology used, the energy savings possibilities etc.

Plant No	Unit Process [Wh kg^{-1}]		
	Mixing	Forming	Compressor (Air)
1	28	139	106
2		804	43
3		1330	148

Table 2. A Unit Process - Plant matrix for some plastics industries.

Results of this kind raises such questions as:
- Are the differences due to treating different material qualities?
- What is the maintenance status of the plants?
- Are there differences in idle time of the plants?
- Are there old and worn-out equipment?

Comparisons could be made in the same way for branches. Parallell questions will then be raised and additional questions related to the technology transfer possibilities between branches.

THE UNIT PROCESS AS A FORECASTING TOOL

The process presented here is based on the recognized need for in depth information on industrial energy to make a forecast on industrial energy use. [1]

In the forecasting of the individual plant two kinds of development must be taken into account: equipment development and customer development. Such change information could be found when questioning the management of the company, from previous knowledge of other customers, from previous experience etc. The basic structure of forecasting is depicted in Fig. 2. The starting point of forecasting is the present energy situation shown as Industrial Plant Energy Use. Unit Processes develop when a customer adopts to changing energy prices and other changing boundary conditions. This is done either by improving existing technology or by installing new technology. This natural process may be accelerated using various strategies such as DSM.

Equipment development is divided in Technology Improvement and Saturation Change. Technology Improvement is the technical possibilities to improve the efficiency of the Unit Process and the Saturation Change is if the individual plant is going to invest in such new technology or not.

The customer development is divided in change in Power Utilization Factor, i.e. how much of the installed power that will be used, and the change in number of Annual Running Hours.

The Forecasted Industrial Plant Energy Use is calculated in the following way for every Unit Process:

Forecasted Industrial Plant Energy Use Unit Process 01 =
= Industrial Plant Energy Use Unit Process 01 * TI01* SAT01 * PUH01, where

TI01 is the Technology Improvement factor,

SAT01 is the Saturation change factor,

PUH01 is the Power Utilization and annual Hours change factors.

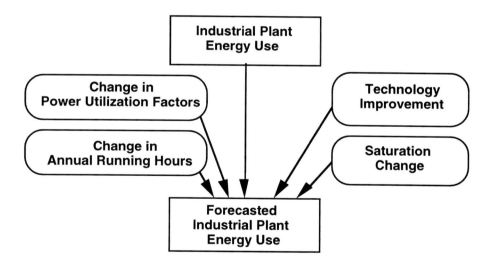

Figure 2. From input data to the Forecasted Industrial Plant Energy Use.

The forecasting result could be saved partly or in its entirety. To simplify the running of many alternative forecasts, the calculations should be made in such a way that it will only be necessary to save the change information itself for the Forecast Year.

AGGREGATION

The aggregation of individual plants to branches and further on to the entire industry must be based on a large number of audited plants and well structured division of their energy use. For every branch the Type Plant, defined at quite a high level of detail according to economic activity, is created. These Type Plants may then be forecasted in the same way as was the individual plant above. The results from Type Plants may the be further aggregated to the branch level using information on the growth and decline of industrial branches.

INFORMATION COLLECTION, STORAGE AND RETRIEVAL

Industrial audit information and data for use in forecasting should be collected using uniform questionnaires and result tables suited for direct data base treatment. This data base could then be used to compare plants, branches etc. if the information is structured, preferrably according to Unit Processes. Information from other sources such as literature or manufacturers of production equipment could be stored in the same manner to enable exchange of information.

DATABASE CONSTRUCTION

Experience from the application of the unit Process concept point at a need for structured storage of the information acquired. The data base suggested consists of two parallell parts, one for industrial plant data and one for literature data. The two parts are constructed in a similar way and it is possible to search for information from one parti n the other. A relational database is used.

The Industrial Plant Data Base is built from a number of individual tables related to each other in a way shown in Fig. 3. The relational structure means that a piece of information is stored at one place only in the database and could be combined with all other pieces of information using key fields (bold in the figure) in the individual tables.

The Literature Data Base is constructed to contain the parallell information but the structure is not so complicated.

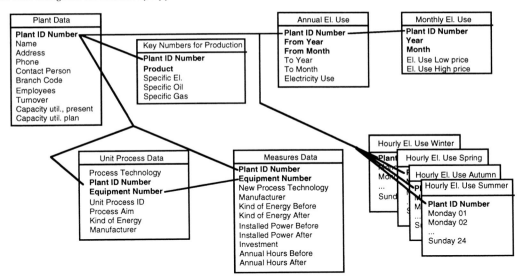

Figure 3. Tables and relations in the Industrial Plant Data Base

CONCLUSIONS

One way of studying the possibilities for improvement in individual processes and entire production systems is to divide the energy use in relation to the aims of production.

Through this structured information with the high degree of detail a number of phenomena could be studied such as the effects of Demand Side Management and the effects of increased production in certain processes. Furthermore the knowledge on the actions and reactions of the industrial electricity customer increases, which may be of paramount importance in the competition on the energy market.

The Results from forecasting based on the Unit Processes could be compared with results from other forecasting models e.g. econometric models at the national level

REFERENCES

1. A. Kahane. "Technological Change and Industrial Electricity Use", *"Electricity, effiient end-use and new generation technologies, and their planning implications"*, pp. 489.502, T.B. Johansson, B. Bodlund and R.H. Williams Eds. , Lund University Press, 1989.

UIE XIII Congress on Electricity Applications 1996   EE 85

Overall Energy Management in Small and Medium Scale Industry

PEKKA EEROLA, MSc; SEPPO LEHIKOINEN, MSc; JUSSI SIPPOLA, MSc

Research Department, Imatran Voima Oy, FIN-01019 IVO, VANTAA, FINLAND
Tel +358 0 85611, Fax +358 0 5636 823

1 ABSTRACT

This presentation covers the possibilities and methods of total energy (electricity) management in the small and medium scale industries. The standard practice in the small and medium scale industries is to optimise each sector (unit) individually (ventilation, space heating, production energy). Each unit operates <u>alone without any interaction</u> with others and even imperfectly, at times, in view of <u>total energy management</u>.

Through the use of electric heating and the utilisation of free heat given off by the production machinery and by applying modern technical equipment and careful planning, Imatran Voima Oy's experiments show that in the small and medium scale industrial halls the actual specific consumption of heating has been ranging between 5 and 15 $kWhm^{-3}$ per annum. The corresponding fuel-heated units, e.g., oil-heated systems, consume an average of 40 $kWhm^{-3}$ per annum.

1 SOMMAIRE

Cette présentation traite les possibilités et les méthodes de gestion totale de l'énergie (l'électricité) dans les petites et moyennes industries. Dans les petites et moyennes industries, il est courant d'optimiser chaque secteur (unité) individuellement (ventilation, chauffage, énergie de production). Chaque unité fonctionne <u>seul sans aucune interaction</u> avec les autres et parfois même imparfaitement étant donné <u>la gestion totale de l'énergie</u>.

Les expériences faites par Imatran Voima Oy prouvent qu'en utilisant le chauffage électrique et la chaleur libre émise par les machines de production, en appliquant l'outillage technique moderne et en planifiant les travaux soigneusement, la consommation spécifique réelle du chauffage des salles industrielles des petites et moyennes industries a varié entre 5 et 15 $kWhm^{-3}$ par an. Les unités correspondantes chauffées aux combustibles, par exemple les systèmes chauffés au mazout, consomment une moyenne de 40 $kWhm^{-3}$ par an.

CONCEPT OF TOTAL ENERGY MANAGEMENT

The total management is a concept of different methods as presented in Figure 1.

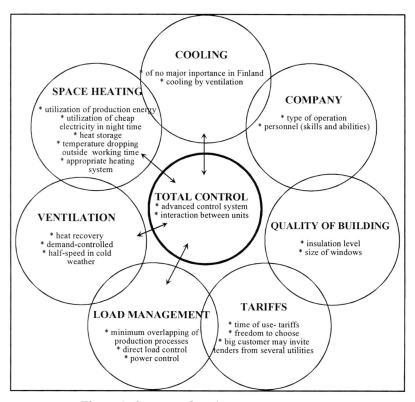

Figure 1. Concept of total energy management.

WHAT ARE THE METHODS OF TOTAL ENERGY MANAGEMENT?

- minimising the overlapping of all the electrical production processes so as to cause as small a power peak as possible
- choosing the appropriate tariff to match a company's operations
- using power regulator to prevent power peaks and to keep the electric connection small
- using appropriate heating systems and efficient heat recovery systems
- using an advanced control system to total management and preventing conflicts between separated systems (ventilation, heating)

BASE ELEMENTS

QUALITY OF BUILDINGS

Due to Finland's northerly location, its construction technology is advanced and insulation levels good. Modern industrial halls built with prefabricated sections are tight, and their calorific value is reasonably low. Also, the windows are generally small.

Therefore, the need for heating power required to replace building conductivity and leak losses is less than 10 Wm^{-3}. In industrial production processes, the amounts of free heat released are considerably higher.

PRICING OF ELECTRICITY

The exact pricing (tariffs) depends on the local utility. However, there are similar features among utilities in the pricing of electricity for industry. Typically, tariffs are composed of four types of fees: connecting, base, energy and power.

The connection fee is based on the cost of constructing the power line from the nearest supply point to the building. It is paid only once. The base fee generally depends on the size of the main fuses in the building. It is paid regularly once a month or once a year. The energy fee depends on the time of use of electricity. The power fee is split up into two parts: active and reactive power charges.

In Finland, effective 1st of November 1995, companies with power peaks of over 500 kW have the opportunity to buy their electricity also from another utility than the local one. The company is also at liberty to choose the preferred tariff, but often there are just very few real alternatives.

COMPANY

In larger companies with the personnel answering for the maintenance of facilities or plant services, the matters are normally well taken care of. Especially at large industrial plants, there is usually at least one person who is fully familiar with the company and its operations. On the other hand, in smaller companies, the energy know-how is often minimal.

LOAD MANAGEMENT

An important fact is to minimise overlapping the electric production processes so that all of them together cause as low a power peak as possible during the power measuring hours. Especially the starting up or warming up periods of machines are often problematic. One should always keep in mind, however, that production is the most important thing in making money, not the peak control.

Figure 2 presents the 24-hour use of electricity (power) in diagram of an actual factory. In this case, the warming up of three large electric machines was carried out without observing the influence of the machines on the peak power.

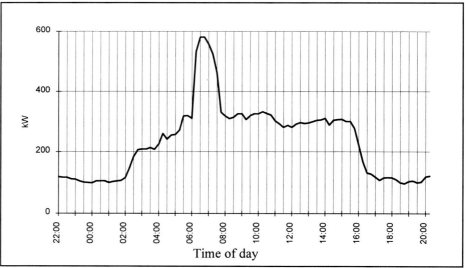

Figure 2. *The power peak caused by three large production machines. Power curve presents 15 minutes average power.*

In this case, the production would not suffer, although warming up periods of devices take place outside the power peak measuring hours (e.g. before 7 a.m.). If the entire power peak of these machines could be transferred, just the power peak fee would be decreased by over FIM 100,000 per year.

Another possibility is to control the load by a power regulator. The load control may have a notable effect on the size of the electric connection in an electrically heated industrial building. A power regulator limits the electric power by disconnecting predefined electric loads, when necessary. Controlling the heating power has no effect on the inside temperature, because production machinery emits heat.

One condition of power regulation is that there are loads which can be controlled. Space heating is this kind of load. The direct control of production machines and processes is often difficult without interfering in the normal production. In an indirect control, for example, a warning lamp indicates to the personnel of the company that there is a need to cut loads.

VENTILATION

The need of ventilation is based on the characteristic of the company production and the government regulation. Ventilation is needed to remove impurities and to cool down the working space during summer months. Ventilation also plays an important role in the balancing of temperature differences in high industrial spaces by moving the warm air at the ceiling level back to the working level with the help of recirculation or heat recovery systems.

In electrically heated industrial halls, the ventilation system is normally based on air displacement, which allows for the elimination of impurities from the air. So-called mixing ventilation is only used only in very clean production spaces, such as assembly halls, offices and staff amenities.

When electric heating is used, the ventilation system is nearly always equipped with a heat recovery unit, which recovers most of the energy included in the discharge air and heats the intake air. This saves energy. Feasibility studies of heat recovery units should not be based on mere energy costs, but savings in connection fees and power fees must also be considered.

The efficiency of heat recovery units ranges between 40 % and 80 %, in some cases even more. In production spaces, it is advisable to use heat recovery units with high efficiency; these are capable of making full use of the free energy from the machinery.

There is also a possibility, by government regulation, to switch ventilation to half speed if the outdoor temperature is 15 degrees or less below the design temperature of that area. The design temperature near Helsinki is about -26 °C. The use of half speed reduces the power peak and thus saves energy. The effect of half speed decreases if there is a heat recovery system in the ventilation.

SPACE HEATING

The larger the number of machines and devices used, the more heat they emit to the working space.

The most important issue in the planning of electric heating is the full utilisation of free heat generated by the process. An energy system is economical if electricity is put into use twice. The first time, electricity is used in production and, the second time, the free heat generated heats the building.

In electric heating, heat adjustment has a significant impact on the heating costs, due to the different pricing of electricity during night and day. It is sensible to consume more heating energy during the night. In heat-storing solutions this is normal, but also in the so-called direct electric heating systems the structures can be utilised.

The need of electric heating can be optimised by allowing the indoor temperature drop outside the working hours and by bringing the temperature back to the desired level, before the working day starts (figure 3).

Optimisation can be carried out by using three different temperature levels:

- temperature during working hours
- lower temperature outside working hours
- raised night-time temperature before working hours in proportion to the outdoor temperature.

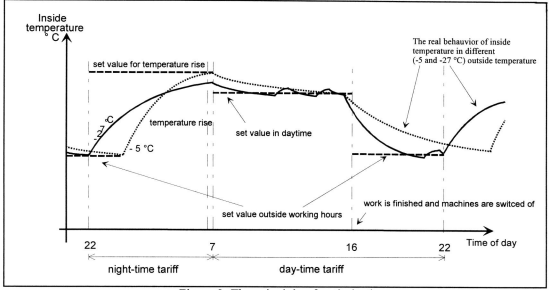

Figure 3. The principle of optimisation.

By raising the indoor temperature above the normal set value in proportion to the outdoor temperature just before the working day commences, night-time electricity is used to store heat in the structures and machines. The time to start raising the temperature is defined on the basis of outdoor temperature in such a way that the set temperature is reached just before the working day starts. While the stored heat is discharged in the course of the morning, the production machinery reaches its final operating temperature and starts releasing large amounts of free heat.

SPACE HEATING SYSTEMS

The system most used in the small and medium scale industrial buildings comprises radiators and heat blowers. Other alternatives are various heat storage and hot air in tube systems.

The heating systems by radiators start heating the surface of the construction materials and the production machines. After that, the heat starts to heat the air. The radiator heating systems apply very well to optimise heating, due to their partial heat storage property. The heat storing capacity makes it possible to utilise the cheap energy in the night-time.

Heating by heat blowers is technically simple, and the investment costs are low. The air temperature is raised quickly, but there are clear temperature layers.

ADVANCED CONTROL SYSTEM

A standard practice in the small and medium scale industries is to optimise each sector (unit) individually (ventilation, space heating, production energy). Each unit operates <u>alone without any interaction</u> with others and even imperfectly, at times, in view of <u>total energy management</u>.

One example about the need of co-operation is the interaction between the space heating and the heat recovery system of ventilation. If there is overheat in the production space, the first thing to do is to switch off space heating and, if that is not enough, only then to adjust lower the coefficient of efficiency. Or, vice versa, if it is too cold, the first thing is to adjust heat recovery to its maximum coefficient of efficiency and only after that to switch space heating on.

The conflicts and the lack of interaction can be avoided by controlling space heating, ventilation, heat recovery and load regulator by one advanced control system. The control system optimises heating by utilising maximum amounts of waste heat, by adjusting ventilation and heat recovery and, if necessary, by using load control in a reasonable way to total management.

There are already such systems on the market, but they are often expensive, typically over FIM 50,000. There is a need for an inexpensive control system which may be used in small industrial buildings.

One must always keep in mind that the control system must tolerate the electric interference, which often occur in industrial buildings. For example, a temporary malfunction in the load controlling device may counteract the success of the total control for the whole year, due to the nature of tariffs in question.

LOAD MANAGEMENT WITH NEURAL NETWORKS

DR. RALF KLÖCKNER, DR. REINER THOMAS
RWE Energie AG

ABSTRACT

Consumers have to be supplied with mains-borne energies, especially electricity, at exactly the point in time when the energy is demanded. This requires sufficient capacity to be available at all times. This is why customers in industry, for instance, are not only charged an energy rate for each kilowatthour, but additionally a demand rate for each kilowatt of the demand set up. Because of that, automatic load management systems are often used for load levelling.

Systems with "trend calculation", often also called optimization computers, are regarded as most advanced among the currently available systems for automatic load management. The trend calculation gives a forecast for the likely quarter-hour demand as early as the beginning of the measuring period.

The weak spot of all the systems, even the most modern equipment, is the highly inaccurate forecast for the mean quarter-hour demand of the current measuring period especially if loads fluctuate heavily. This results in unnecessary switching operations which unduly interfere with operational processes. In a research project carried out at RWE Energie it is investigated, how the use of neural networks can improve the load forecast.
First results have shown that neural notworks can in principle forecast loads and that they are generally significantly more accurate than the classical methods.

RÉSUMÉ

Ce qui est important au niveau de l'approvisionnement de consommateurs en énergies liées à la puissance, et en particulier en électricité, c'est de mettre l'énergie concernée à dispositon exactement au moment où elle est requise. Cela exige toujours une fourniture suffisante de puissance. Pour cette raison, on ne facture pas seulement un prix de travail pour chaque kilowatt-heure aux clients industriels, par exemple, mais aussi, en plus, un prix de puissance pour chaque kilowatt de la puissance sollicitée. Dans le secteur industriel, on emploie donc souvent des systèmes de contrôle de charge pour étalonner la puissance absorbée.

Les appareils dotés d'un dispositif de "calcul de tendance", que l'on appele également souvent calculatuers optimalisants, sont considérés comme les plus évolués parmi les appareils proposés actuellement sur le marché. Le calcul de tendance se caractérise par l'établissement d'un pronostic pour la puissance moyenne prévue, dès le début de la période de mesure déteminée par la compagnie d'électricité correspondante. La faiblesse de tous ces appareils, et même des plus modernes d'entre eux, est la grande imprécision du pronostic de puissance moyenne de la période de mesure actuelle pour des charges subissant des variations relativement importantes. Il en résulte des opérations de commutation superflues qui perturbent le déroulement de l'exploitation de manière inacceptable. Dans le cadre d'un projet de recherche mis en oeuvre chez RWE Energie, on étudie dans quelle mesure l'utilisation de réseaux neuronaux peut améliorer le pronostic de puissance moyenne. Les premiers résultats montrent qu'un pronostic de charge avec réseaux neuronaux est fondamentalement possible et dépasse en général sensiblement la précision des méthodes classiques.

TERMS OF REFERENCE

Consumers have to be supplied with mains-borne energies, especially electricity, at exactly the point in time when the energy is demanded. This requires sufficient capacity to be available at all times. Both the power plant capacities and the transmission and distribution systems have to be designed for maximum demand situations. This is why customers in industry, for instance, are not only charged an energy rate, which varies depending on the tariff time-of-day, for each kilowatthour, but additionally a demand rate for each kilowatt of the demand set up. In simplified terms, it is fair to say that an energy consumption as even as possible results in lower electricity costs than a highly uneven consumption with pronounced load peaks when the amount of energy is the same.

The electricity utilities usually measure the mean quarter-hour demand. Since an industrial enterprise is interested in minimizing the electricity costs, it will strive to keep the highest mean quarter-hour demand occurring within a given billing period (usually a year) as low as possible because this determines the demand-related costs.

There are countless different ways of achieving a certain mean quarter-hour demand (fig. 1).

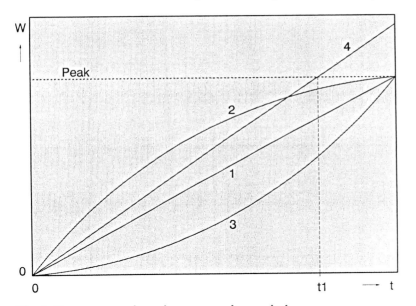

Fig. 1: Energy curve in a given measuring period

In the first three cases, the mean quarter-hour demand is the same. The fourth curve shows too high an energy consumption in the current measuring period. All consumers should have been switched off at the point in time t_1 at the latest in order not to exceed the defined maximum. Automatic load management systems serve to comply with given load limits (fig. 2).

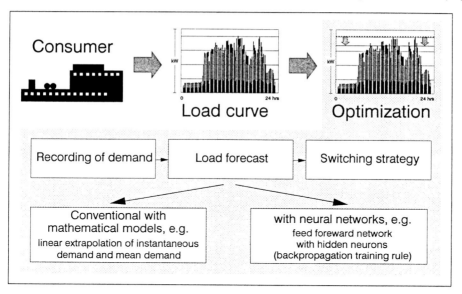

Fig. 2: Automatic load management

STATE OF THE ART

Systems with "trend calculation", often also called optimization computers, are regarded as most advanced (among the currently available systems for automatic load management). The trend calculation gives a forecast for the likely quarter-hour demand as early as the beginning of the measuring period. This forecast is based on very simple assumptions, however, and is therefore usually not accurate enough. Examples of this classical, model-based approach are:

- a) linear extrapolation of instantaneous demand to the end of the measuring period,

- b) linear extrapolation of the previous mean period demand to the end of measuring period,

- c) method according to a) or b) with additional consideration of the known time-related demand as set up by various consumers.

The weak spot of all the systems, even the most modern equipment, is the highly inaccurate forecast for the mean quarter-hour demand of the current measuring period especially if loads fluctuate heavily. This results in unnecessary switching operations (both on and off), which unduly interfere with operational processes and according to experience frequently lead to a situation where the maximum to be complied with by the load management system is set extremely high, which in turn reduces the savings effect of the system. The quality of the forecast for the mean quarter-hour demand and the switching strategy based thereupon is therefore pivotal for the performance of the load management systems.

IMPROVING THE LOAD FORECAST

The mean quarter-hour demand of the current measuring period can be forecast by neural networks. The neural network can be implemented by powerful microprocessors, e. g. signal processors.

Compared with the conventional procedures, the forecast of the current quarter-hour demand mean value is improved because a neural network "learns" to record the time-related demand set up by individual consumers, entire consumer groups and the overall demand set up by a given plant over a representative period of time. The previous mean quarter-hour demands and instantaneous demand of individual consumers, consumer groups and the entire plant are input as training data. Other relevant data, which have an impact on the set-up demand, are also provided, e. g. pressure in compressed-air networks, melt temperatures, filling level in tanks etc. Learning patterns at the back end are the associated actual current and future mean quarter-hour demands (fig. 3). Various possible switching operations are to be carried out deliberately at this stage so that their effects can also be learned. The learning phase can be continued during normal operation. In this way, the system can also become adapted to new conditions.

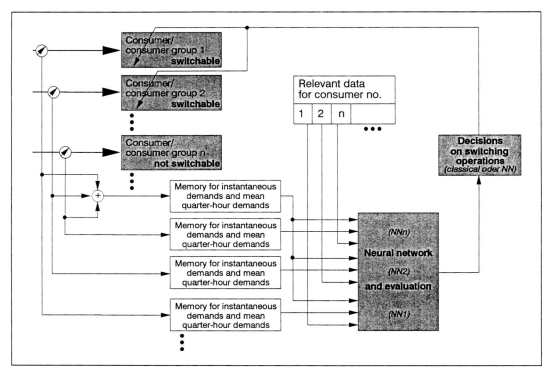

Fig. 3: Integration of NN in a load management system

CURRENT STATUS OF RESEARCH AND OUTLOOK

A programme package is currently being prepared enabling the demand set up by an industrial enterprise, controlled by a load management system, to be simulated.

This software can be used to test both classical and neural-network-based forecast methods and switching strategies with a view to their effectiveness. Pure feed-forward network structures are used first, with the back propagation algorithm adjusting the weights.

First results have shown that neural networks can in principle forecast loads and that they are generally significantly more accurate than the classical methods.

Short-time demand mean values of the current measuring period are for the time being used as input data. Area coding is used for the output data. The data available for the forecast algorithm are therefore very similar to those of the classical methods so that the various forecast methods can be well compared. Moreover, the forecast of the neural network can be improved by providing additional input data without having to resort to complex mathematical models.

The preprocessing of input data and the optimization of the network structure require further research so as to ensure versatility as is the case with the conventional methods.

UIE XIII Congress on Electricity Applications 1996

EE 101

Efficiency Monitoring Of Induction Motors Without Torque Transducers

Controle de l'efficacite des moteurs a induction sans transducteurs de couple

LEWIS, Chris - *Electricity Corporation of New Zealand Ltd*
WALTON, Simon J. and PENNY, John R. - *Innovative Developments Ltd*
NEW ZEALAND

ABSTRACT

It is well known that much of the electrical energy used in industry today is consumed by loads connected to induction motors. Therefore the efficiency with which these motors can convert energy is of major concern to users. Other side effects of poor conversion, apart from cost, can also create major problems and poor efficiency itself is often an indicator of a deteriorating or damaged motor. However it is difficult to measure motor efficiency especially when the motor is insitu. The common methods use torque transducers but these are difficult to install and calibrate to the required accuracy, especially for larger motors. This paper describes a Motor Efficiency Monitor which is based on direct estimates of motor losses and torque using terminal measured quantities and motor parameters derived from an identification process. The meter uses identification algorithms which enable it to give efficiency readings in real time or for prescribed load conditions. The influence of harmonics in the supply, time varying loads, temperature effects, and frequency and voltage variations on the estimator are discussed. The performance of the system is verified by experimental results.

RESUME

Il est bien connu que la plupart de l'énergie électrique utilisée aujourd'hui dans l'industrie est consommée par des charges connectées aux moteurs à induction. C'est la raison pour laquelle l'efficacité avec laquelle ces moteurs peuvent convertir l'énergie est au souci important chez les utilisateurs. Les autres effets secondaires d'une mauvaise conversion, hormis les coûts, peuvent

EE 102

également conduire à d'importants problèmes de maintenance et une médiocre efficacité est souvent le signe d'un moteur endommagé ou en train de se détériorer. Il est cependant difficile de mesurer l'efficacité du moteur notamment lorsque celui-ci est in-situ. Les méthodes les plus fréquemment rencontrées utilisent des transducteurs de couple, mais ceux-ci sont difficiles à installer et à calibrer au degré d'exactitude requis notamment par les gros moteurs. Cet article décrit un système de contrôle d'efficacité d'un moteur basé sur des estimations directes des pertes et du couple du moteur en utilisant des quantités mesurées au niveau de la boîte de connexions et des paramètres du moteur découlant d'un processus d'identification. Le mètre utilise des algorithmes d'identification qui lui permettent de faire des relevés d'efficacité en temps réel ou pour des conditions de charge prescrites. L'article aborde l'influence des harmoniques dans l'approvisionnement, des charges variables en temps, des effets de la température, des variations de fréquence et de tension sur l'estimateur. Les performances du système sont vérifiées par des résultats expérimentaux.

INTRODUCTION

AC induction motors are major users of electric power in world industry (estimated to exceed 50% of all industrial consumption). Major energy savings can be made through the adoption of high efficiency motor designs and the identification of poorly performing motors for replacement. The achievement of such energy savings has been hindered, world wide, by the lack of practical and easy to use test equipment to perform motor efficiency measurements and comparisons in the industrial environment.

The need and benefits of such an instrument was identified by the Energy Services group in the Electricity Corporation of New Zealand (ECNZ) and the directors of Innovative Development Ltd (a private New Zealand electronics development company). In April 1994 a joint venture between ECNZ and Innovative Development was formed to facilitate the development of a practical instrument.

This instrument has been named Motor Monitor and the development is now entering its commercialisation phase. Challenging performance specifications have been verified through direct comparison in professional motor laboratory tests and industrial trials.

ENERGY EFFICIENCY IMPLICATIONS

In a world where electric motors consume a great proportion of the electric power used by industry, the widespread use of high efficiency motors has potential for massive energy savings. Unfortunately, the adoption of such induction motors into industry has been slow due to the unavailability of reliable and quantitative information regarding the motors presently in operation, and hence of the benefits to be gained. Motor Monitor is one of the first portable instrument's available to provide the means to identify the efficiency of existing motors in industry, without the requirement to fit a torque transducer onto the motor shaft.

In the past the lack of ability to quantify the efficiency of existing motors has presented difficulties in justifying the additional capital cost of their replacement with high performance induction motors. Further, in some instances questionable claims of efficiency for new motors have been made. Motor Monitor is intended to be used to check the efficiency of both existing and new motors. Motor Monitor will provide the information to allow the calculation of payback periods which can then be used to make informed decisions. This new ease of availability of directly comparative information has the potential to have a major impact on both the local and the international motor markets.

Most induction motors consume many times their capital value in energy use annually. Thus even small efficiency gains can give rapid payback.

THE INSTRUMENT TECHNOLOGY

Today induction motor efficiency measurement is the exclusive domain of motor test houses which utilise laboratory standard dynamometers and power measuring equipment. Employing sophisticated microprocessor controlled instrumentation, Motor Monitor achieves near laboratory standard accuracy in the diverse industrial environments in which AC motors are found.

The Motor Monitor embodies specialised motor knowledge accumulated by the development team at Innovative Development over many years of developing world leading AC motor speed controllers, together with the practical requirements of a useful industrial test instrument. Considerable technical difficulties, particularly related to achieving precise measurement accuracy in non-ideal industrial environments, and complex motor modeling, had to be overcome during the course of the development.

The instrument carries out a number of measurements upon the operating motor, to acquire the motor parameters which affect loss and efficiency. This information is then incorporated into a mathematical loss model of the motor allowing motor loss components, and thus efficiency, to be calculated under any load condition.

Considerable emphasis has been placed in the development of a user friendly operator interface and the availability of results in a format that facilitates easy interpretation.

The value of the measurements cannot be realised until communicated to the motors owner. An automatic report writer has therefore also been developed as part of the Motor Monitor package. The report writer professionally presents thorough details of the motor test results in both tabular and graphical form many times faster than could be possible manually.

UIE XIII Congress on Electricity Applications 1996　　　　　　　　　　　　EE 105

MOTOR MONITORS OPERATING PRINCIPLES

An induction motor is commonly modelled using the single phase equivalent circuit as shown in figure 1.

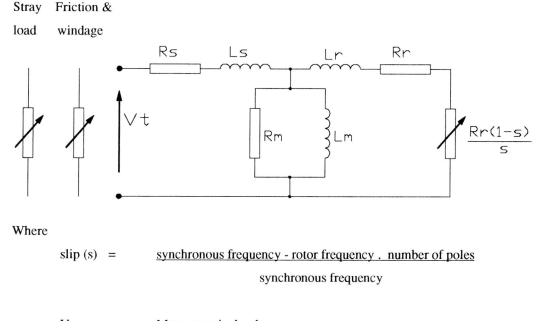

Where

slip (s) = (synchronous frequency - rotor frequency . number of poles) / synchronous frequency

Vt	=	Motor terminal voltage
Rs	=	Stator winding resistance
Ls	=	Stator leakage reactance
Rm	=	Resistance representing magnetising circuit loss
Lm	=	Magnetising reactance
Rr	=	Rotor resistance
Lr	=	Rotor leakage reactance

Figure 1: Single phase equivalent circuit of the induction motor.

Motor efficiency can be simply calculated by measuring and calculating the individual losses in each of the motor loss components and input power using equation 1.

$$\text{Motor Efficiency } (\mu) = \frac{\text{Pin} - \Sigma \text{ Individual loss components}}{\text{Pin}}$$

Where

    Pin = The motors input electrical power measured on the motor terminals

Individual loss components are:-

Stator loss	=	Loss in Rs
Magnetising loss	=	Loss in Rm
Rotor loss	=	Loss in Rr
Friction and windage loss	=	Mechanical losses through friction and windage
Stray load loss	=	Load dependent loss components which includes additional eddy current and other load dependent losses in the motor

All of the above loss components, with the exception of stray load loss, can be acquired from the motor through measurement of winding resistance, shaft speed and motor electrical input parameters. Both light (ideally uncoupled shaft) and high load (ideally greater than half load) tests are required to be carried out. Stray load loss is not able to be acquired in this way and is instead estimated using the technique defined in IEC 34.2 [1].

Using the observed loss parameters motor efficiency can be calculated for a range of operating conditions or at the actual operating point of the motor.

To verify the accuracy of the Motor Monitor instrument parallel tests were carried out using Motor Monitor and conventional separated loss techniques at an accredited motor test laboratory at Pope Motors in Australia. Comparative results for a 90 kW motors tested follow as figure 2 and show agreement in results to within the instruments claimed accuracy.

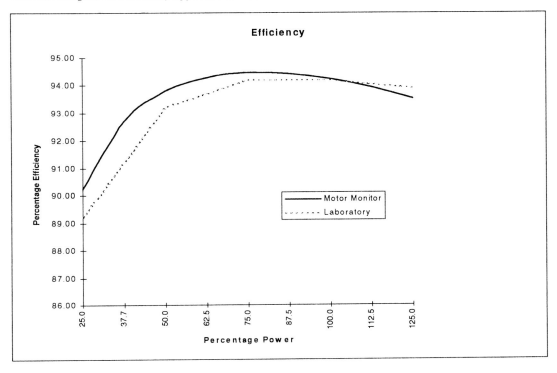

Figure 2: Comparative motor efficiency results for a 90 kW motor using Monitor and conventional separated loss measurement procedure.

PRACTICAL USE OF THE INSTRUMENT IN INDUSTRY

A primary focus taken by the developers of the Motor Monitor test instrument was practicality of its use in an industrial environment. Induction motors operating in industry are influenced by harmonics in the supply, temperature effects, supply frequency and supply voltage variations. It is also common for the motor to be subject to significant load variations when operating.

The Motor Monitor instrument contains a powerful 16 bit micro-computer that simultaneously measures all electrical parameters and shaft speed during a high or low load test. Software algorithms have been designed to provide the accurate parameter identification in conditions where there may be significant variations in the measured parameters.

The instrument digitally samples all electrical parameters at 1 kHz which means higher order power system harmonics are ignored. Real time instrument readings take into account root mean square (rms) stator loss effects of harmonic currents. Calculated efficiency curves however assume that the motors supply voltage is sinusoidal, balanced and at the magnitude of the motors name plate rating. This enables motor performance to be compared without the effects of a non-ideal supply influencing the results.

CONCLUSION

A new test instrument has been developed that allows the practical measurement of induction motor efficiency insitu. Information acquired by the instrument is available in the form of a comprehensive motor test report that can be used to make sensible decisions about motor replacement and to identify problems with motor health.

The Motor Monitor instrument has considerable commercial importance for use in quantifying the motor efficiency and identifying poorly performing induction motors in industry. Once identified information is then available to allow financially based decisions to be made on the sensibility of motor replacement with new higher efficiency motor designs. The instrument also has the significant application potential for motor rewinders who can use the instrument to verify motor performance post rewind.

REFERENCES

1. Rotating Electrical Machines - Part 2: Methods for determining losses and efficiency of rotating electrical machinery from tests (excluding machines for traction vehicles), International Electrochemical Commission, 1972.

2. W. LEONARD: 'Control of Electrical Drives',
   Springer-Verlag Berlin Heidelberg, New York, Tokyo, 1974.

3. A.E. FITZGERALD, C. KINGSLEY and A. KUSKO 'Electrical Machinery',
   McGraw - Hill, 1971.

Improving Electrical Efficiency

In Electric Arc Furnace Steelmaking

R.D. McKELLAR*, B. O'ROURKE* and R.A. REESOR**
* Ontario Hydro, Energy Services, Toronto
**Ontario Hydro Technologies, Toronto

ABSTRACT

The economic conditions of the last several years have forced industry to improve the efficiency of their operations to better their competitive position. Electrical utilities have faced the same threats as the cost of building new power plants became excessive. This was the situation in Ontario in the late 1980's. In 1990, Ontario Hydro undertook a major initiative in partnership with the steel industry to improve the electrical efficiency of electric arc steelmaking furnaces. Over the next five years, the Ontario steel industry achieved an electrical energy saving of 20 percent in the existing mills through a series of audits, improvements and the use of new technology. The electrical utility supported this work by providing technical support and information, financial incentives and rate options.

Les conditions économiques des durent plusieurs d'années ont forcé l'industrie à améliorer l'efficacité de leurs opérations à améliorer leur position compétitive. Les utilités électriques ont affronté les mêmes menaces comme le coût de bâtir nouveaux centrales électriques devenait excessif. C'était la situation dans Ontario dans le tard 1980. En 1990, Ontario Hydro entreprenait une initiative majeure dans l'association avec l'industrie d'acier à améliorer l'efficacité électrique d'arc électrique fournaises. Sur les prochain cinq années, L'Ontario industrie d'acier réalisait une énergie électrique sauvant de 20% dans les fabriques actuelles en réalisant une série de vérifications, améliorations et l'emploi de nouvelle technologie. L'utilité électrique soutenait ce travail en fournissant l'information et soutien technique, incitations financières et estiment des options.

INTRODUCTION

For many years, the steel industry in North America was considered to be in decline as part of the old industrial economy. The future appeared to be bleak as competitive pressures from the new global economy reduced or eliminated profits. However, the situation has been improved and the steel industry is now revitalized and once again competitive in the global marketplace.

The utility business has also been affected by these same global market forces. Many utilities in North America are now facing the prospect of deregulation that will end their monopoly and allow competitors to serve their traditional customers. The utility industry must be prepared to provide customers with competitively priced power along with a high level of customer service.

This situation has created an opportunity for industry and utilities to work together to their mutual benefit. When utilities work to help make their customers more competitive then it is possible for those companies to maintain or increase their market share. This in turn will ensure the long term viability of the utility. One cannot survive without the other. This has been the approach taken by Ontario Hydro in partnership with the steel industry over the past five years.

In Ontario, the electric arc furnace (EAF) is used for melting and refining in five mini-mills and for refining at two integrated steel mills. The total connected load from these furnaces is over 300 MW. At the start of this program, the average electrical efficiency of melting furnaces in Ontario was in the order of 500 kWh/ton. During the past five years, the specific electrical energy consumption of these furnaces has been reduced by almost 20 percent and productivity has been increased.

These improvements have been achieved through a continuing process involving **audits, feasibility studies, research and development support, financial incentives and alternative rates**. This paper will review the process that was used and the results achieved.

THE ARC FURNACE AUDIT

The starting point for improving the operating efficiency of electric arc furnace meltshops is to first determine the electrical energy use patterns within the meltshop. This involves electronically recording the average load on each individual furnace and piece of auxillary equipment during each 15 minute interval over an entire month. A record of production rates and any unusual occurances is simultaneously kept during the monitoring period. Important observations can be made from this information on the way electricity is consummed in relation to the production requirements.

This preliminary analysis led many of the mills to realize that economies could be achieved by operating the furnace with greater utilization. This resulted in most of the mills installing ladle refining stations allowing the main arc furnace to operate as a high powered melter. This in turn led to higher productivity from individual furnaces allowing most shops to become one furnace operations.

The next stage of the audit process involved the optimization of the electrical system. In the industry, this is commonly referred to as a `tune-up'. The audit or `tune-up' of the electric arc furnace is based on the premise that most melt shop operators do not appreciate the significance of electrode regulation. The regulation system controls the power input to the furnace by moving the electrodes up and down as the scrap is melted. This is often accomplished by maintaining a fixed current level in each electrode by controlling arc length between the electrode and the scrap charge or the molten bath. In the past, little attention was paid to this control system but recent developments have created a need for improved control. These developments include:

- the move to higher powered furnaces;
- increasing costs of electricity and electrodes;
- increased productivity of the arc furnace.

Due to these developments, a small change in regulator set point and response can cost several thousands of dollars each day in electrical losses, refractory wear, electrode consumption and production rate. To realize savings in these areas, it is necessary that a complete system analysis be done on each furnace. The savings are achieved first by calibrating the existing electrode regulating systems for the best response, secondly by balancing the power input at the tip of each electrode and thirdly, by optimizing the set points through computer techniques.

The main element of the above procedure is the computer simulation of each arc furnace and its associated utility transmission system. The computer program has been developed to identify the imbalances in the power system and generate detailed operating curves for the furnace. This information also allows the furnace `hot spots' to be moved around the furnace to allow an even meltdown of the scrap and protect the furnace walls from premature wear. These audits have been conducted on both melting furnaces and ladle refining furnaces with good results. Table I shows the average results that have been obtained on melting furnaces at Ontario mills.

Table I - Tune Up/ Optimization Results for Average Melting Furnace

	Tons/ Month	Electrodes (#/ton)	Energy (kWh/ton)	Electrodes ($/ton)	Energy ($/ton)	Monthly Savings (000's$)
BEFORE OPTIMIZATION						
On-Peak	145,000	6.89	434.2	9.03	21.71	-
Off-Peak	173,000	6.89	436.6	9.03	21.83	-
AFTER OPTIMIZATION						
On-Peak	145,000	6.31	422.3	8.27	21.12	196
Off-Peak	173,000	6.54	425.6	8.57	21.28	175

Note: Costs based on unit prices of electrodes at $1.31/pound and electricity at $0.05/kWh

FEASIBILITY STUDIES

The audits described above were conducted at the full expense of Ontario Hydro with the main objective being to identify areas for energy saving or load shifting. This has been necessary since most companies do not have the time, personnel or money to seek out these opportunities. In many cases, the recommendations can be implemented directly with little or no capital cost. For larger projects, an engineering study or feasibility study may be required. In this case, Ontario Hydro will initially pay half the cost of the study and if the project proceeds to implementation, the entire cost will be paid.

RESEARCH AND DEVELOPMENT

As audits and feasibility studies identify and evaluate more and more opportunities many ideas are developed that require additional study. Often these involve technologies or processes that are not yet used in Ontario industry or that need minor modification or demonstration to suit a particular customer need. In these cases, Ontario Hydro provides facilities and manpower to provide technical support as required. This support may make use of Ontario Hydro's Research Division (Ontario Hydro Technologies), outside research facilities or take place in the customers plant.

At Ontario Hydro Technologies, a wide range of facilities is available but in this case, the most useful is a 700 kW electric arc furnace facility. The facility provides an ideal setting for mini pilot-scale work. The overall facility is designed for maximum flexibility such as capability to operate in either AC or DC mode and it is possible to accommodate a variety of projects with only minor modifications.

ALTERNATIVE RATES

Since the start of this program, Ontario Hydro has developed several new rates that give the steelmaker greater flexibility in improving efficiency and controlling costs. The most significant was the introduction of the Time of Use rate providing much lower prices in the off peak period. The off peak period consists of weekends, holidays and an eight hour overnight period each weeknight. This rate alone resulted in a major energy shift to the offpeak as shown in Figure 1 and proved to be a very effective demand management tool. In the last two years, several other rates have been designed to utilize spare generating capacity at a lower rate whenever available on short notice. These rates have further optimized the load profile for the utility while providing cost savings to the steel mill.

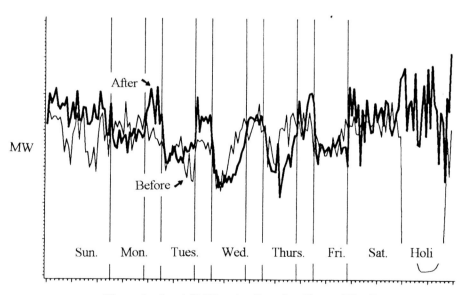

Figure 1 - Load Shifting Attributed to Time of Use Rates

FINANCIAL INCENTIVES

In many cases, the audits, feasibility studies and technical support identify an energy saving or load shifting opportunity that is technically attractive but economically requires an unacceptably long payback period to the customer. This situation was helped by financial incentive programs to facilitate implementation:

- Accelerated Paybacks - an incentive equal to $0.10 for each kWh saved during the first year of operation as the result of equipment modification or the amount needed to reduce the payback period to 1.5 years, whichever is less;
- Load Shifting - an incentive to be used for holding or storage equipment to allow shifting to the off-peak period. The incentive is equal to $400/kW shifted or the amount needed to reduce the payback period to 1.5 years, whichever is less.

IMPLEMENTED TECHNOLOGIES

During the past five years, the steel industry and Ontario Hydro have jointly invested millions of dollars in new EAF technology. The trend has been to a one furnace meltshop where a single high powered furnace is totally dedicated to melting with maximum productivity. The ladle furnace can then perform the required chemical and temperature adjustments without any delay. This change alone has improved energy efficiency by almost 25 kWh/ton. The furnaces themselves have also been modified with several add-on technologies including;

Oxy-fuel Burners

Oxy-fuel burners have greatly improved the tap-to-tap time by assisting with the scrap meltdown process. The burners can be positioned through the door, the sidewalls or the roof panel to provide heat in the furnace cold spots. The use of oxy-fuel burners can save as much as 50 kWh/ton in some cases.

Oxygen Injection

The injection of oxygen directly into the melt can improve the electrical efficiency of the furnace. Although oxygen is produced in an electrically intensive process, the exothermic energy produced from the reaction with impurities in the steel or with added carbon can more than compensate for the energy used in oxygen production. Oxygen injection also provides for the production of a foamy slag that can allow a longer arc to be used resulting in more effective energy transfer to the melt. Asian and European steelmakers use much more oxygen than North American steelmakers but the gap is narrowing. Electrical efficiency gains in the order of 25-35 kWh/ton are easily achievable.

Bottom Stirring

Several mills have experimented with bottom stirring. Bottom stirring requires the installation of a porous plug in the floor of the furnace through which a small volume of gas flows. The major benefit is a more homogeneous melt with fewer hot spots. This saves on refractory costs. A secondary benefit is electrical energy savings of 10-12 kWh/ton. The net benefit achieved at one mill where a full evaluation was done was about $0.50/ton.

Current Conducting Electrode Arms

The high currents used in the arc furnace can result in electrical losses through the Joule Effect. In the past few years, some mills have changed from the conventional copper bus bar design to the current conducting electrode arm. In these new arm designs, the current carrying area is much greater resulting in fewer losses. This can be done by using copper clad steel in the arm fabrication or by constructing the arm entirely out of aluminium. The aluminium arms not only conduct the current with low losses, they weigh much less allowing easier arc regulation.

EMERGING TECHNOLOGIES

The above measures have helped improve the specific electrical energy consumption of Ontario arc furnace meltshops by almost 20 percent but much more work needs to be done. Not long ago, a highly efficient furnace was considerd to be one with an electrical efficiency of 400 kWh/ton. In recent years, that level has been reduced to 300 kWh/ton and further improvements are expected. At this time, several technologies are under consideration for implementation in Ontario steel mills.

Substitute Fuels

Electric arc furnaces are engaged in a continuous search for the lowest cost energy supply to their furnaces. Many companies consider that the emerging furnace of the future will be a combination of electric arc and chemically derived heat usually involving higher oxygen consumption and carbon based fuels.

There are already a lot of arc furnaces directly charging anthracitic type coal and with the adoption of the foamy slag practice, the use of coal will become even more favourable. It would appear that the resulting cost saving from energy efficiency will have to be balanced against the extra emission control required for the added off-gas created by the coal additions. It appears that the future of a hybrid style furnace is promising, and that local power costs and individual steel plant environmental capabilities (baghouse capacities) will be the main factors affecting the rate of its adoption.

Post Combustion and Oxygen Injection

The increased level of interest in the use of large amounts of chemical energy in the EAF has led to advances in oxygen injection technology. Lancing technology has evolved over the years from straight oxygen injection through a steel pipe and oxygen-fuel burners to submerged sidewall tuyeres and even bottom injection. The direct injection of oxygen results in a highly exothermic reaction maximizing the energy derived from carbon, and therefore, the trend to post combustion will continue. Total electrical savings of 25-50 kWh/ton are expected.

Electromagnetic Shielding

While a great deal of money and effort has been directed at research on the process side of electric arc furnace operation, only a limited amount of research and development has been dedicated to the electrical side of the operation. One such area is the loss of electrical efficiency due to stray current losses. The high currents used in electric arc furnaces induce currents in the steel structure supporting the electrode arms and in the structural steel in the immediate area of the furnace significantly reducing the electrical efficiency of the furnace and thus increase operating costs. Recently, models for 3 dimensional steel structures have been developed to estimate the stray current losses in the electrode arm supports and the steel structure surrounding the furnace. The furnace designer can then develop solutions such as shielding of the structure or isolating loops in the structural steel to eliminate a current path.

SUMMARY

The steel industry is part of the industrial technological revolution that is taking place in Canada. Through a series of initiatives conducted jointly by the steel industry and Ontario Hydro, significant cost savings and energy savings have been achieved. During the past five years, the steel industry in Ontario has seen the average electrical energy efficiency of arc furnaces improve by almost 20 percent. This represents a decrease in specific energy consumption from over 500 kWh/ton to almost 400 kWh/ton. For the steelmaker, this has reduced steelmaking costs by over $5.00/ton and reduced average demand on the utility system by more than 25 MW. This effort has also resulted in a significant amount of load shifting to off peak periods providing additional benefit to Ontario Hydro and cost savings to industry.

There is still much work to be done as competitive pressures continue. There are many technologies under consideration for future implementation and it is expected that these technologies will again result in significant savings. The pressures facing utilities are also increasing and deregulation will require utilities to maximize the use of existing assets while also improving customer service and value. By working together, the steel industry and electrical utilities can both achieve their objectives.

Energy Saving in Industry by Efficient Use of Electricity

DE HOE JEAN-MARIE

LABORELEC

Rodestraat 125

B1630 Belgium

This text presents the brochure written by the UIE "**Energy Saving in Industry by Efficient Use of Electricity**" working group.

The 50 case studies presented in the work clearly demonstrate the variety of electrotechnologies that are available and their ability to compete in all industrial sectors on the terms of profitability as well as primary energy saving and reduction of CO_2 emissions. The Primary Energy Ratio (PER) provides a basis for comparison of competing electrotechnologies and fossil fuelled technologies through an assessment of their impact on primary energy consumption.

Ce texte présente la brochure écrite par le groupe de travail UIE "**Economie d'Energie dans l'Industrie par l'Utilisation Rationnelle de l'Electricité**".

Les 50 études de cas présentées dans la brochure montrent clairement la variété des électro-technologies et leur compétitivité dans tous les secteurs industriels, tant en termes de rentabilité que d'économies d'énergie primaire et de réduction des émissions de CO_2. Le "Primary Energy Ratio" (PER) permet la comparaison des électrotechnologies et des technologies utilisant les combustibles fossiles du point de vue de leur consommation d'énergie primaire.

Chairman: G. Claesen (B) - Secretary: J.M. De Hoe (B)

Members: D. Baggs (UK) - A. Colombo (I) - B. Nord (S) - J. Gaulon (F) - H. Hamelin (CND) - W.J.L. Jansen (NL) - R.D. Mc Kellar (CND) - T. Motohashi (J) - M.A. Porto Fonseca (Br)

EE 118

SCOPE OF THE BROCHURE

Reducing CO_2 emissions has become one of the most important and pressing challenges for energy policy. Improved generating efficiency of thermal power plants and promoting more efficient energy technologies are the key solutions offered by electricity to meet that challenge.

Avoiding wasteful use of energy is a widely recognized way of reducing CO_2 emissions. It can be achieved for example by better insulation of domestic, commercial and industrial buildings, installation of energy efficient lighting and by good house keeping in industry.

However within industry the greatest opportunities for energy saving are likely to come from the adoption of more efficient technologies. This includes making greater use of electricity to reduce CO_2 emissions.

Electricity is the most refined form of energy and can be produced from the widest variety of primary energy sources. It can be applied more efficiently at the point of use, than fossil fuels, and this often more than offsets energy losses incurred in its production. Electricity is renowned for its cleanliness and control, and offers the widest choice of techniques.

These strengths allow electricity to compete with fossil fuels by reducing industry's operating costs, even its energy costs and primary energy requirements. In some cases, the investment for an electrotechnology is also cheaper than the one for the alternative with fossil fuel.

The brochure takes a look at some of the issues that need to be considered in putting the evaluation process into practice. It shows how the energy efficiency of alternative technologies can be compared and how to assess the impact on CO_2 emissions. The choice of electrical techniques is discussed and examples, drawn from the member countries of UIE, demonstrate how changing from a fossil fuel to electricity can benefit both industry and the environment. Reconciling the needs of industry and society doesn't have to be a painful process.

COMPARING PRIMARY ENERGY CONSUMPTIONS

All countries have a different mix of generating plants, which is influenced by available energy resources and many other techno-economic constraints such as capital, labour cost and plant reliability. As a result, the primary energy needed to produce 1 kWh of electricity varies from one country to another. A generation efficiency of 38% - generation efficiency of thermal power plants - has been assumed for all the case studies presented in the brochure, in order to allow for a uniform presentation. This means that all these cases are at least examples of primary energy saving on this basis.

PRIMARY ENERGY RATIO

An important value for comparing the efficiency of electrically powered processes with those supplied with fossil fuel is the **Primary Energy Ratio (PER)**. This is the ratio of the primary energy consumption of a process requirement met with fossil fuels to the primary energy consumption of the same requirement partially or totally met from electricity.

$$PER = \frac{W \text{ (MJ}_{thermal})}{3.6 \, E \text{ (kWh}_{electric})/\eta_{\text{power station}}} = \frac{W \text{ (MJ}_{thermal})}{E \text{ (MJ}_{thermal})}$$

Figure 1 shows the primary energy ratio for several industrial processes. All processes on the line PER = 1 have the same primary energy consumption whether supplied by a fuel based technology or by an electrotechnology. Processes having a PER > 1, (P1, P2, P4), have a lower primary energy consumption if they use electricity. Whereas those with a PER < 1, (P3, P5), are more efficient from a primary energy stand point if they use fossil fuels.

The value of the PER ratio is dependent on the overall efficiency of the mix of electric power generation stations used ($\eta_{\text{power station}}$). If this improves, for instance by using more CCGT instead of coal fired plant, then the PER points move to the right and more processes (P5 for instance) can show a primary energy saving by using electricity.

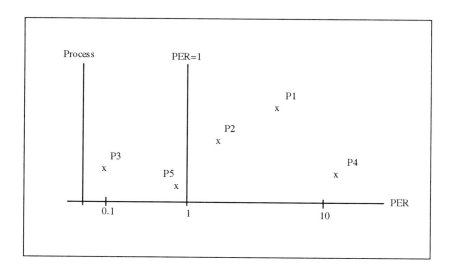

Figure 1 : The Primary Energy Ratio (PER) of processes P1, P2, P3, P4 and P5

OVERVIEW OF THE CASE STUDIES

It cannot be claimed that the use of electricity as an energy source in an industrial process always leads to immediate savings on primary fuels. However, the potential for indirect energy savings, due to the 'economising' effect of electricity, should not be underestimated.

Perhaps of equal, if not more, importance to industry are the other benefits that can be achieved by switching to electricity: increases in quality, reduction in products rejects and waste... These can often produce cost savings and improvements that more than justify the investment.

50 cases of successful implementation of electrical techniques are reported in a uniform format. These cases are drawn from 9 countries and are representative of electrical techniques allowing for primary energy saving and profitability. Although the cases are far from exhaustive, they are representative of a broad range industrial sectors as well as most electrical techniques.

The term electroheat is often used to refer to a number of electrotechnologies including resistance, infra-red, dielectric and induction heating as well as ultra-violet curing, although the last of these does not strictly involve heat transfer. These technologies may not always appear to be the first choice as an energy saving technology, but savings are possible because the process can be better controlled and high energy densities can be applied selectively. Other electrotechnologies, employing electricity for motive power, include mechanical vapour recompression, heat pumps, air knives and variable speed drives. These can all be very effective for energy saving. Membrane techniques like reverse osmosis and ultrafiltration are still more used in effluent treatment and in food and drink processes. They have a low power consumption when compared with traditional separation and concentration techniques based on evaporation.

50 case studies give an illustration of the possible benefits of all these electrical techniques; they cover the major industrial sectors:
- pulp, paper and wood (7 cases)
- textiles (3 cases)
- chemistry, plastics, pharmaceutics (5 cases)
- agro-food (8 cases)
- waste treatment, water (5 cases)
- mechanics, electronics (8 cases)
- metallurgy (9 cases)
- minerals, glass, ceramics (5 cases)

In summary this brochure illustrates how efficient applications of electricity contributes to energy savings in industry. It shows how the potential conflict of the need for profit with the wish to save energy and reduce CO_2 emissions, can be successfully reconciled to the benefit of both industry and society.

Research & Education

Contents

The teaching fellowship scheme in British universities 1
Oliver, T.N. (Aston University), Metaxas, A.C. (Cambridge University) UK

Expert system for teaching electrotechnologies – intelligent learning aid 9
Neascu, D.O., Lakhsasi, A., Xu, J., Yao, Z., Rajagopalan, V., Skorek, A. (Université du Quebec à Trois-Rivières) CANADA

Software complex for teaching and design of induction crucible furnaces 17
Bukanin, V., Nemkov, V., Koutchmassov, D., Zenkov, A. (St Petersburg Electrotechnical University), Smolnikov, L. (Moscow Commercial University) RUSSIA

Use of hypermedia in teaching electroheat 25
Le Goff, Y. (ENSAM) FRANCE

Experimental and numerical investigation of industrial induction heating processes 33
Mühlbauer, A., Lessmann, H-J. (University of Hannover) GERMANY

High performance switched mode power supply for microwave heating application 41
Petkov, R. (Swichtec Power Systems) NEW ZEALAND, Hobson, L. (University of Glamorgan) UK

High temperature resistance heating in semi-conductor production – new processes control system 49
Michalski, L., Sankowski, D. (Technical University of Lodz), Lobodzinski, W., Orzylowski, M. (Industrial Electronic Institute, Warsaw) POLAND

Artificial neural networks: drying process modelling and control . 57
Jay, S., Oliver, T.N. (Aston University) UK

Radial temperature differences and thermal stresses in arc furnace electrodes 65
Sundberg, Y. (Konsult AB) SWEDEN

La modélisation des arcs électriques: des équations théoriques aux applications industrielles 75
Delalondre, C., Simonin, O., Mineau, L. (EDF-DER), Roth, J-L., Gevers, C. (IRSID) FRANCE

Etude thermique du chauffage d'acier liquide par plasma d'arc tranféré **83**
Poilleaux, P., Bouvier, A., Trenty, L., Mineau, L., Debourcq, J., Delalondre, C. (EDF-DER), Leclercq, A. (SOLLAC), Roth, J-L. (IRSID), Heanley, C.P. (Tetronics R & D Ltd) FRANCE

A handbook of eletrotechnologies **93**
Peña, J. (Comité Español de Electrotermia) SPAIN

UIE presentation **101**
Van Dommelen, D. UIE Study Committee (ERE)

The Teaching Fellowship Scheme in British Universities

T. N. OLIVER* and A. C. METAXAS**

*Aston University, Birmingham, UK
**Cambridge University, Cambridge, UK

ABSTRACT

In the 1970's the Electricity Industry in the UK formed the Electroproduction Education Initiative. One of the key aspects of the programme was the setting up, in selected universities, of an industry supported academic presence or Teaching Fellow, to promote and develop electroheat related teaching material. This paper reviews the current state of this scheme in light of recent developments in the UK electricity industry. Highlighted are the many and varied links that have been developed between the industry and the Universities of Aston and Cambridge. One of the successes of the scheme is the number of undergraduates at both universities who now study electricity utilisation to second and final year degree level as well as the establishment of active areas of research in specific topics in electroheat.

Dans les années 1970, l'industrie électrique britannique lança l'Initiative pour l'Enseignement de l'Electroproduction. Un des principaux aspects de ce programme concernait la mise en place, dans des universités désignées, d'une présence académique parrainée par l'industrie, soit un "Teaching Fellow", dont le rôle serait de promouvoir et de développer les programmes d'enseignement sur l'énergie électrique. Cet article fait le bilan du stade actuel du projet face aux récents développement réalises par l'industrie électrique britannique. Il souligne les liens nombreux et variés que l'industrie a créés avec les universités d'Aston et de Cambridge. Le succès du projet est entre autres reflété par la nombre d'étudiants qui dans ces deux universités ont choisi d'étudier l'utilisation de l'énergie dans leurs seconde et dernière années de licence.

INTRODUCTION

The Teaching Fellowships in Electricity Utilisation at selected British Universities were established as a co-operative venture between the electricity industry and the academic institutions. The first Teaching Fellows were appointed in 1977. These first Fellows paved the way to establishing formal activities in electroheat education many of which continue to the present, albeit in modified form. The scheme was originally funded by the then Electricity Council as part of the Electroproduction Education Initiative and therefore was funded collectively by the electricity industry. In this context the term *electroproduction* refers to the technologies of electricity utilisation and not generation. In the early 1980's the structure of the initiative was formalised to provide support for a Teaching Fellow with a support technician at both Aston and Cambridge Universities plus a visiting professorship at Aston. In addition support was given to a Masters programme at Loughborough University. At its inception electroheat was not generally perceived as an important component of undergraduate electrical engineering programmes. The more usual emphasis in the power modules of such degrees was on machines, power electronics and power system analysis. As a result of the Electroproduction Education Initiative the Electricity Council has estimated[1] that during the period 1980 to 1988 the annual flow of undergraduates through courses with an electroheat input increased from around 100 to 1400.

Following electricity supply privatisation the structure of the industry has changed considerably. As a result the Fellowships, whilst still having a similar remit, are funded somewhat differently. At Cambridge the electroheat group's activities are supported by Eastern Electricity plc and National Power plc. At Aston all electroheat activity is currently supported by a single sponsor, Midlands Electricity plc(MEB).

The rationale of the Teaching Fellowships remains the same as at their creation: the promotion and better understanding of electroheat processes through teaching and research. This paper describes the present activity at both Cambridge and Aston and reviews the development of course content. In the case of the Aston Teaching Fellowship it will be shown that the involvement of a single sponsor has led to a mutually beneficial integration of a range of utilisation activities.

COURSE DEVELOPMENT

The genesis of an electricity utilisation course has been reported by Metaxas[2]. The introduction of electroheat as a specific topic requires careful thought and must be designed to suit the requirements of a particular educational establishment. It is crucial that the course be examinable and of equal standing to other engineering subjects. It is also important at degree level that the course will meet the requirements of accreditation laid down by the professional institutions, for example those of the Institution of Electrical Engineers. It is generally accepted that the subject of electricity utilisation cannot be introduced in any great depth in the early years of a degree course. The basics of the subject, at this level, require an understanding of electromagnetics, electrostatics, thermodynamics and heat transfer. Electricity Utilisation provides an academically challenging area of study to the student who having studied Maxwell's equations and the skin depth concept asks the question "to what eventual use do I put this?".

It has been proposed[2] that the subject matter in a utilisation course can be based on a classification of electricity utilisation processes (Figure 1). Here the term 'utilisation' is preferred to electroheat because motive power technologies such as heat pumps and air knives are included. From Figure 1 it is possible to establish several common themes which enable material to be offered in different contexts. For example, the topic area of *electromagnetic heating* can be developed from the technologies of direct resistance, induction, radio-frequency and microwave heating. The result is a unified approach to electromagnetic heating[3] of both metals and dielectrics over a frequency range from 50Hz to 2450MHz.

A second unifying theme evident from Figure 1 could be entitled *the ionised state* and includes the laser, arc furnace, plasma processes and glow discharges. The laser is included because the two most important industrial lasers, the CO_2 and Nd:YAG, derive their energy from electricity through the ionised state in the form of either a glow or flashbulb discharge.

A third mechanism inherent in most of the processes is that of heat transfer. Whether the problem is heat distribution in a steel billet during induction heating or moisture profiling of paper the temperature distribution is governed by a generic heat flow equation[4]. The study of utilisation from the heat transfer viewpoint provides material suitable for courses other than electrical engineering, for example mechanical, production or chemical engineering. At Aston plans are underway to incorporate electricity utilisation case study material into an environmental studies module.

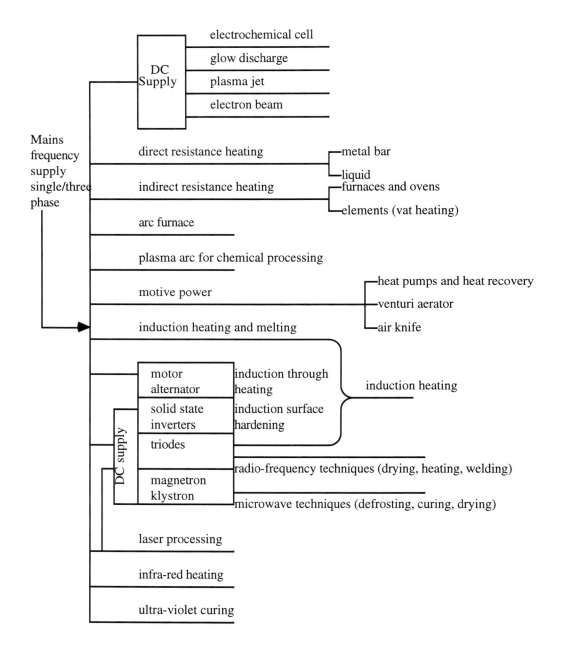

Figure 1 Classification of Electricity Utilisation Processes (After Metaxas, 1990)

ELECTRICITY UTILISATION ACTIVITIES

CAMBRIDGE UNIVERSITY

An effective means for introducing electroheat topics is through experiments and projects. At Cambridge a number of laboratory units have been developed for set experiments, project work or demonstration. One example is an electrically driven heat pump drying plant for particulates. The unit at Cambridge was funded as well as designed in collaboration with industry. Such a unit can be introduced at second year level before a more detailed lecture series on electroheat is offered as a final year option. The heat pump experiment is based on fundamental principles relating to the first and second law of thermodynamics. As well as being designed as a set undergraduate experiment at second year level (and hence ensuring a large audience of about 80 students per year) it also enables realistic trials to be carried out in feasibility studies involving energy recovery. This experimental unit has recently been awarded the PowerGen prize for innovation and development in education.

Recent final year undergraduate projects, which last 300 hours, include microwave heating of ceramics, microwave plasmas, radio frequency moisture levelling, computer controlled laser engraving, infrared curing and expert systems for electromagnetic drying. At final year level the students have the choice of 12 subjects of specialisation, each of 16 hours duration, one of which is based on the unified approach for electroheat discussed earlier. It has been estimated that a student who answers a full compliment of electroheat examination questions and undertakes a final year electroheat project will graduate with nearly 29% of an overall mark due to electroheat topics.

The Electricity Utilisation Group undertake a number of research projects, many of which are funded by industry and/or EPSRC studentships. Recent work includes the use of numerical techniques to model the electromagnetic field distribution in radio frequency and microwave heating cavities. Various applicators and material configurations are under investigation where coupled electric and thermal fields are considered. Another area of current work is associated with the problem of arcing in industrial applications of radio frequency heating. Arcing can present a major hazard and be harmful to many applicator/material configurations and a pre-arc detector has been developed and licensed to industry which has a much faster response than the "line of sight" infrared detectors used previously.

ASTON UNIVERSITY

During the 1980's the activity at Aston mirrored that at Cambridge except that a second year undergraduate course, *Electricity Utilisation*, was developed and operated as a core module in electrical engineering courses. The course provides one third of a 50 hour electrical power and machines module and comprises material giving an introduction to many of the heating techniques of Figure 1. Some of the topics are considered in greater detail in a final year optional module of 25 hours, *Electroheat*, which develops the unifying theme of electromagnetic heating as well as introducing detailed studies of drying applications, both dielectric and heat pump. Two experiments are undertaken at final year level, infrared paper drying and RF plastics welding.

Final year students undertake a project of 120 hours timetabled duration. Recent electroheat projects include an infrared heat flux sensor, PC based induction heater monitor, magnetostriction in steel undergoing electromagnetic heating, an analysis of a 1MW resistance heater for steel bar, energy distribution in a domestic microwave oven and the performance of molybdenum disilicide heating elements. There are other means for introducing electroheat topics. For example all final year students carry out a group based design project. Recently one group undertook the design of a fast response infrared emitter. By appropriate choice of options and projects it is possible for a final year electromechanical engineering (elmech) student to spend about 170 out of 450 timetabled hours studying electroheat material. Typically a total of 30 to 40 students per annum are examined on electroheat related topics. A new course, Electricity (Environment and Economics), was offered as a full 50 hour final year option in 1994/5. This has given the opportunity for topics such as electrochemical cells, the venturi aerator and heat pumps to be studied as potential solutions to environmental problems. Electroheat material also has great potential as the source for relevant and up-to-date case study material.

Since 1990 the Aston Teaching Fellow has been supported by Midlands Electricity plc (MEB). As a result substantial links have been developed which have resulted in increased industrial collaboration. MEB aim to sponsor two undergraduates on each year of the elmech course as well as a rolling programme of three postgraduates undertaking utilisation research to doctorate level. Current research projects include vibrations in electromagnetically heated steel, the use of neural networks for the control of drying processes and demand side measures in industrial electroheat processes.

MEB's Energy Technology Centre at Halesowen is made available to second year students for industrial visits which include a number of short electroheat investigations. Final year students visit the centre to study heat pump technology. Project students make use of technical facilities and the expertise of MEB staff. The Aston Teaching Fellow takes an active part in activities of the sponsoring company. These include project support in areas of technology development and courses for engineering staff.

EXTERNAL ACTIVITIES

Both Teaching Fellows continue to promote utilisation activities to as wide an audience as possible. For example, the British National Committee for Electroheat (BNCE) through its education and training committee organises short courses for academics and publishes teaching notes in the form of monographs for lecturers, recent examples being Induction Heating[5] and Electric Discharges for Heating[6]. Currently the Teaching Fellows are responsible for promoting two newsleters, one for BNCE edited at Aston and one for AMPERE (Association for Microwave Power in Europe for Education and Research) edited at Cambridge. These newsletters detailing electroheat activities in academia and industry are distributed to a wide readership in further and higher education as well as in industry. Through the BNCE route electroheat courses have been introduced into other universities. For example at Kingston University the MSc in Advanced Manufacturing Systems includes a module of one week's duration entitled *Electro-manufacturing Processes*. Kingston also benefits from the availability of a Teaching Fellow supported by SEEBOARD plc and a well equipped electroheat laboratory in which multimedia PC based experiments and demonstrations are being developed.

BNCE has funded the development of small scale demonstration equipment initially to support its short courses but generally to be made available to lecturers to assist in the teaching of utilisation topics. A recent example is a 1kW heat pump dehumidifier[7] which, together with the necessary instrumentation which has also been funded by BNCE, can demonstrate the practical aspects of coefficient of performance and the nature of refrigeration cycles.

CONCLUSIONS

It is possible to structure an undergraduate course on topics in electroheat which is topical, marketable and examined at a level of equal standing with other engineering or physics subjects. Industrial sponsorship has proven to be paramount in support of these activities especially at the embryonic stage. Courses developed with this support in the 1980's now attract a considerable number of undergraduates and where optional modules are provided student take up is high.

Recent changes in the structure of the electricity industry have, to date, resulted in continued support of the Teaching Fellowship scheme as well as a more integrated level of activity between sponsors and Teaching Fellows.

REFERENCES

1. R D Langman: 'Electro Production Education 1980 - 1989', The Electricity Council, Report M140, London, 1989.

2. A C Metaxas: 'Undergraduate Teaching in Electroheat', *Power Engineering Journal*, January 1990, 45 - 50.

3. A C Metaxas: 'Heating with Electromagnetic Fields - A Unified Approach', The Electricity Council, EC5055, London, 1989.

4. A C Metaxas: 'Foundations of Electroheat; a unified approach', John Wiley and Sons, Chichester, 1996.

5. F W Walkden: 'Induction Heating', The Electricity Association, EA1013, London, 1991.

6. J E Harry: ''Electric Discharges for Heating', BNCE, London, 1993.

7. T N Oliver: 'Heat Pumps for Industrial Drying - A Demonstration Unit', BNCE Electroheat Education Newsletter, 1993, **18**, 7.

Expert System for Teaching Electrotechnologies - Intelligent Learning Aid

D.O.NEACSU, A.LAKHSASI, J.XU, Z.YAO, V.RAJAGOPALAN, A.SKOREK

Chaire de Recherche Industrielle CRSNG/Hydro-Québec, Université du Québec à Trois Rivières, CP 500, Trois Rivières, Québec, Canada

Abstract. This paper outlines the major requirements and concepts of a computer-aided learning aid for electrotechnologies and gives the characteristics of a prototype of an expert system named **CATELECT**. It provides the basic information for beginners as well as all the required tools for specialists (access to simulation of electrical drives or design of other electrotechnologies). The usefulness of this software is illustrated by a case study. Finally, the future directions of development are presented.

Résume. Cet article ressort les conceptes majeurs requis par un tuteur intelligent pour l'apprentissage des électrotechnologies et donne les caractéristiques du prototype du systéme expert nommé **CATELECT**. Il donne les informations de base aussi bien pour les débutants que pour les specialists (ayant accès à des simulations des systèmes d'entrainements ou le design des autres électrotechnologies). L'utilisation de ce logiciel est illustrée à l'aide d'une étude de cas. Finalement, les directions future de développement sont présentées.

1. INTRODUCTION

An engineer is often confronted with the vastness of the available electrotechnologies and is required to learn or update his/her knowledge usually outside the normal working hours of work and most often outside conventional classroom depending on the availability of time. Computer-aided tutor or learning aid will be a very useful tool for such applications which will help the user to understand a subject from basic to advanced level which could involve sophisticated computer-aided analysis and design tools.

This paper outlines the major requirements and concepts of a computer-aided learning tutor for electrotechnologies and gives the characteristics of a prototype of an expert system named *CATELECT*.

CATELECT gives a modern introduction to various electrotechnologies[1] and their industrial applications involving specialized knowledge of

- electrical power conversion aspects such as AC-AC, AC-DC and DC-AC converters,
- electrothermal technologies such as resistance heating, induction, infrared radiation, dielectric hysteresis, electric arc, plasma, laser and electron beam heating;
- electromechanical aspects such as variable speed drives, mechanical vapour recompression and heat pumps.

CATELECT provides not only the basic information for beginners and all the required tools for specialists but also permits a user to make some simulation or to obtain a fast design of desired applications.

Recently, several computer-aided analysis tools have been available for the simulation of power electronic systems.[2] Expert system techniques have been successfully applied for the fault diagnosis of sophisticated drive systems.[3-5] Our group has been working for the past five years to design an expert system learning aid for teaching power electronics and some of the results of a preliminary prototype have been presented.[6-8] This paper explores further the methodology of development of an intelligent tutor for teaching electrotechnologies. Section 2 gives the characteristics of a prototype of an expert system called *CATELECT*. Two examples are chosen for illustration: one on the electrical drives and another on electrothermal analysis. Section 3 gives some future directions of development.

2. CATELECT TUTOR

Availability of affordable portable computers offers a new environment for the dynamic learning of a subject that may comprise an electronic manual and other simulation tools integrated in an intelligent tutoring environment. The teaching aid should be available for a portable computer and must have most of the required notes and manuals similar to a textbook. Such an interface needs to make accessible the necessary text and software tools in a user-friendly manner and as far as possible, provide graphic environment, taking into account the availability of affordable expert system shells, hypermedia techniques.

CATELECT has been developed in ***ToolBook*** environment[9] which is an object-oriented development environment that provides drawing tools for creating objects; it uses a programming language (***OpenScript***) for programming object behavior. The applications created by ***ToolBook*** are directly executable under Windows. It also provides full-featured animation tools, introduction of linked or embedded objects interactive training and the possibility of running other Windows applications. This opportunity allows the interaction with simulation packages such as ***SIMUPELS*** or ***ATOSEC5*** in order to obtain simulation results. All of these features make ***ToolBook*** an excellent tool for computer aided training applications. Moreover, an application developed in ***ToolBook*** can be easily introduced in another learning environment ***CATPELS***.

CATELECT is structured into two parts. The first one corresponds to the teaching of electrical drives (***CATELEC***) while the second part is a presentation of electrothermal aspects (***CATTHERM***).

2.1. VARIABLE SPEED DRIVES: CATELEC

An adjustable speed drive (ASD) is a device which is used to provide continuous range process speed control (as compared to discrete speed control as in gearboxes or multi-speed or pole changing motors). Our computer aid in teaching electrical drives gives the features and the operating modes of the most known converter fed drives. The cases of DC machine, asynchronous machine, synchronous machine are presented both with the converter and machine description. In each case, a brief description of the drive system introduces the reader into the field and more information is obtained by request. Moreover, one can observe the desired waveforms by accessing the ***SIMUPELS*** library[2]. This library has been developed in ***MATLAB-SIMULINK*** environment and provides models for a number of electronic converters and electrical drives.

The following electrical drives are available in the ***CATELECT*** system:
- DC drives, unidirectional and bidirectional topologies;
- VSI-Fed Variable Speed Induction Motor Drive;
- CSI-Fed variable Speed Induction Motor Drive
- CSI-Fed Variable Speed Synchronous Motor Drive
- Slip Power Recovery Scheme for Wound Rotor Induction Motor
- PWM Inverter-Fed Induction Motor Drive
- Comparison of Variable Speed Drives

- Applications

In order to emphasize the advantages of this computer aided tutor for electrical drives, let us take a simple example: the case of a bidirectional DC drive. In the appropriate screen we can get some information about this application and a schematic diagram of the system. For more details, one can activate the "hotwords" corresponding to the drive characteristics or can activate one block from the schematic diagram representing the circuit. Moreover, it is easy to get direct access to SIMUPELS in order to see some waveforms or to analyse the drive behavior. Figure 1 presents some screen captions.

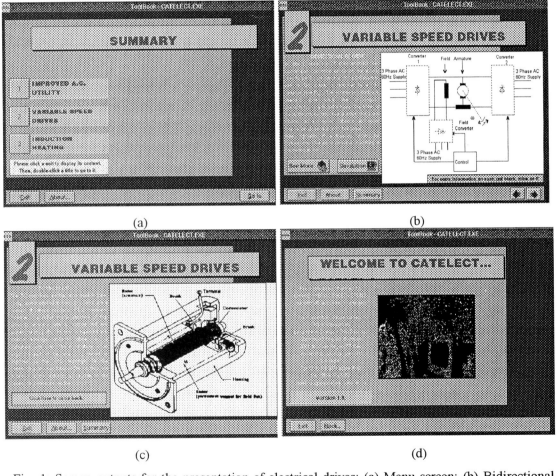

Fig. 1: Screen outputs for the presentation of electrical drives: (a) Menu screen; (b) Bidirectional DC drive; (c) DC machine; (d) Welcome screen of CATELECT

Finally, for more information, *CATELECT* gives some precise references on each topic. Details or basic definitions are delivered by some viewers that are activated by clicking on the desired "hotwords".

2.2 ELECTROTHERMICAL ASPECTS: CATTHERM

The main objective of the *CATTHERM* that concerns the electrotechnologies is the presentation of electrical heating for industry used by electrothermal processes.[10-11] A brief description of this field is given below.

A direct resistance heating ocurrs where a medium is heated directly by a current passing through it. In turn this implies intimate contact between the medium or workpiece and the power source.

Indirect resistance heating invariably requires the use of a resistance element which produces heat which is transferred to the workpiece in an oven or furnace, where radiation and convection are the main heat transfer mechanisms.

In induction heating, the material is heated by a current passsing through it. This current is induced from a separate source. Whatever the configuration, induction heating always involves a varying magnetic field. The metal to be heated is the secondary circuit of the transformer.

In its simplest form, dielectric heating may be represented as the exploitation of loss mechanism in a capacitive circuit. The application of an electrical field across a material causes distortion of the alignment of electrons in the molecules of material. When the field is reversed, the distortion is also reversed, and work involved in this distortion is apparent as heat given off in the material.

At this point, having covered direct, indirect and induction heating of conductors and dielectric heating, it is now possible to link these various mechanisms together to consider metal melting.

The most common arc furnace is a metal melting furnace, often melting scrap steel. Three consumable electrodes are used in all but very small furnaces. Three-phase power is connected to the electrodes, and the metal becomes the neutral point of the system. An arc is initiated by inserting the electrodes to touch the load, and arc current is controlled by extrating the electrodes to lengthen the arc.

Some captions corresponding to the presentation of the electrotechnologies are shown in Fig.2. The ToolBook features of making presentation by inserting animation are well utilized in order to describe the physical aspects of the electrothermal processes.

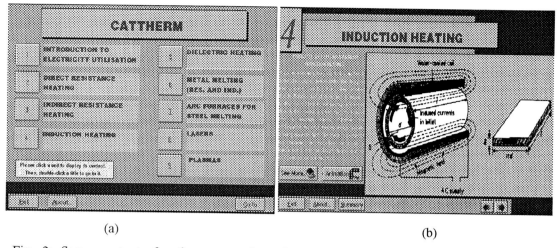

(a)   (b)

Fig. 2: Screen outputs for the presentation of the electrotechnologies: (a) Menu Screen of *CATTHERM*, (b) Sample screen presenting the induction heating.

3. CONCLUSION

This version of *CATELECT* comprises a number of completed chapters with excellent opportunities for teaching electrotechnologies. The text is integrated in the form of short explanations concerning the subject. More information is available by viewers that are activated by mouse. A set of figures and tables and some animation features enrich the presentation. Moreover, *CATELECT* ensures the access to some simulation tools such as *ATOSEC5* and *SIMUPELS*.

The structure of *CATELECT* comprises two parts: one part for teaching electrical drives and the second part for electrothermical applications. In the paper, both cases are illustrated with some examples.

ACKNOWLEDGMENTS

This work is supported by the grants received from the Hydro-Québec and NSERC Canada through the Industrial Research Chair at the Université du Québec à Trois-Rivières.

REFERENCES

1. V. RAJAGOPALAN: 'Principles and Applications of Electrotechnologies', CCE, Les Editions de la Chenelière Inc, 1993

2. V. RAJAGOPALAN and Z. YAO: 'SIMUPELS User Manual', Groupe de Recherche en Electronique Industrielle, Département d'ingénierie, Université du Québec à Trois-Rivières, April 1994

3. K. DEBEBE, V. RAJAGOPALAN, and T.S. SANKAR: 'Diagnosis and Monitoring for AC Drives', *IEEE IAS Conference Record,* 92-CH 3146-8, 1992, pp. 370-377.

4. D. SHEN and B.K. BOSE: 'Expert System Based Automated Selection of Industrial AC Drives', *IEEE IAS Conference Record*, 92-CH 3146-8, 1992, pp. 387-392.

5. K. DEBEBE, V. RAJAGOPALAN, and T.S. SANKAR: 'Expert System for Fault Diagnosis of VSI-Fed AC Drives', *IEEE IAS Conference Record*, 91-CH3077-5, 1991, pp. 368-373.

6. K. DEBEBE, and V. RAJAGOPALAN: 'Expert System Learning Aid for Power Electronics' *IEEE Power Electronics Society, Proceedings of the Third Workshop on Computers in Power Electronics*, 92TH0504-1, 1992, pp.7-17.

7. V. RAJAGOPALAN, D. BRILLON, Z. YAO and M.L. DOUMBIA: 'Computer-Aided Teaching of Power Electronics- Classroom of the Future', *The 10th ISPE/IFAC International Conference on CAD/CAM, Robotics and Factories of the Future,* 1994, pp.197-202

8. K.DEBEBE and V.RAJAGOPALAN: 'A Learning Aid for Power Electronics with Knowledge Based Components', *IEEE Trans. on Education*, 1995, **38**(2). 171-176

9. 'ToolBook User Manual', *Asymetrix Company*, 1994.

10. C. J. ERICKSON: 'Handbook of Electrical Heating for Industry', IEEE PRESS Marketing, New York, 1995.

11. R. D. LANGMAN: 'Worked Exemples in Electroheat', ASTON/Cambridge 1987.

Software Complex For Teaching And Design Of Induction Crucible Furnaces

V. BUKANIN, V. NEMKOV, D. KOUTCHMASSOV, A. ZENKOV

Saint Petersburg Electrotechnical University

L. SMOLNIKOV

Moscow Commercial University, RUSSIA

ABSTRACT

The first multipurpose software complex CADIFTB is developed for teaching, learning, study and design of induction crucible furnaces (ICF). Educational part permits teaching/learning of basics, design, equipment, economics, safety, environment protection and maintenance of ICF. This part is created in a visual media Asymetrix ToolBook for Windows. It contains texts, pictures of actual equipment, facilities for calculation of components parameters, computer created graphs and animation. The second part containing a code for ICF study and CAD may be used independently or in combination with the first part. It allows to calculate the thermal and electromagnetic parameters of ICF, to choose materials and components from data-bases and so on. The complex is intended for teaching students of different curricula, for induction heating schools, courses and seminars, as well as for ICF design, sales management and in-plant technical service.

Le premier ensemble logiciel à usages multiples CADIFTB a été développé pour l'enseignement et l'apprentissage des techniques relatives aux fours-creusets à induction (ICF), ainsi que pour l'étude et la conception de ces équipements. La partie didacticielle est destinée aux enseignements et apprentissages des principes de base, des méthodes de conception, d'instrumentation et d'entretien des ICF ainsi que de la sécurité et de la protection de l'environnement. Cette partie est réalisée à partir de Asymetrix ToolBook pour Windows; elle comporte des textes, des vues d'équipements réels, des outils de dimensionnement de composants, des graphiques et des animations. La deuxième partie comporte un code de calcul destiné à l'étude et CAO des ICF. Cette partie peut être utilisée indépendamment ou en relation avec la première; elle permet la détermination des paramètres thermiques et électromagnétiques des ICF, le choix des matériaux et des composants dans des bases de données etc.

INTRODUCTION

Electronic educational tools built on a base of personal computers and modern Hypermedia packages are of special interest for such particular sectors as induction heating and melting. Development and appropriate application of induction melting equipment requires special knowledge in electrical engineering, electromagnetism, thermal processes, materials and components. At the same time preparation and publication of traditional course books is rather expensive and non-economic due to the relatively small market.

The electronic books have a set of advantages compared to traditional papers books:

- the user receives not only "static" information, but also animation, facilities for real "on line" calculations, additional sounds effects, etc.;
- possibility to adapt educational process to any particular demand or situation, including translation to other language;
- possibility of editing, updating and further development of the book;
- economical reproduction of copies.

Compared to books, video or slide projection Hypermedia programmes allow for the user to have a feedback, to control the presentation of material and to be an active participant of the educational process. The aim of this work is to create a flexible multi-purpose tool for teaching and self-education in induction crucible furnaces with more profound study and calculation of real furnaces in a wide range of types and sizes.

The complex is intended for users with different special knowledge and general background. Professor, student or engineer can choose a desired level of material presentation beginning from a general survey of the problem and up to profound study of particular features of this type of equipment and even to CAD of induction furnace installation. The complex is designed as an open system for further development, modification and adaptation.

CODE DESCRIPTION

The complex consists of two coupled parts. An instructive, part contains general and special information in the theory, design and main industrial applications of induction furnaces in the form of texts, tables, graphs, computer-made pictures and animation. Photographs, charts and drawings of real equipment are scanned and inserted in the programme. Examples of on-line calculations are provided to illustrate the theoretical considerations. The second part of complex is a code for study and Computer Aided Design (CAD) of real ICF installations.

INSTRUCTIVE PART

The ToolBook software allows to arrange all the instructive material in the form of a computerized book. As a paper book it contains parts, chapters and pages devoted to different topics of induction melting technique. Any parts of the electronic book may be printed to receive a hard copy or projected to a screen for demonstration during a lecture, seminar or other presentation.

The instructive part contains two systems, providing friendly user interface. The first system organized all the control operations (i.e. movements, show, calculations, printing, etc.). The second system EditText is developed to facilitate editing of texts in the frames of the code structure.

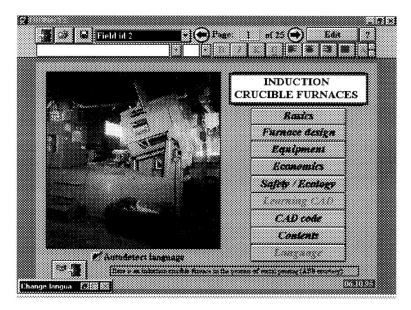

Figure 1

Basics

The chapter *"Basics"* contains compressed information about types, typical applications, theoretical fundamentals, physical phenomena, advantages and weaknesses of ICF. Not only electromagnetic and thermal phenomena but hydrodynamic processes are under consideration too. Magnetic field pressure on the melt causes a meniscus and excites vortexes of molten metal. Animation is used to show the patterns of metal movement for different conditions of coil energizing. Upper, lower or both parts of coil may be supplied by current. Real on-line calculations of hydrodynamic processes are rather complicated and qualitative predictions are used for animation.

Furnace Design

This chapter deals with the furnace assembly and its components. Hot words are in the text to help for the user to identify the components on the figure. On the contrary you can ask for a name of a component by clicking its image on the picture. *Induction coil, crucible, inactive coils, magnetic yokes, lid, tilting mechanism and water cooling system* may be identified by hot words.

The following pages show different types of windings and coil tubings used in ICF; types of crucibles: *Conductive, Non-conductive* and *Special*, such as metallic cold crucible; types of resonant circuits: parallel, series, parallel-series, series-parallel and "autotransformer" coil connection. Inherent properties of different circuits may be studied in general form and on some examples.

Economics

This part contains general consideration of economics in application to ICF (investments, maintenance, energy, labor and material costs, quality of products, etc.) in comparison to other types of melting furnaces. The ways of efficacy improvement are considered too. Some illustrative cases are given, for example diagrams, demonstrating power losses in each of the components of ICF installation, beginning from the main line and right up to the melt (Fig. 2).

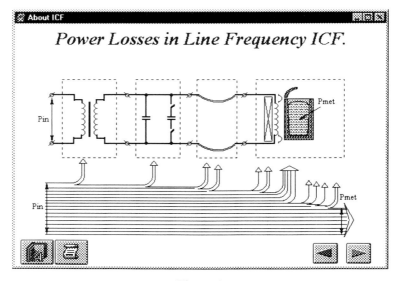

Figure 2

Safety and Ecology

Safety and Ecology chapter (Fig. 3) deals with the problems of environmental protection and human safety. The process of metal melting in ICF is much more environment friendly than melting in other types of mass production furnaces.

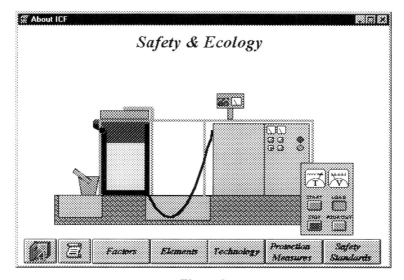

Figure 3

Nevertheless, there is a set of dangerous or harmful factors general for melting furnaces such as high temperature, molten metal, mechanical dangers, exhausting gases, etc. and specific for ICF (toned sound (Fig. 4), magnetic field in close environment) takes place in this case too.

The user of this programme can become a furnace operator and carry out a technological process of metal melting, from scrap loading to melt pouring in the a tundish. Unproper design and maintenance of equipment or process may lead to unpermissible levels of industrial environment pollution or harmful impact on operators. Main dangers such as electric shocks, burns, mechanical impacts, sound and electromagnetic field exposure as well as safety standards and protective measures are studied here.

EditText

Typically, educational programmes are being developed by professors or scientists (not by programmers) and the process of their composition must be simplified as mush as possible. Inspite of relatively simple technology of teaching tools development provided by modern Multimedia systems, such as ToolBook, it remains complicated for non-professional user. It is especially difficult to add new language to or make other changes in already existing programmes because usually they are protected from unauthorized or non-intended changes. At the same time a possibility to change language is very important for educational tools. To solve this problem a special programme EditText was developed which includes the very editor and a programme for it's installation. Using EditText any user can easily update or translate texts into other languages. When translation is completed a proper language may be chosen from option menu in block "Language". The author's version of text is saved in "protected" file in other format. EditText may be used with most part of ToolBook programmes.

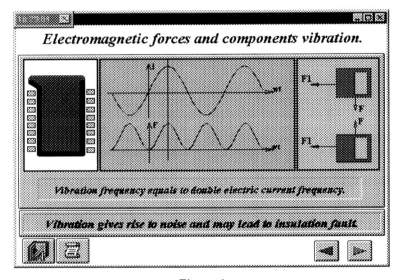

Figure 4

Now the main part of complex for teaching induction furnaces is developed in Russian and English but some parts are available in French too. The standard text files are used for text saving in the process of translation or editing, so automated translators may be applied for draft translation.

EditText contains also an opportunity to autodetect the user language. When started at the programme tends to install the text language corresponding to the language of it's own system. The user can keep autodetection or switch it off.

CAD CODE

This part is intended mainly for professional use, but it may be rather effective for more profound education in ICF. A special chapter *"Learning CAD"* is inserted in the complex in order to help to non professional user to understand the problems of ICF design and whole installation projecting and to facilitate the use of *CAD code*. It includes two parts: detailed description of objectives and procedures of design and demonstration of calculations on some simple examples. Two modes of demonstration are provided -- autoshow and on-line operation. In fact, it is CADIF tutorial.

CAD code is an independent programme's complex. This intelligent dialog system places at your disposal all necessary facilities for calculation and design of installation. It includes the blocks for thermal, electromagnetic and coils cooling calculations, as well as resonant circuit and supplying line choice and calculation.

CADIF permits to: choose the design of furnace, dimensions and types of its elements; study the influence of furnace elements, their dimensions and properties on furnace parameters and hence to optimize its design; study the evolution of furnace parameters during the operating cycle when the mass of metal changes; choose the optimal power source, supplying line and resonant circuit configuration; display and documentate the furnace layout, data tables, temperature distributions, etc.; create and modify the data bases of elements and materials; create the archive of furnace designs.

The input data may be taken from archive or data bases, corrected or introduced by user (Fig. 5). The inherent checking of input data compatibility permits to avoid the errors of this type. The validity of calculations and the effectivity of CADIF are proved by experience of its use.

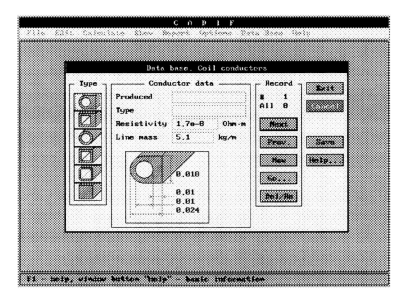

Figure 5

CONCLUSION

Complex CADIFTB is developed on the basis of long-term experience of some of the authors in study, design and teaching of induction furnaces in St. Petersburg and Moscow (Russia). It's essential that this new educational and design tool was created in close cooperation of professors and students. The authors strongly believe that this complex will be widely used for both educational and industrial applications due to its flexibility, friendly interface, possibility to pass easily from general consideration to real calculations and design, user's ability to obtain new information non existing in explicit form and to other advantages. Minimum configuration for software installation is IBM PC AT-386, 4MB RAM with Windows 3.1 or higher.

ACKNOWLEDGEMENT. The authors thank professor Yves Le Goff (ENSAM, Paris) for support and useful discussion.

REFERENCES

1. V. NEMKOV, L. SMOLNIKOV, V. BUKANIN, D. KUCHMASOV, A. ZENKOV: 'Hypermedia course plus CAD complex - a new tool to teach induction crucible furnaces', *Proceeding of the Seminar "Using Hypermedia for Education in Electricity Utilization"*, Leuven, 27-28 April 1995.

2. V. BUKANIN, D. KUCHMASOV, V. NEMKOV, A. ZENKOV: 'Multi-media Learning Program for Induction Crucible Furnaces', *40. Internationales Wissenschaftliches Kolloquium, Technische Universität* Ilmenau, 18-21.09. 1995

Use of Hypermedia in Teaching Electroheat

Yves LE GOFF.

ENSAM - ENG/LMF - 151, Bd de l'Hôpital - 75013 PARIS - FRANCE

D'une part, les outils informatiques que sont les hypertextes, les interfaces graphiques et les langages de programmation orientés objet ouvrent des voies nouvelles pour lier de façon dynamique les multiples aspects d'une réalité donnée ; les logiciels auteurs permettent notamment à tout intervenant, même non informaticien, de présenter des informations sous une forme détaillée et accessible à un public diversifié, que ce soit en vue d'une présentation publique ou à l'intention d'apprenants ou de clients individuels. D'autre part, les techniques électrothermiques mettent en oeuvre des technologies nouvelles et mobilisent souvent des connaissances scientifiques de haut niveau ; de plus, leur apprentissage s'adresse à des personnes d'origines diverses. Pour ces raisons, la réalisation et l'utilisation de documents interactifs rédigés en logiciel auteur permettent de démontrer la pertinence de choix technologiques en liaison permanente avec des aspects économiques, techniques ou scientifiques ; elles devraient intéresser les chercheurs, ingénieurs, enseignants, agents commerciaux. Des exemples, inspirés de techniques électrothermiques diverses, seront présentés en séance, et l'auteur s'efforcera d'en dégager des orientations pour de nouveaux modes d'accès à la connaissance scientifique et technique.

On one hand, computer techniques such as hypertexts, graphical user interfaces and object oriented programming yield new ways to link in a dynamic way several aspects of a given reality: authoring softwares allow especially any author, even not computer science specialist, to present information in a detailed and accessible form to a wide variety of audiences, either for public presentation or for individual learners or customers. On the other hand, electroheat techniques involve recent technology and often high level scientific knowledge. Moreover, teaching those techniques is aimed at people from very varied backgrounds. Therefore, production and use of Interactive documents through authoring softwares make it possible to demonstrate the best technological choices in connection with economical, technical and scientific aspects,, and should interest scientists, engineers, teachers, sales managers. Examples relying on various electroheat devices, will be presented at the congress, and the author will endeavour to show some directions in new ways of access to scientific and technical knowledge.

STAKES AND ASSETS

Electroheat covers all the studies and research work intended to improve industrial processes by using resources related to electrical energy.

The improvements targeted concern the reduction of energy consumption or of combustibles or waste flow, as well as the times of response of facilities as regards energy transfer. These results are to be obtained by means of the relative complexity of electro-heating systems and of the continuous gathering of required knowledge for the research into the electrical, thermal or mechanical behaviours of very varied materials. This is to be done in context with the natural competition with conventional and well established technology.

Any document dedicated to electro-heating techniques (as to any technical topic) is aimed at people from very varied backgrounds. The requests they make are different according to their initial training (institutes of technology, schools, universities), whether they need continuous training (self-teaching, work placements, skills up-grading, follow-up), scientific or technical information (factory or laboratory visits), technological transfer (design offices), commercial approaches (presentation to clients, after-sales service)

The spread of knowledge and know-how in electroheat, the access to new markets and the establishment of new skills could all take advantage of the new techniques offered by hypermedia to explain, demonstrate, calculate and make comparisons.

HYPERMEDIA AND MULTIMEDIA

It would be wrong to expect hypermedia techniques to transform into monomedia techniques i.e. that they could be a substitute for all other information circulation means : lessons or conferences, paper documents, analogue video cassettes. The obstacles to *all-digital* are :

- psychological - the receptivity of a group is sometimes better than that of an individual
- cultural - reading a good work will always be a pleasure
- technological - the techniques of compression/decompression of digitised audio-visual sequences (by means of peripheral devices and software) are still rarely diffused in professional environments and are faced with problems of compatibility and standardisation. The passing from simple hypermedia to real multimedia (which integrates digital audio-video, transfer of information by digital networks, and giving ample media coverage via 'information highways') should be envisaged as the final step in the production of digital documents on electroheat. This would widen the experience acquired by the 'electroheat community' in the hypermedia domain.

Here as elsewhere, step by step innovation is a reasonable approach. It should consist in integrating the hypermedia documents within the wide range of circulation aids which already exist for scientific and technical information; on one hand as a specific aid, on the other hand as an addition to conventional aids (use of projection during presentations, conventional video-cassettes or even to give help to find a text within a voluminous paper document, etc.).

Our own experience leads us to focus especially on what hypermedia, unlike any other media (paper, analog video), are able to do

- very high level interactivity, quantitative simulations, very low cost animation
- choice of several approaches (deductive, inductive, ...), branching off, digressions on demand, indexing, case researching, removals/insertions of specific information
- very easy far distance transmission

Authoring softwares which are nowadays widely available on PC platforms meet all those features and requirements. Those softwares are the result of the fusion of three kinds of digital techniques: hypertexts, graphical interfaces, object oriented programming languages. They allow mixing of hypertexts, interactive calculations and graphics and powerful linking between different items.

OBJECT ORIENTED ACCESS TO KNOWLEDGE

Authoring softwares make easier what can be called the *mapping of knowledge* : knowledge is easily mapped in dictionaries (alphabetically) or atlas (geographically), but the mapping of knowledge in physics, technology, economics, etc... seems to be a big challenge : any description of any real object needs texts, drawings, theories, models, etc... and belongs therefore to *virtual reality*: the contribution of modern computers and softwares is to put virtual reality much closer to (real) reality than it has never been before, mainly thanks to Object Oriented Programming (OOP).

COMPLEXITY	OOP
•many categories of many elements	•many categories of many (virtual) objects,
•hierarchical levels	•heritage, embedded objects
•many categories of links	•linked objects
•non-linear behaviour	•non-linear navigation

In a somewhat metaphoric way, it is possible to define *Object Oriented Access* (OOA) to study of technical systems: such systems have a complex structure, because they involve several devices, each of which includes several components, each of which is made of several materials: the table above lists a correspondence of some characteristics of complexity with some features of OOP.

Examples below show Object Oriented description of Electro-heating devices which start with an hypertext linked to an interactive picture: this (vectorised) picture displays at a more or less detailed scale several parts with some own characteristics which can be changed on demand; programmed models yield interactive simulation and allow learner or teacher to answer in real time questions like how is it built ? how does it work ? what happens if ? how to explain what happens ?

This *virtual experimentation* is the first step of *inductive knowledge acquisition*; the second step is the displaying of the involved *models* : numerical coefficients, equations, computations; the third step is the access to higher level *concepts*, definitions and laws. A well-designed package on Electroheat techniques must endeavour to offer a wide variety of examples (galvanisation vessel, microwave applicator), all of which are related to coherent models (energy balance, penetration depth, heat transfer), all of which are related to a small number of basic concepts (first an second law, Maxwell equations)

THREE STAGES

Taking into account the current conditions of development and rate of spreading of micro-computing facilities it is possible to define the following schedule :

FIRST STAGE

An inventory of what already exists. A good start could be to deeply examine the achievements in electroheat by the institutes present at the conference and from there to identify common objectives, compare currently available and known hypermedia/authoring softwares which allow the development of documents in WINDOWS or WINDOWS + Mac format integrating, produce texts and hypertexts in several languages, drawings and animation, digitised photographs, interactive calculations, classify the work which can now be linked together in a register, construct the architecture of 'prototype' documents already produced and to copy these onto disc for international (non-commercial) circulation and evaluation, the setting up of a 'reading committee' working via a network (Internet or other).

SECOND STAGE

Development of an international and commercial publishing project to construct a compilation on CD-ROM (and CD-I ?) of documents duly tested, integrating audio-video multimedia elements, consideration of questions posed as regards the legal, financial and commercial frameworks in order to put in place a permanent structure to ensure that document up-dates are carried out and recent developments recorded.

THIRD STAGE

The development of a knowledge base according to a client-server frame-work in addition to the compilation on CD-ROM/CD-I.

THE CEE/CFE EXPERIENCE

Electro-heating techniques are at an ideal meeting point for university and industry. This represents a large potential for training, research and development. Certain meetings have taken place in France via the 'CLUB ELECTROTHERMIE ENSEIGNEMENT' (an electroheat teaching association, related to Centre Français de l'Electricité, CFE). This association carries out many activities, among them and one which they initiated is the production of hypermedia documents in electroheat within the framework of summer courses which take place in August of every year. This was made possible with the help of CFE, DOPEE publications and NOVELECT network in conjunction with university institutes in Marseilles (ESIM), Nantes (ISITEM), Paris (ENSAM), Saint-Nazaire (LRTI), Toulouse (ENSEEIHT). This initiative first took the form of a feasibility trial i.e. the writing of an interactive document related to techniques of inductive heating, and then of the development of a data-base involving students on industrial placements or final year projects. The subjects currently treated cover membrane techniques, and infra-red and RF and MicroWaves heating. A survey is currently in progress in parallel involving about a thousand potential users.

The experience gained over the past two years allows us to make several quantitative estimations :

• The presentation of an electro-heating technique can quite reasonably form the subject of an interactive document comprising around 50 views of screens.

• The achievement of each screen view, i.e. the putting into an interactive form of a document available as a series of texts, diagrams and photographs represents 5 to 10 hours of work for someone who has mastered the authoring-software.

A navigation system seems to be quite acceptable to users. It would consist of a first choice from the following list (there is no hierarchy shown in the list) : presentations of case studies, elements of classification of techniques relative to a given domain, different technologies, quantitative simulations which mainly present the criteria for dimensioning facilities, basic theoretical elements, research subjects

The different screen views are linked either sequentially, or in a logical way by means of hyper words, with memorisation of random routes (dynamic branching).

INTERNATIONAL OPPORTUNITIES

The ERE conference held at CNIT on May 17, 1994 was the opportunity for ENSAM, the St Petersburg Electrotechnical University (SPETU, represented by V.NEMKOV) and the Université du Québec à Trois-Rivières (represented by V.RAJAGOPALAN) to consider student exchange programmes for work devised from developments in Hypermedia. These exchanges have already been undertaken, including the registration at the conference in Brussels, April 1995, of two SPETU students and two ENSAM students.

Moreover, French ADEME (Agence de l'Environnement et de la Maîtrise de l'Energie) supports students and teachers exchanges with eastern Europe in the frame of European Union Programs such as TEMPUS and COPERNICUS : beside ENSAM and SPETU, TUS (Technical University of SOFIA), MPEI (Moscow Power Engineering Institute), LRTI (Laboratoire de Recherches en Techniques Inductives) are to be involved in common production of hypermedia materials on the basis of energy savings, environmental protection, renewable energies and electroheat techniques promotion.

THE CONTRIBUTION OF THE UIE

The experience shown up to now is proof of the crucial role of the UIE in hypermedia production in electroheat, which could not grow in the long term otherwise than to an international scale. This is taking into account of the large number of tasks that need to be accomplished in the domain and also of the necessary opening of the market for electroheat equipment constructors. We trust that the present conference will open other perspectives for profitable meetings between industrial partners and universities in all the represented countries.

REFERENCES

J. de ROSNAY: 'Le Macroscope', Editions du Seuil, Paris 1975

J.P. MEINADIER:'L'Interface Utilisateur', *Pour une informatique plus conviviale*, Informatique et Strategie, DUNOD Paris 1991.

P. WHITE: The Idea Factory, *Learning to think at the MIT*, PLUME-PENGUIN Books USA, 1992

C. McKNIGHT, A.DILLON and J. RICHARDSON: HYPERTEXT, *a psychological perspective*, ELLIS HORWOOD SERIES IN INTERACTIVE INFORMATION SYSTEMS, London 1993.

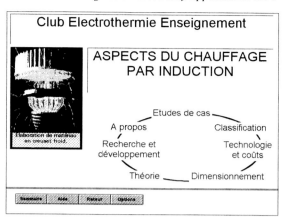
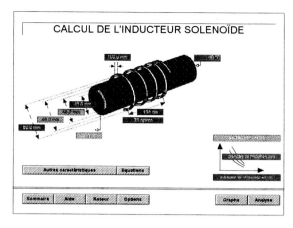
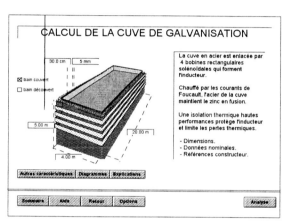
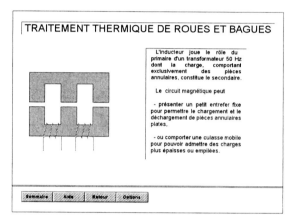
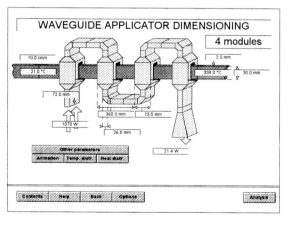
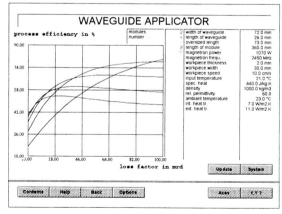

RE 32

UIE XIII Congress on Electricity Applications 1996

TEMPERATURE DISTRIBUTION

Workpiece temperature in °C

modules number	4 modules
width of waveguide	72.0 mm
length of waveguide	26.0 mm
oversized length	73.0 mm
length of module	360.0 mm
magnetron power	1070 W
magnetron frequ	2450 MHz
workpiece thickness	2.0 mm
workpiece width	30.0 mm
workpiece speed	10.0 cm/s
input temperature	21.0 °C
spec. heat	440.0 J/kg K
density	1000.0 kg/m3
rel. permittivity	60.0
loss factor	100.0 mrd
ambient temperature	23.0 °C
int. heat tr.	7.0 W/m2 K
ext. heat tr.	11.0 W/m2 K

covered distance by workpiece in m

HEAT DISTRIBUTION

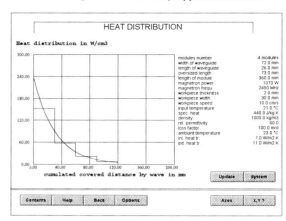

Heat distribution in W/cm3

cumulated covered distance by wave in mm

WATER BEHAVIOUR AT RF AND UHF FREQUENCIES

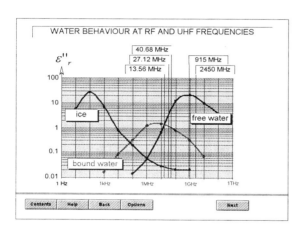

40.68 MHz, 27.12 MHz, 13.56 MHz, 915 MHz, 2450 MHz — ice, free water, bound water

1D PROPAGATION AND ABSORPTION

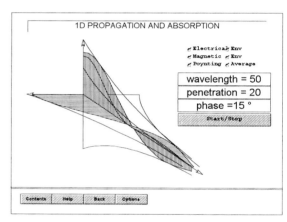

- Electrical Env
- Magnetic Env
- Poynting Average

wavelength = 50
penetration = 20
phase = 15 °

SPLITTED WAVEGUIDE IN TE10 MODE

$$\frac{1}{\lambda_0^2} - \frac{1}{\lambda_g^2} = \frac{1}{4a^2} \Rightarrow \lambda_g = \frac{\lambda_0}{\sqrt{1 - \frac{\lambda_0^2}{4a^2}}}$$

$$\vec{E}_{10} = E_M \cos\frac{\pi x}{a} e^{kz} \cdot \vec{y}$$

penetration $\quad p = \frac{\lambda_0}{2\pi \varepsilon''_r} \frac{a}{e} \sqrt{1 - \frac{\lambda_0^2}{4a^2}}$

THE PRINCIPLE OF MAGNETRON

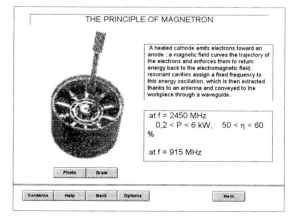

A heated cathode emits electrons toward an anode; a magnetic field curves the trajectory of the electrons and enforces them to return energy back to the electromagnetic field; resonant cavities assign a fixed frequency to this energy oscillation, which is then extracted thanks to an antenna and conveyed to the workpiece through a waveguide.

at f = 2450 MHz
   0,2 < P < 6 kW,   50 < η < 60 %

at f = 915 MHz

Experimental and Numerical Investigation of Industrial Induction Heating Processes

A. MÜHLBAUER AND H.-J. LESSMANN
Institute for Electroheat, University of Hannover, Germany

Traditionally, the research at the Institute for Electroheat in Hannover, Germany, is closely connected with the development of electrothermal processes in industry and is therefore carried out in cooperation with the companies concerned. The different projects described in this paper are examples of successfull and productive cooperation between industry and university, between application and research. Accordingly the main emphasis of research at the Institute for Electroheat the projects belong to the field of induction heating and induction melting, i.e. the float zone silicon single crystal growth, the transverse flux heating of strips, the induction tube welding and the induction furnace with cold crucible. In all cases the necessity of an efficient further development of the industrial processes requires the use of powerful numerical tools combined with selected experiments for verification. The computer programs have been especially designed for the specific application leading to short calculation time and a high suitability for the parameter studies to be carried out.

Traditionnellement, la recherche à l'Institut de chaleur électrique de Hanovre, Allemagne, est étroitement liée au développement de procédés électrothermiques dans l'industrie et est, par conséquent, menée en coopération avec les compagnies concernées. Les différents projets décrits dans ce document sont des exemples d'une coopération réussie et fructueuse entre l'industrie et l'université, entre l'application et la recherche. En conséquence, à l'Institut, l'accent est principalement mis sur la recherche. Les projets appartiennent au domaine du chauffage et de la fusion par induction, c'est à dire de la zone flottante du tirage de monocristaux de silicium, du chauffage diélectrique transverse par ruban chauffant, de la soudure de tubes par induction et du four à induction à creuset froid. Dans tous les cas, la nécessité d'un développement ultérieur efficace de procédés industriels exige l'utilisation d'instruments numériques puissants associée à des expériences sélectionnées en vue de vérification. Les programmes informatiques ont été spécialement conçus pour l'application spécifique aboutissant à un temps de calcul réduit ainsi qu'à une haute aptitude concernant les études de paramètre devant être menées.

FLOAT ZONE SILICON SINGLE CRYSTAL GROWTH

The float zone (FZ) method is used for the growth of silicon single crystals of large diameters. The electric current is induced to melt the feed rod (fig. 1), which is pulled from above. The molten silicon forms the liquid zone with the free surface and crystalizes at the growing interface. The obtained crystal is pulled down. The interface shapes are strongly coupled with the distribution of the electromagnetic (EM) field, and as a consequence with the shape of the inductor. The quality of the growing crystal depends on the shape of the growth interface, the temperature gradients and on the fluid flow near this interface. Especially the fluid flow influences the shape of the growth interface and thus determines the resistivity distribution in the grown crystal. As it is very difficult to investigate and develop the FZ method experimentally, numerical simulation is necessary.

The main objectives of the investigations carried out in cooperation with the industrial partner is to study the interface shape and the flow velocity field depending on the process parameter and to find the limits of the process as well as a set of optimal growth parameters to improve the crystal properties[1]. The EM field, the temperature field and the melt motion are simulated. The unknown shape of the molten zone is calculated as a coupled EM-thermal-hydrodynamic problem. All driving forces in the melt, i.e. EM-, buoyancy, thermocapillary and centrifugal, are taken into consideration. Fig. 1 shows the calculated EM-, temperature and flow velo-city fields for a typical FZ system.

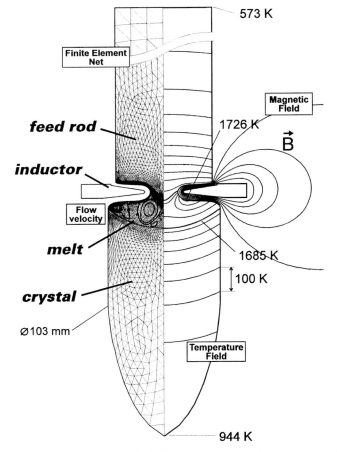

Fig. 1: Geometry of the FZ system, finite element net and calculated EM-, temperature and flow velocity fields

The calculation of the melt flow is usually started with zero velocity field. In the case of instability, after some transient periods the quasisteady oscillation regimes are searched. An example of the typical structure of the unsteady melt motion at three different moments of time is shown in fig. 2. The vortex at the outer region of the melt is rather steady but noticable oscillations of the flow are observed at the central region.

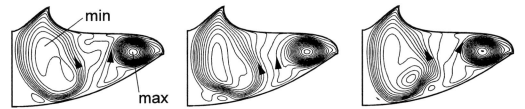

Fig. 2: Meridional melt flow velocity field at 3 different moments of time; line direction corresponds to direction of melt motion, density of lines corresponds to magnitude of velocity; crystal diameter 100 mm, growth rate 3.3 mm/min, rotation rates 5 rpm (single crystal) and 20 rpm (feed rod, opposite direction)

The calculated distributions of the oscillation amplitude of the tangential velocity at the growth front correspond to the radial resistivity variations measured in the grown crystal by the photo-scanning method[1].

TRANSVERSE FLUX HEATING OF STRIPS

In recent years, induction heating installations have increasingly been used for heating plates, strips and thin slabs. Induction heating offers significant advantages over conventional gas-fired furnaces because the heat is generated in the workpiece itself. This results in less requirement for space, reduced heating times and lower heat and oxidation losses.

A program package has been developed, which enables the numerical simulation of the current and temperature distributions in transverse flux induction heating arrangements, to investigate the physical correlations and to find guidelines for an optimum design[2]. For the experimental verification of the numerical model and the calculation results the temperature distribution in the strip during its passage through an industrial pilot plant is recorded with the aid of an infrared measuring system.

In a transverse flux induction heating arrangement like that shown in fig. 3, the inductor coils are separately arranged on both sides of the strip to be heated which runs through the installation. A magnetic core is arranged above each inductor; fig. 3 shows only the lower cores. In rough approximation, the currents induced in the strip are a projection of the inductor current onto the strip surface. This results generally in an inhomogeneous temperature distribution. The arrangement in this example is characterised by the fact that the inductors are laterally offset against the longitudinal axis of the strip, whereby a protruding inductor is formed at one strip edge leading to an overheating and a receding inductor at the other leading to a temperature minimum. The resulting temperature distribution is shown in fig. 4. A great variety of different temperature profiles can be obtained by various combinations of the two edge effects.

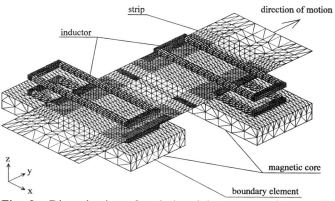

Fig. 3: Discretization of an industrial transverse flux configuration

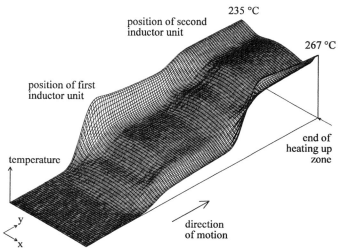

Fig. 4: Calculated distribution of the steady state temperature field in the strip

Beside the investigation of the configuration shown in fig. 3 parameter studies have been carried out in order to determine the influence of the most important geometrical quantities on the power sources and the temperature distribution. The aim of the research is to define guidelines for an optimum design depending on the strip temperature distribution demanded at the heater outlet. The first parameter studies with the program developed are basing on a simple configuration consisting of the strip and a pair of one turn inductors which are arranged symmetrically to both sides of the strip and to the strip axis.

INDUCTION TUBE WELDING

Beside the quality of the tube the requirement of power is one of the most important aspects for the operating authority of the tube-welding facility. The requirement of energy per meter of tube should be as small as possible. To achieve this by a certain design, the influence of the operational and geometrical parameters on the power have to be known. Because of the complicated threedimensional (3D) current distribution in the tube and the coupling between the field of heat sources and temperature, the knowledge mainly based on empirical investigations. Until now, exact quantitative correlations were hardly known.

At the Institute for Electroheat a program package has been developed to determine the 3D current distribution in arbitrarily shaped configurations[3,4]. Basing on the Boundary Element Method the 3D-program is also suitable to analyze the electromagnetic processes in induction tube welding installations. As an example, fig. 5 shows the current distribution in the tube. In a close cooperation with an

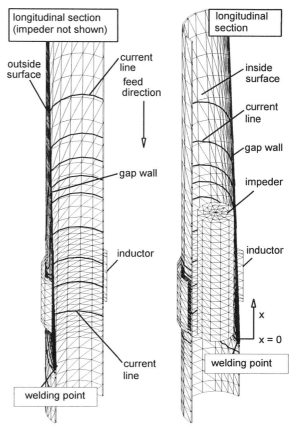

Fig. 5: 3d-current distribution in a tube with cylindrical impeder

industrial partner another computer program has been developed especially for the investigation of tube-welding configurations[4]. It is made for a configuration consisting of a one-turn cylindrical inductor, a tube with the vee-gap and an impeder. Among other things the program calculates integral quantities, such as the total power, the active resistance, the inductance and the inductor current necessary to reach the required temperature in the welding point. Besides, the distributions of the heat sources and the temperature in the gap walls are determined (fig. 6). The practical suitability has been verified by measurements at industrial welding installations.

The parameter investigations carried out with the program for low carbon steel tubes describe how it is possible to reduce the requirement for power[4]. Among the free parameters, the length of the impeder and the distance between the inductor and the welding point are the most important ones. The welding speed also shows a distinct influence on the requirement for energy. However, usually the speed is not completely optional, because it determines the whole tube production. Anyway, the speed should be chosen as high as possible. The impeder should extend from the welding point downstream at least two tube-diameters beyond the inductor. Compared with a configuration without an impeder the total power is reduced by about 50 %.

The results correspond well with the known qualitative statements from the practice. However, these qualitative statements are not sufficient for an energetical optimization of tube-welding configurations. The program "*Rohr*" shows for the first time the quantitative correlations for practical configurations. It allows to minimize the requirement for energy per meter of tube taking into consideration the effect of all important parameters, which are influencing each other.

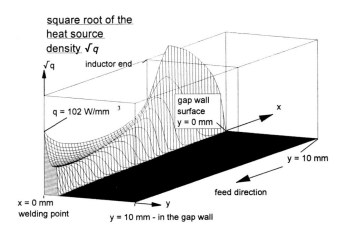

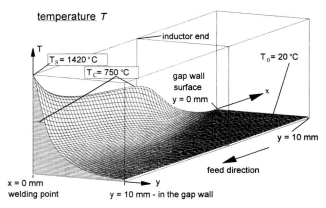

Fig. 6: Calculated heat source density and temperature distribution in the gap walls for low carbon steel

INDUCTION FURNACE WITH COLD CRUCIBLE

Components subject to extreme thermal and mechanical exposure, such as valves, are a focal point in the further development of internal combustion engines. Titanium alloys make excellent materials for such components. With the exception of aerospace, however, they are little used in industry. The reason can be found in the production process currently used for the titanium alloys, which is extremely complex and cost-intensive, since the reactivity of the melt precludes melting in a conventional induction furnaces with a ceramic lining.

Production at a rational price level becomes possible with a new manufacturing process using the induction furnace with cold crucible (IFCC). It makes it possible to mix, superheat and cast the required alloy in a single operation. The water-cooled copper crucible features slits in order to permit the transmission of the electrical energy from the inductor to the material to be melted (fig. 7 and fig. 8).

The development status achieved up to now demonstrates that the cold wall induction crucible furnace will be capable of achieving an industrial breakthrough for titanium alloys in the automotive sector. This objective is now to be realized in the form of a joint project[5]. The parties involved include basic research establishments, plant engineering companies and users.

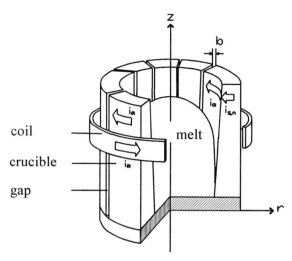

Fig. 7: Schematic of an induction furnace with cold crucible

At the Institute for Electroheat the melting process in an IFCC is investigated theoretically and experimentally as well. Due to the nature of the energy transmission the 3D-EM-field distribution determines the process and therefore has to be analysed in order to find an optimal design of the IFCC. For this reason a 3D (fig. 9) and a 2D numerical model have been developed. They are now used to investigate the influence of different parameters on the EM-field and the power induced in the charge. First results confirm the distinct influence of the number of slits, the width of the slits and the penetration depth. The results of the numerical calculations are compared with measurements at an industrial IFCC and with results of an integral analytical model.

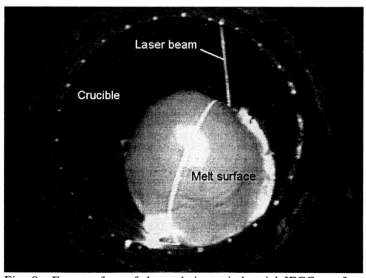

Fig. 8: Free surface of the melt in an industrial IFCC; surface shape detected by a laser beam

REFERENCES

1. A. MUEHLBAUER, A. MUIZNIEKS, J. VIRBULIS, A. LUEDGE and H. RIEMANN: ´Interface Shape, Heat Transfer and Fluid Flow in the Floating Zone Growth of Large Silicon Crystals With the Needle-Eye Technique´, *J. Crystal Growth*, 1995, **151** .66-79.

2. J.-U. MOHRING, H.-J. LESSMANN, A. MUEHLBAUER and B. NACKE: ´Numerical and Experimental Investigations into Transverse Flux Induction Heating´, *ETEP Eur. Trans. on Electr. Power Engng.*, 1996, **6** in print.

3. A. MUEHLBAUER, A. MUIZNIEKS and H.-J. LESSMANN: ´The Calculation of 3D High-Frequency Electromagnetic Fields During Induction Heating Using the BEM´, *IEEE Trans. on Magn.*, 1993, **29** (3).1566-1569.

4. H.-J. LESSMANN: ´Elektromagnetische und thermische Vorgänge beim induktiven Längsnahtrohrschweißen´, *Fortschr.-Ber. VDI Reihe 2 No.325*, VDI-Verlag, Düsseldorf, 1994.

5. A. MUEHLBAUER, E. WESTPHAL, A. CHOUDHURY, M. BLUM: ´Neues Fertigungsverfahren für Automobilventile aus Titan-Aluminium-Legierungen´, *elektrowärme international*, 1995, **53** (1).13-17.

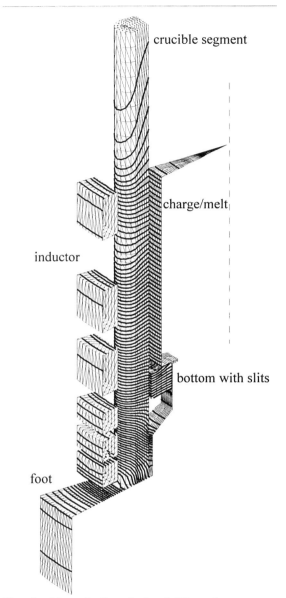

Fig. 9: Numerically calculated 3D surface current lines in one segment of the industrial IFCC; 26 segments in total

High Performance Switched Mode Power Supply for Microwave Heating Application

R. PETKOV   and   Prof. L. HOBSON

Swichtec Power Systems   *University of Glamorgan*
PO Box 11-188   *Pontypridd*
Christchurch, New Zealand   *Mid Glamorgan CF37 1DL, UK*

This paper explores the implementation of the switched mode power conversion technique in power supplies for industrial microwave heating. This leads to a significant improvement of power density, input power factor, control characteristics and efficiency of the conventional power supplies. Switched mode power supplies (SMPS) provide also very favourable operating conditions for the magnetron leading to extended life expectancy.

Power conversion stage of the SMPS was based on a quasi-resonant power converter employing output transformer parasitics in the resonant network. Power factor corrector was employed as a preconverter stage of the SMPS to shape the input current and satisfy power factor requirements. The later stage also provides very low ripple of the magnetron voltage which is essential for some specific microwave heating applications. A 2.5kW, 4kV, 40 kHz prototype of the SMPS was built and tested and practical results showed that SMPS efficiency and input power factor at full load were as high as 0.92 and 0.995 respectively. Output power control is very broad and precise and the control range matches power control capability of the magnetron used. Prototype power density achieved was $350 \, W/dm^3$.

INTRODUCTION

Conventional, mains frequency power supplies currently used to supply the magnetrons consist of a single phase half-wave voltage doubler with a high voltage, high leakage inductance transformer[1]. It is a very simple and reliable but quite bulky and not precisely controlled unit.

Comparatively small in size and precisely controlled switched mode power supplies (SMPS) for use in microwave ovens have been developed [4, 5, 6, 7, 8] and introduced as an alternative to conventional, mains frequency power supplies. They comprise a high frequency DC/DC converter utilising an IGBT, GTO or BJT as the main switching device, a high frequency ferrite transformer and an output rectifier. Schematically they are more complicated and still more expensive, but due to their excellent volumetric parameters, both easy and precise output power control and high input power factor, they tend to replace the conventional, mains frequency power supplies.

This paper reports the achievements of an investigation into a high performance SMPS for industrial microwave heating.

SMPS SPECIFICATIONS AND SCHEMATIC

The specifications of the magnetron SMPS required by the sponsoring organisation EA Technology Ltd, UK were as shown in Table 1.

Table 1. SMPS specifications

PAPAMETER	VALUE
Maximum output power [W]	2500
Output power control range	10 - 100% of the maximum power
Output voltage (dc) [V]	3900
Output voltage ripple [%]	< 1
Input voltage (ac) [V]	187 - 274
Input power factor	> 0.95
Magnetron type	NL10250
Filament current [A]	20A, controlled with the anode current

BLOCK SCHEMATIC

The Block-schematic of this magnetron power supply is shown in Fig. 1 and comprises an EMI filter, full-bridge rectifier, preconverter, double-forward quasi-resonant converter, filament power supply and a control circuitry. The latter can be split into two sections: a)preconverter section, driving and protecting the preconverter and b)converter section, driving and protecting the converter. Most of control and protection functions of these two sections, except overvoltage protection (O.V.PROT) and current limit (CURR. LIMIT), are available in the TK84819 combo-controller [3] which is designed to drive simultaneously in a PWM mode both the preconverter and converter. This simplifies greatly the control system of the SMPS and increases its reliability.

POWER CIRCUIT

A boost converter topology is implemented as the SMPS preconverter (Fig. 1) which is driven in power factor correction mode and features high input power factor. The nondissipative snubber (SNUBBER) connected in parallel with the switching transistor reduces its turn-off loss and provides high preconverter efficiency. The preconverter output voltage is constant, low ripple and independent of input voltage variations leading to a very low crest factor (peak value/average value) of the magnetron current (Table 2) and in turn extended life expectancy of the magnetron.

A novel, hybrid, quasi-resonant power converter topology has been used for the converter stage of the SMPS. It combines the favourable switching conditions for the active devices with the effectiveness of PWM control method and therefore it has been patented [2].

CONTROL CIRCUIT

Preconverter Section of the Control Circuit

The "peak-current" control strategy is employed in the TK84819 combo-controller for achieving high power factor operation and the control blocks involved (Multiplier, Adder, Voltage Divider) form a sinusoidal reference signal to define transistor switching instants. The reference signal produced by the adder is applied to a pulse width modulator comprising a comparator (comp) and a ramp generator (RAMP GEN.). In addition to input current shaping, the current transformer monitoring transistor current is used by the control system (CURR. LIMIT block) for over current protection as well.

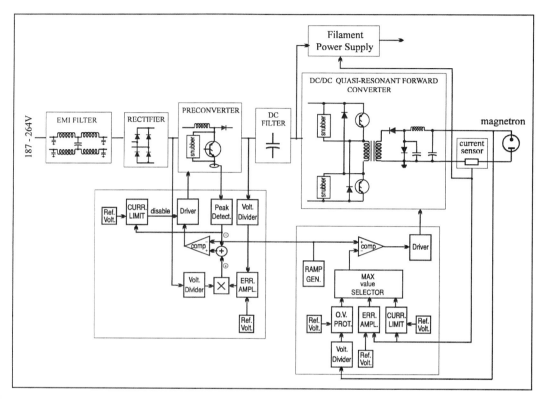

Fig. 1. Block-schematic of a 2500W magnetron SMPS prototype

Converter Section of the Control Circuit

This section comprises three Error amplifiers, three Reference voltage blocks, a Voltage divider, a Maximum value Selector, a pulse width modulator (comprising a comparator COMP and a ramp generator RAMP GEN.) and a Driver. The first error amplifier (ERR. AMPL.) controls the magnetron anode current and in turn magnetron power, the second one (CURR. LIMIT) limits the output current in the case of an overload and the third one (O.V. PROT) monitors the output voltage and limits it to an acceptable value defined by the magnetron. The maximum value selector chooses the active (controlling the pulse width) error amplifier according to the highest output signal magnitude. This block gives control priority to the protection error amplifiers in the case of abnormal load conditions (overvoltage or overcurrent).

The operating frequency of the converter (and preconverter) is defined by the stabilised (40kHz) ramp generator (RAMP GEN.). This generator and the comparator (comp) form a pulse width modulator which controls the turn-on time of the switching transistors in relation to the MAX value SELECTOR output signal. This operating principle is well known and employed in most PWM controllers. The driver (Driver) amplifies and shapes the control pulses for both switching transistors.

POWER CHARACTERISTICS

Fig. 3 presents power characteristics measured of the overall microwave heating system comprising the developed SMPS and NL10250 magnetron. As can be seen, the microwave power can be controlled down to 10% of its maximum value and preconverter + converter system efficiency is 85% at low power and 92% at maximum power. Microwave system efficiency, which accounts also for the filament power and magnetron efficiency, is good. Input power factor has a minimum value of 0.96 at low power and approaches 0.995 at maximum power. Experimental waveforms of SMPS input voltage and current at full power are shown in Fig. 4 and one can see that the input current is nearly sinusoidal and in phase with the input voltage.

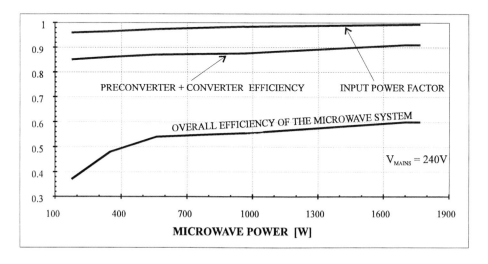

Fig. 3. Power characteristics of the 2500W magnetron SMPS prototype

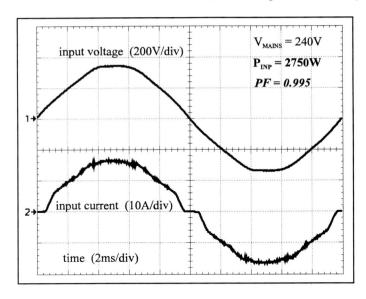

Fig. 4. SMPS input voltage and current at nominal conditions

CONCLUSION

The practical results of the magnetron SMPS are summarised in Table 2. As can be seen, the developed prototype meets all target specifications and the results achieved in terms of the efficiency, power density, output power control range and input power factor are very good.

The weight and volume of the prototype are approximately 20% and 25% respectively of those achieved by an equivalent mains frequency power supply.

Input power factor value is approaching unity due to input current shaping (see the preconverter stage). Output power control is very precise and the control range matches power control capabilities of the magnetron used.

Filament voltage is controlled according to the specified correlation between magnetron anode current and filament voltage.

Anode current crest factor value is approaching unity and therefore this SMPS is suitable for some specific applications of microwave heating (microwave plasma) requiring very low anode current ripple.

Converter efficiency varies between 0.85 - 0.92 over the whole output power control range. The overall efficiency of the microwave system is good as well.

Overvoltage and overcurrent protection of the converter output is provided which increases SMPS reliability.

In general, the characteristics of the prototype are very close to the characteristics of the "ideal" magnetron power supply and therefore it is currently being prepared for commercial production. In addition, a patent application [2] has been made for the power converter stage of the SMPS.

Table 2. Practical results of the magnetron SMPS prototype

PRACTICAL RESULTS	
Maximum output power [W]	2500
Output power control range [%]	10 - 100
Input voltage (RMS) [V]	187 - 264
Output voltage (DC) [V]	3900
Output voltage ripple [%]	< 1
Efficiency at full power [%]	92
Power density [W/dm^3]	350
Input power factor:	
at full power	0.995
at low power	0.95
Anode current crest factor	1.02
Filament current	specified

REFERENCES

1. J. PLATTS: "Microwave Ovens", Peter Peregrinus Ltd, 1991.

2. R. PETKOV: "Magnetron power supply" Patent pending No. GB 9418392.8.

3. TOKO AMERICA: "TK84819 power factor controller" *TOKO AMERICA application note,* 1992.

4. A. BONNET and J. DANNER: "Resonant mode monoswitch converters with corrected power factor, reduced harmonics and limited inrush current", *Proceedings PCIM'92, Nurnberg 1992, 373-383.*

5. B. TAYLOR: "A new converter topology uniquely suited to high output voltage applications" *Proceedings PCIM'93, Nurnberg 1993, 1-10.*

6. K. WADHIA: "Using IGBT's for switching mode power supplies", *TOSHIBA application note* Ref. No: X3502, April 1990.

7. C. HAMMERTON: "The GTO as a fast thyristor", *Proceedings EPE'89, Aachen 1989, 81-85.*

8. I. TAKASHI: "IGBT for microwave oven.", *Fuji Electric Review,* No 151, 1990.

High Temperature Resistance Heating in Semiconductor Production. New Process Control System.

L. MICHALSKI; D. SANKOWSKI.

Technical University of Lodz, Department of Electroheat, Al. Politechniki 11, 90-924 Lodz, Poland

W. LOBODZINSKI; M. ORZYLOWSKI.

Industrial Electronic Institute, ul. Dluga 44/50, 00-240 Warsaw, Poland

ABSTRACT

The paper presents an introduction to the technology of semiconductor production in which extremely precise temperature control is necessary. The highly specialised electric diffusion furnace and its properties are described and the principles of applied adaptive, cascade temperature control are discussed. Experimental results confirm the usefulness of described installation.

RÉSUMÉ

Le rapport presente une introduction a la technologie de la production de semiconducteurs, dans laquelle le contrôle de temperature d'une precision trés poussée et necessaire. On a decrit un four a' diffusion et ses proprietes ainsi que les principes du contrôle de température utilisé, adaptative a cascade.

Les resultats experimentaux ont confirmé l' utilité de l'installation presentée.

INTRODUCTION

The electric furnaces for diffusion and oxidation are listed among the most important technological equipment in the complete lines for production of VLSI semiconductor structures. For example the production of VLSI memory chips of Mega- or Gigabites capacity or of microprocessors, demands many (up to 20) thermal treatment operations. Table 1 presents an outlook of actual trends in this production for the years 1995-2005[1]

Table 1. Trends in production of VLSI semiconductor structures						
Parameters	Unit	Year			
		1995	1998	2001	2005
Path width	μm	0.35	0.25	0.18	0.13
DRAM	Mb/Gb	64 M	256 M	1 G	4 G
Diameter of Si-wafers	mm	200	200	300	300
Area of VLSI structure	mm^2	190-450	280-660	420-750	640-900

The following operations are executed in the diffusion furnaces: doping prediffusion (N or P), doping redistribution, oxidising (isolating layers), contacts inserting.

A complete production equipment is composed of:
1. multisection tubular furnace with quartz or SIC reactor,
2. technological gas feeder,
3. hermetic loading chamber equipped with automatic loading and unloading arrangement of Si-wafers,
4. microprocessor based process control.

The above mentioned technologies demand very precise temperature control of the batch placed in the central part of a tubular furnace. The control accuracy should be ± 0.2° C at a level of about 900 °C to 1200 °C for dwell period and ± 1 °C for ramp period. The uniformity of temperature distribution along the central part of rector tube should be better than 0.5 °C

High cost of the furnace batch, comparable with the cost of the whole equipment and the fact that even short-time deviation from the pre-imposed temperature limits may cause the irreversible destruction of the batch, demand very precise and reliable control and a special construction of the furnace.

FURNACE DESCRIPTION

The cross-section of a typical tubular furnace applied in semiconductor production is shown in Fig.1. The furnace is equipped with 5 sections of heating elements and a quartz reactor of 200 mm diameter and 2 m long. The three central sections of heating elements ensure an uniform temperature distribution in 1 m long central part of the reactor in which the batch is placed. The two end sections, called also active thermal insulation, are used to limit the axial heat losses from central sections and help in assuring along them the linear temperature distribution.

UIE XIII Congress on Electricity Applications 1996 **RE 51**

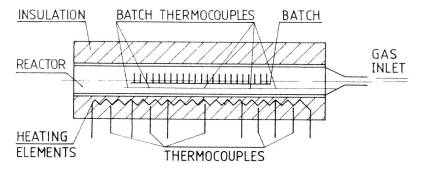

Fig.1. Cross-section of the diffusion furnace.

The structural model of the furnace[2] is shown in Fig.2 where the following denotations are used: u_{hi} - input power, y_{hi} - temperature close to heating elements, y_{bi} - batch temperature, i=1...5. The signals y_{bi} are measured at locations as shown in Fig.1.

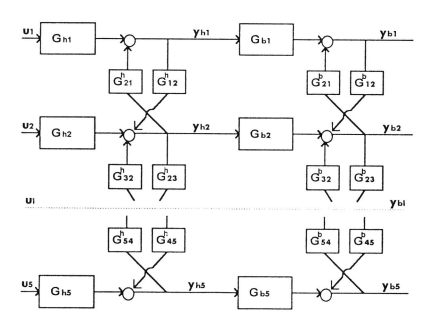

Fig.2. Structural model of the furnace

In the model the transfer function $G_{hi}(s)$ and $G_{bi}(s)$ are relevant for heat flow in each section of the heating elements and batch respectively; $G_{ik}^h(s)$ and $G_{ik}^b(s)$ where $i \neq k$, are relevant for cross-section interaction. The functions $G_{hi}(s)$ and $G_{bi}(s)$ can be described by first order inertia with time delay[3]. As the cross-section interaction is relatively weak[4], the transfer functions $G_{ik}^h(s)$ and $G_{ik}^b(s)$ can be reduced to first order inertia elements.

TECHNOLOGICAL PROCESS

The sequence of technological operations is as follows (Fig. 3):

1. The support carrying the cold Si-wafers is pushed into the reactor, maintained at the stable starting temperature $\vartheta_s \approx 700\ °C$.
2. Temperature of the batch is equalised,
3. Operator initiates the automatic identification of system parameters by MBS test signals, for preliminary setting of temperature controllers,
4. Programmed linear temperature increase (ramping) starts,
5. Approaching the preset temperature ϑ_o (1000 °C to 1250 °C) MBS identification of system parameters is automatically initiated, resulting in updating of controller settings.
6. Thermostatic temperature control starts,
7. Programmed flow of technological gases, during which prediffusion, diffusion and oxidation occurs.
8. Programmed cooling down of batch to the starting temperature ϑ_s
9. Unloading of the ready batch.

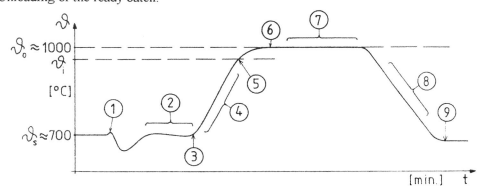

Fig. 3. Sequence of technological operations

PROCESS CONTROL

A simplified block diagram of process control of semiconductor production showing only one section and no section interactions is presented in Fig. 4. $G_{h1}(s)$ and the corresponding $G_{b1}(s)$ are as shown before in Fig. 2.

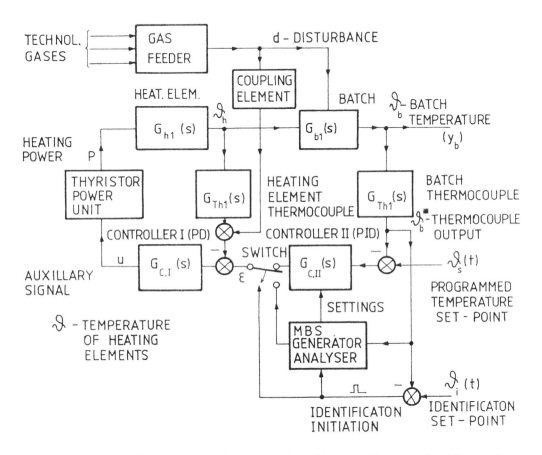

Fig. 4 Block diagram of process control of one section of heating elements of a diffusion furnace

The control algorithm[2,5] is a multiloop adaptive, cascade control with feed-forward disturbance correction. In each section there are two control loops. The output signal of the controller II determines the set-point of the controller I, which is simultaneously corrected by the output signal of the thermocouple placed close to heating elements of the section.

To shorten the settling time after any disturbance caused by the changing flow of technological gases, the disturbance signal is also fed to the input side of controller I. The settings of controller II, operating in a closed loop comprising important lags, are crucial for the stability and dynamics of the system. Therefore regarding the furnace parameters changing with temperature, they have to be on-line identified at least at the level of starting temperature ϑ_s ($\vartheta_s \approx 700\,°C$) and approaching the final operating temperature ϑ_0 (marked 3 and 5 in Fig. 3).

To this aim Multifrequency Binary Signals generator-analyser is applied. Multifrequency Binary Signals (MBS) constitute a versatile sub-group of the important class of binary signals. Their main power is concentrated in a few chosen harmonics, so that a few points of the frequency response of the system can be obtained in one short on-line experiment. The MBS have excellent signal-to-noise ratio for dominant harmonics. Based on step functions they are ideally suited for microprocessor environment. The short identification procedure is automatically switched on when reaching the identification set-point $\vartheta_i(t)$ (Fig. 3, Fig. 4).

A detailed theory and applications of MBS are presented elsewhere[6,7,8].

In each section all the functions of both controllers as well as MBS generation, identification and adaptive setting of controller parameters are implemented in microprocessor environment. In practical realisations the described method of control can be enriched by variable structure control leading to suboptimal control[2] with accuracy better than $\pm 0.2\,°C$ in steady state at the temperature level of $1000\,°C$

The whole system operates in adjustable 10 to 30 s long, control steps.

EXPERIMENTAL RESULTS

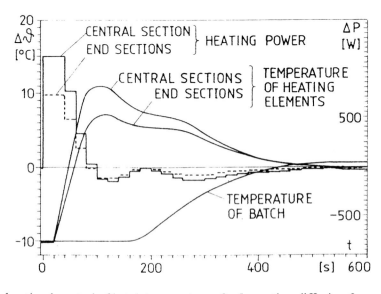

Fig. 5. Suboptimal control of batch temperature of a 5 - section diffusion furnace

The set-point step input response of a closed-loop temperature control of a 5-section diffusion furnace is shown in Fig. 5. The presented response is relevant for a suboptimal control based on a simplified gain matrix of reduced order.

The value of linear, quadratic performance index of the suboptimal control is 1.4 % bigger than that of the optimal control. Nevertheless the results of suboptimal control are well within the permissible values.

CONCLUSIONS

The cascade temperature controller implemented and applied for the diffusion furnace grants high accuracy - better than $\pm 0.2^{0}C$ in steady state at the temperature level of $1000^{0}C$

REFERENCES

1. S. WOLF: 'The Future Fab' *Solid State Technology*, 1995, 48-55.
2. M.ORZYLOWSKI:'Temperature Control of Multisection Electric Furnace; *Elektrowärme International*, 1995, (**53,3**). 169-170.
3. W. LOBODZINSKI, D. SANKOWSKI, J. KUCHARSKI: 'Monitoring the parameters of an electric resistance furnace during the start-up period'. *5th International Congress on Condition Monitoring and Diagnostic Engineering Management COMADEM'93*, 1993, Bristol, UK.
4. D.SANKOWSKI: 'Frequency Domain Diffusion Furnace Model', *Elektrowärme International*, 1989, **47,B**, 131-135.
5. T.FURNIJAKA and M.ARAKI:'Discrete-time Optimal Control of Systems with Unilateral Time-delays', *Automatica*, 1987,23(6).763-765.
6. A.VAN DEN BOS: 'Construction of Binary Multifrequency Test Signals'. *IFAC Symp.* Prague, 1968, Paper 4-6.
7. J. McGHEE et al.: 'Recursive data measurement for multifrequency analysis by the correlation method'. *10th International Conference on System Engineering, Coventry, UK*, 1994, pp.817-822.
8. D.SANKOWSKI et al.:'Application of Multifrequency Binary Signals for Identification of Electric Resistance Furnaces in *'Perturbation Signals for System Identification'*, Prentice Hall, University Press, Cambridge, 1993, Chapter 8.

Artificial Neural Networks:
Dryer Process Modelling And Control

S.JAY, T.N.OLIVER

Aston University, Birmingham, UK.

ABSTRACT

It has been shown that improved understanding and control of drying processes can lead to savings in energy, together with increased production throughput and enhanced product quality. Drying is still not well understood and normal computational modelling techniques have been hampered by the sheer complexity of the processes. Artificial neural networks offer a convenient platform for capturing the non-linearity and dynamics inherent in drying systems. This paper discusses the neural network modelling of the microwave drying of gypsum cove material. The experimental analysis is described together with development of the model architecture. The model's ability to predict moisture loss characteristics, including a measure of product quality, are revealed highlighting the accuracy of the applied technique. Comments are made concerning application of the model within the fields of control and of equipment evaluation.

Il a été démontré que la compréhension et le contrôle des processus de séchage peut permettre de réaliser des économies d'énergie et d'optimiser la productivité et la qualité des produits fabriqués. Le séchage n'est pas entièrement résolu et le développement des techniques ordinaires de modelage informatique a été ralenti par la complexité intrinsèque des processus. Les réseaux neuronaux artificiels offrent une plan d'analyse idéal de la non linéarité et des caractéristiques dynamiques inhéréntes aux systèmes de séchage. Cet article commente le modelage par réseau neuronal du processus de séchage par micro-ondes des corniches en gypse. Il décrit l'analyse expérimentale ainsi que le développement de l'architecture de modelage. La capacité de prédiction des caractéristiques de perte d'humidité par modelage et la méthode de détermination de la qualité du produit y sont révélées, soulignant le précision de la technique employée. L'article commente l'application du modelage aux domaines du contrôle et de l'evaluation de l'équipement.

INTRODUCTION

INDUSTRIAL DRYING

Industrial drying is an energy intensive operation, consuming an estimated 3×10^{11} MJ of energy annually in the UK.[1] In recent years advances in drying have been achieved.[2] However drying is still not well understood (meaning that good phenomenological models may not be readily available), due mainly to the complexity involved in the simultaneous transfers of heat, mass and momentum during the process. Modelling has increased our knowledge, but many models are complex and in many cases require data that is either unavailable or outdated. Escalating energy costs and more intense competition provide the impetus for continued efforts in improving drying efficiency.

CONTROL AND MODELLING OF INDUSTRIAL DRYING

Until the advent of process computers, manual and automatic feedback systems were the most commonly used methods for drying control.[3] Many of these techniques are still in operation in older plants. The availability of process computers and advances in sensor technology have increased the use of control systems combining feedforward and feedback loops. Control of drying processes is an important factor which is directly related to an understanding of the drying system. A poorly controlled process is likely to be wasteful, both in terms of energy and inferior quality product. An understanding of the process will enable both the control and design of a dryer to be optimised, enhancing product quality, maximising throughput rates and minimising energy costs. Many systems are still based upon empirical data obtained from a vast number of experimental tests, scaling up of small scale tests and past design experience.

Feedforward control techniques have superseded feedback systems where there is a significant time lag between a control change being made to an input and its effect being felt on the outputs. Feedforward control depends upon having an accurate mathematical model of the system. Drying systems, like most industrial processes are very complex and sufficiently accurate models do not exist.[3] Due to the non-linear dynamic nature of drying processes linear based modelling and control methods are unsuitable. The characteristics of drying may vary because of changing raw materials properties, gradual changes in process equipment and changes in environmental variables.

NEURAL NETWORKS FOR MODELLING AND CONTROL

An approach that has evolved as a powerful computational technique in the past few years is artificial neural networks.[4,5] In chemical engineering neural networks have been applied to solve problems of fault diagnosis, sensor data interpretation, prediction of dryer performance and process control.[4,5,6]

A variety of neural network configurations and training algorithms have been developed over the past 20 years.[5] Multilayer perception architectures have been successfully applied to many modelling processes. These networks have one input layer, two or more hidden layers and an output layer. The number of neurons used in the hidden layers is one of the architectural parameters. Many articles have been published describing neural network configurations and architectures.[5,8]

Pao, Bhat and McAvoy, Psichogios, and Ydstie, are among many researchers who have investigated the use of neural networks in process control.[4] Several types of neurocontrol architectures have been employed by researchers in the past and many of these require on-line training. Such training is feasible if it is fast and inexpensive, but in many cases on-line training will disrupt the normal operating environment and may even raise some safety concerns.

Over the past 12 years, model predictive control has become the standard procedure for control using a process model. In this architecture, an on-line model of the process is used to predict the outcome of future control actions. This paper describes work aimed at developing and evaluating a neural model of a drying system, with the possible applications to control and investigation of the process.

APPLICATION OF NEURAL NETWORKS TO A MICROWAVE DRYING SYSTEM

An industrial 4kW 1m^3 multimode microwave unit was selected as the drying equipment. Tests were carried out using 100mm lengths of Gyproc cove. Although presently not commercially processed using microwaves, paper coated gypsum cove was chosen due to its availability, drying characteristics (including quality terms) and ability to be laden with a measurable quantity of water.

The dielectric properties and thermal-physical characteristics of the cove material have been examined demonstrating that it is very susceptible to microwaves, especially at high moisture contents.[9] Furthermore, careful control of the drying process (even using conventional methods) is essential due to the thermal sensitivity of the gypsum material. Excessive drying will result in the loss of the water of hydration forming the hemihyrate; a material with a marked change in structure and overall strength characteristics.

Optimising the energy efficiency of a machine and maximising product throughput may adversely affect product quality. It was essential that the model analysing the drying system should include some form of quality terms in its output. In an industrial situation there may be many parameters that effect the characteristics of the system, however a three input model was selected for initial investigation. The objectives were to evaluate the ability of a neural network to model a drying system, including a representation of product quality. Figure 1 shows the parameters studied.

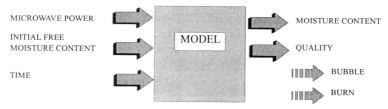

Fig. 1 - Model parameters

The cavity was modified and an electronic data-logging balance installed allowing the product weight to be recorded throughout the drying process. Samples of gypsum were soaked in water and allowed to reach an equilibrium moisture content. Drying characteristics were analysed for samples with moisture contents between 0g and 60g of added water, subjected to input magnetron powers of between 0 and 100% (4kW). Product quality defects were noted using a video camera with digital timing facility. Results were transferred directly to a PC version of MATLAB, where a third of the data was retained for model verification, and the neural networks developed. A backpropagation architecture network using a Levenberg-Marquardt approximation was utilised. Three separate networks were produced modelling moisture loss, paper bubbling and burning characteristics.

RESULTS AND DISCUSSION

MOISTURE LOSS

The drying curves (mass loss with time) obtained when paper coated gypsum cove was subjected to a microwave field showed typical drying characteristics, i.e. heating phase, constant drying phase, and falling rate period. The exact shape of the curve was found to be dependant upon the initial moisture content and magnetron power setting. Figures 2 and 3 show a comparison of network output to non-training test data for high and low power levels respectively.

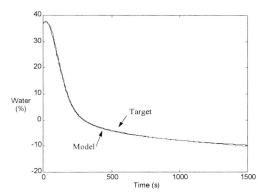

Fig. 2 - *Comparison between model and experimental data for low power*

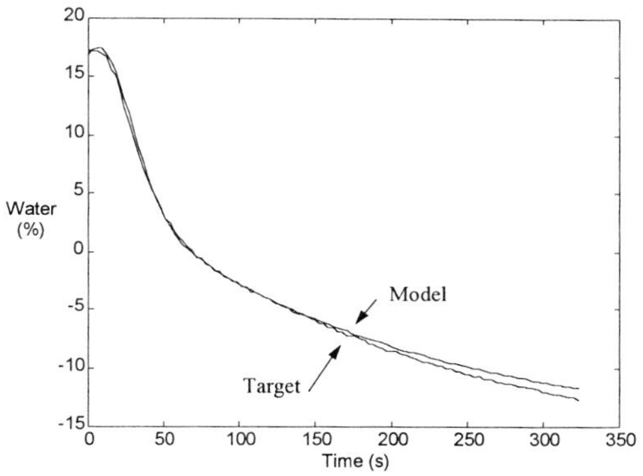

Fig. 3 - *Comparison between model and experimental data for high power*

Figures 2 and 3 show that the neural model constructed was capable of simulating the water loss drying curves of paper coated gypsum cove subjected to microwave heating within the parameter range analysed. Although small errors were encountered during the initial and final stages of drying the model could easily predict the point of zero free moisture content.

PRODUCT QUALITY

Paper bubbling occurred during microwave heating trials and was attributed to the paper coating being less porous than the gypsum core. The water was removed from the core due to internal pressure and temperature gradients being established during heating. As this water was removed, especially during the liquid movement phase that proceeds the constant drying period, the porosity of the paper was insufficient to allow the volume of moisture to migrate through the coating. Paper deformation due to the coating being forced away from the gypsum core thus occurred. Furthermore, due to the bubbling occurring during the initial stages of heating, the time for deformation to occur was not analysed. Paper shrinkage may have also contributed to the movement of the paper away from the central gypsum body. Since any form of paper deformation cannot be tolerated on a standard production line a model was produced to predict if any bubbling will occur. Figure 4 shows a plot of the experimental data obtained.

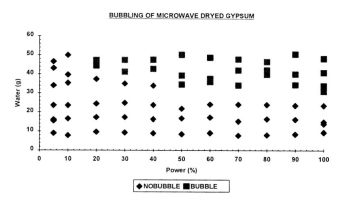

Fig. 4 - *Experimental data showing bubble deformation as a function of magnetron power and added water*

Various networks were assessed. A two layer structure with tangent and logsigmoid transfer functions was the most successful. Eight neurons formed the hidden layer and a sumsquared error (indication of the fit of the network output to training data) of 0.00037 was obtained after only 55 training epochs. Testing of the network with non-training data proved that the model could predict the likelyhood of paper deformation due to bubbling to a high degree of accuracy. Tables 1 and 2 show some typical results.

Table 2 shows that the model is capable of predicting the bubble characteristics for non-training data.

Training Data	Network Output
0	0.0000
0	0.0003
1.0000	1.0000
1.0000	1.0000
0	0.0000
1.0000	0.9807

Table 1 - *Comparison of the network output for modelling the training data*

Test Data Target	Network Output
0	0.0000
0	0.0002
1.0000	0.9249
1.0000	1.0000
0	0.0000
1.0000	0.9987

Table 2 - *Comparison of the network output for modelling of non-training data*

Burning of the paper coating the gypsum core occurred during microwave heating. The time to burn and dependence on the power level was attributed to the rate of removal of moisture from the body. Figure 5 shows a scatter diagram of the water content verses the time for initial burning for various microwave power levels.

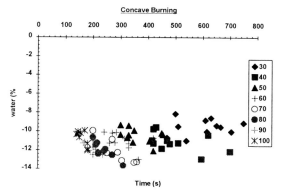

Fig. 5 - *Time and moisture group plot for concave burning*

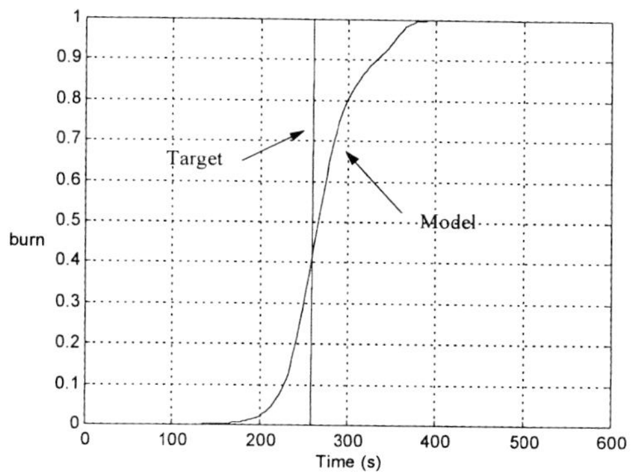

Fig.6 - *Comparison between model and non-training experimental data*

A three layer architecture (Tan-Log-Log) with 8 and 10 hidden neurons respectively was implemented. The network underwent 225 training epochs. Figure 6 shows a comparison of the network output when presented with non-training data, (where 1 represents burning, and 0 no burn). Although the models capability of predicting the time for burning in comparison to non-training experimental data resulted in some errors as high as 15%, most tests were within 4% of the target. The noise level and lack of training data was the most probable cause of the error. Further tests may have resulted in a more accurate representation of the burn phenomena. However, burning only occurred at a moisture content of below zero free-moisture, and hence was not considered important to warrant further investigation.

CONCLUSIONS

Artificial neural networks offer a convenient platform for capturing the non-linearity and dynamics inherent in drying systems. A neural network model was constructed capable of simulating the water loss drying characteristics of paper coated gypsum cove subjected to microwave energy to a high degree of accuracy. Furthermore, two quality terms notably, paper burning and bubbling were also investigated and successfully modelled.

The model could be applied as the 'model' component in a predictive control system, or used by engineers to maximise system performance and design by evaluating the response to input variations, without the need to run full scale tests. Further development could expand the technique to examine other parameters which have a direct influence on the drying characteristics of the process. Although the model is specific in nature, neural networks of this kind can be used in 'adaptive' control systems, where the model is retrained using modified data from the process, thus optimising the model to any changing process parameters.[4]

ACKNOWLEDGEMENT

This work is supported by the United Kingdom Engineering and Physical Science Research Council and Midlands Electricity plc, Halesowen, West Midlands, UK.

REFERENCES

1. S.Jay, T.N.Oliver: 'Energy Consumption for Industrial Drying Processes in the United Kingdom', 9th International Drying Symposium. Gold Coast, Australia. 1-4 August, 1994.

2. J.W.Robinson: 'Improve Dryer Control', Chem.Eng.Procress, December 1992, 28-33.

3. A.Mercer: 'Industrial Drying Technologies', CADDET Analysis Series. No. 12. CADDET, Sittard, Netherlands, July 1994.

4. B.Joseph, F.W.Hanratty: 'Predictive Control of Quality in a Batch Manufacturing Process Using Artificial Neural Nework Models', Ind. Eng. Chem. Res, 1993. 32. 1951-1961.

5. C.M.Bishop: 'Neural Networks and their Applications', Rewiew of Scientific Instruments, 1994, 65(6). 1803-1832.

6. B.Huang, A.S.Mujumdar: 'Use of Neural Networks to Predict Industrial Dryer Performance', Drying Technology, 1993. 11(3), 525-541.

7. D.A.White, D.A.Sofge: 'The Handbook of Intelligent Control', Van Nostrand Reinhold. New York, 1992.

8. B.E.Ydstie: 'Forcasting and Control Using Adaptive Connectionist Networks', Comput. Chem.Eng. 1990, 14(4), 583-599.

9. N.G.Evans, M.G.Hamlyn, S.Jay, T.N.Oliver: 'A Study of the Dielectric Properties of Gypsum and their Relation to Microwave Drying Behaviour'. Microwave and High Frequency Heating. Cambridge, 17-21 September, 1995.

RADIAL TEMPERATURE DIFFERENCES AND THERMAL STRESSES IN ARC FURNACE ELECTRODES

SUNDBERG, Yngve - INDUSTRIELL ELEKTROVÄRME Y SUNDBERG KONSULT AB, SWEDEN

ABSTRACT

Formulae are given for:
- the build-up of radial temperature distribution in an electrode when loading with evenly distributed current
- the temperature drop at the cylindrical surface of the electrode during and after recharging
- thermal stresses resulting from the temperature differences
- bulging of the end surfaces of electrode units because of the internal overtemperature, and resulting slots between joined units.

The formulae are used for calculation of temperatures and thermal stresses in a number of electrode sizes, applying currents which have been used in practice. The experience from operation with these currents gives knowledge on allowable temperatures, and in this way the formulae may be used for precalculation of allowable current in other electrode sizes.

RÉSUMÉ

Formules sont dérivées pour
- la croissance de la température à l'intérieur d'une électrode après l'enclenchement d'un courant également distribué
- la chute de température de la surface cylindrique de l'électrode pendant et après rechargement du four
- tensions thermiques dues à la différence radiale de température
- voussure des pignons des tronçons d'électrode à cause de la surtempérature à l'intérieur, et l'interstice qui en résulte entre les tronçons jointes.

Les formules sont employées pour le calcul des températures et les tensions thermiques d'un nombre de dimensions d'électrode. Les courants appliqués sont utilisés dans la pratique. L'expérience de l'utilisation de ces courants donne une **connaissance** des températures tolérables, et ainsi les formules pourraient être employées pour le calcul du courant tolérable pour d'autres dimensions d'électrode.

BACKGROUND

The first obstacle when loading an electrode with a higher current is that the electrode becomes hotter above the furnace roof. A high surface temperature above the roof means rapid oxidation. An early method to counteract this is to use coated electrodes. A recent method is to spray water on the electrode surface below the holder. A higher current density can then be allowed without an excessive oxidation.

Next obstacles are due to the higher temperature difference which is built up between the interior and the cylindrical surface of the electrode. This implies:
- risk of reaching temperatures where the graphite becomes plastic
- risk of reaching thermal stresses which cause cracks
- risk of reaching so large differences in thermal axial expansion that slots appear between joined electrode units.

It is a good way for getting an understanding of the problems to calculate the temperature distribution radially and its time dependence. A theory will be summarized below for the simplified case that the temperature of the cylindrical surface of the electrode is kept constant. This applies raughly above the furnace roof when water spraying is applied, and inside the furnace when the charge is in molten state.

RADIAL TEMPERATURE DISTRIBUTION AFTER SWITCHING ON FULL CURRENT

The following formula can be derived [1, p. 204]

$$\vartheta(r,t) = \frac{p_v}{4\lambda}\left[\frac{d^2}{4} - r^2\right] - \frac{p_v d^2}{2\lambda}\sum_{n=1}^{\infty} e^{-\frac{t}{T_n}} \frac{J_0\left(\frac{2r}{d}\alpha_n\right)}{\alpha_n^3 J_1(\alpha_n)} \quad (1)$$

where ϑ = overtemperature; p_v = volume power density = ρs^2; ρ = resistivity; s = current density (assumed constant over the cross section); d = diameter; n = a positive integer; t = time; $T_n = \frac{d^2}{4\alpha_n^2}\frac{c}{\lambda}$ = time constant; J_0 = Bessel function of the first kind; J_1 = ditto; α_n = zero points for J_0: 2,405; 5,520; 8,654; 11,792 ... c = specific heat (here expressed in Ws m^{-3}K^{-1}); λ = heat conductivity.

The distribution of $\vartheta(r,t)/\vartheta(r,\infty)$ for some values of t/T_1 is plotted in Fig. 1.

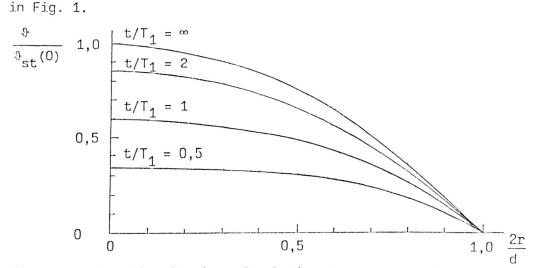

Fig. 1 Radial distribution of relative temperature rise

TEMPERATURE DROP AT THE CYLINDRICAL SURFACE OF THE ELECTRODE AFTER RECHARGING

We can get an idea of the temperature drop at the surface during the first minutes after recharging by introducing a negative surface power density $\bar{p}_\sigma$ equal to the interrupted re-radiation from the walls and the roof of the furnace. On the other side we have to take oxidation heat on the electrode surface into account. The resulting cooling process is superimposed on the heating process going on. We have [5]:

$$\Delta\vartheta = (-\bar{p}_\sigma + \bar{p}_{ox}) \frac{2}{\sqrt{\pi}} \sqrt{\frac{t}{c\lambda}} \qquad (2)$$

The "penetration depth" is illustrated in Fig. 2. It is

$$\delta = \frac{2}{\sqrt{\pi}} \sqrt{\frac{\lambda t}{c}} \qquad (3)$$

Fig. 2 Temperature distribution in an electrode after recharging

As the heating of the charge goes on, the superimposed temperature drop is spread over the cross section of the electrode with a time constant which is shorter than that given on page 3. It is

$$T_1 = \frac{d^2}{3{,}832^2} \frac{c}{4\lambda} \qquad (4)$$

THERMAL STRESSES IN THE ELECTRODE [6]

We shall approximate the temperature distribution in Fig. 1 with the formula

$$\vartheta = \vartheta_0 \left[1 - (2r/d)^m\right] \qquad (5)$$

where $2 < m < \infty$.

The value $m = 2$ corresponds to steady state according to Eq. (1). Higher values correspond to flatter curves, see for instance the curve for $t/T_1 = 0{,}5$ in **Fig. 1**. An idea of the value m for the temperature distribution can be obtained by deriving an expression for "penetration depth", see Fig. 2. We get

$$\delta = \frac{d}{2m} \qquad (6)$$

The following formulae can be derived for the stresses in a long solid circular cylinder with the above temperature distribution:

$$\sigma_r = -\frac{\alpha E}{1-\nu}\frac{\vartheta_o}{m+2}\left\{1-(2r/d)^m\right\} \quad (7)$$

$$\sigma_\varphi = -\frac{\alpha E}{1-\nu}\frac{\vartheta_o}{m+2}\left\{1-(m+1)(2r/d)^m\right\} \quad (8)$$

$$\sigma_z = -\frac{\alpha E}{1-\nu}\frac{\vartheta_o}{m+2}\left\{2-(m+2)(2r/d)^m\right\} \quad (9)$$

where $\sigma_r \; \sigma_\varphi \; \sigma_z$ = stresses in the r φ z - directions (positive for tensile stresses); α = coefficient of thermal expansion; E = modulus of elasticity; ϑ_o = overtemperature in the centre; ν = Poisson's constant.

The radial distribution of the stresses is shown in Fig. 3.

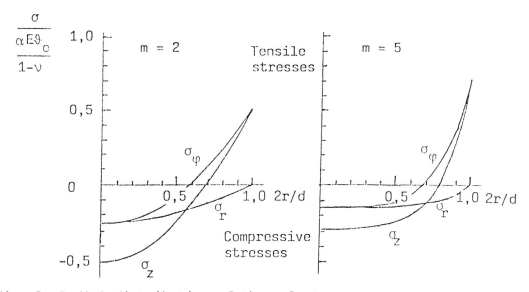

Fig. 3 Radial distribution of thermal stresses

For r = d/2 we get

$$\overline{\sigma}_r = 0 \quad (10)$$

$$\overline{\sigma}_\varphi = \frac{\alpha E \, \vartheta_o}{1-\nu}\frac{m}{m+2} \quad (11)$$

$$\overline{\sigma}_z = \frac{\alpha E \, \vartheta_o}{1-\nu}\frac{m}{m+2} \quad (12)$$

and especially for m → ∞ (corresponding to cooling of a thin layer)

$$\bar{\sigma}_\varphi = \bar{\sigma}_z \to \frac{\alpha E \vartheta_0}{1 - \nu} \quad (13)$$

CALCULATIONS

Table 1. Physical properties of electrode graphite of special grade

Temperature	°C	1400	1700
Resistivity	Ωm		$7{,}44 \times 10^{-6}$
Specific heat	$Ws\ m^{-3}K^{-1}$	$3{,}19 \times 10^6$	$3{,}28 \times 10^6$
Heat conductivity	$W\ m^{-1}K^{-1}$	48	45
Coefficient of thermal expansion	K^{-1}	$3{,}7 \times 10^{-6}$	$4{,}8 \times 10^{-6}$
Modulus of elasticity	Pa	$5{,}6 \times 10^9$	$5{,}7 \times 10^9$
Poisson's constant	-	0,1	0,1

Results during operation with surface temperature 1500 to 1700°C

Electrode currents which have been used in practice are applied.

Table 2. Overtemperature in the centre and thermal stress in the cylindrical surface during operation

Electrode diameter	Current DC	Time const. T_1	Overtemp. in centre After time			Steady	Thermal stress After time			Steady
			$0{,}5T_1$	T_1	$2T_1$		$0{,}5T_1$	T_1	$2T_1$	
mm	kA	sec	K	K	K	K	MPa	MPa	MPa	MPa
406	42	519	61	107	153	179	1,1	1,8	2,4	2,7
0,7 x 406	42	254	124	218	313	366	2,3	3,6	4,8	5,6
457	42	658	48	84	121	141	0,9	1,4	1,9	2,1
457 x 0,7	42	322	98	172	247	289	1,8	2,8	3,8	4,4
610	90	1172	124	217	312	365	2,3	3,6	4,8	5,5
610 x 0.7	90	574	253	443	636	744	4,7	7,3	9,8	11,3
711	120	1593	162	284	408	477	2,9	4,7	6,3	7,3
711 x 0,7	120	780	331	575	833	974	6,1	9,5	12,9	14,8

Comments to Table 2

1. The time constant T_1 is to be compared with the time during which the cylindrical surface of the electrode has a temperature of 1500 to 1700 °C during the later part of melt-down of a part charge. In modern operation the time for melt-down is often about 30 minutes, and the time at 1500 to 1700 °C can be estimated to be about 15 minutes. In view of this one ought to consider

the figures valid at
- steady state for diameter 0,7 x 406 mm and 0,7 x 457 mm
- T_1 " " 0,7 x 610 mm
- $0,5\ T_1$ " " 0,7 x 711 mm.

2. The overtemperature added to the surface temperature gives the centre temperature. This is to be compared with the level where graphite becomes plastic (2500 °C according to [2]). Of course also the quality of the joint between electrode units plays a role.

3. The thermal stress is to be compared with the rupture point for tensile stress which is 5 to 8 MPa for graphite of special grade [3] (tensile strength is about one half of flexural strength according to [2]).

4. There is some experience from operating DC arc furnaces with the currents applied in Table 2:
   - 406 mm: - longitudinal cracks in the cylindrical surface near the tip
     - some "stub end losses"
   - 457 mm: - practically no stub end losses
   - 610 mm: - longitudinal cracks in the cylindrical surface near the tip. Graphite consumption about 1,2 kg/ton
   - 711 mm: - deep longitudinal cracks in the cylindrical surface near the tip. Graphite consumption about 2 kg/ton.

5. The results in Table 2 can explain why cracks appear in the cylindrical surface of an electrode at a high current. Additional stress is caused by the surface cooling during and after recharging, see next chapter.

The results cannot fully explain the stub end losses at high currents, as the temperatures in the centre after the normal melting time are not so high that the graphite becomes plastic. An additional explanation is the bulging of the end surfaces of the electrode units, which gives slots between joined units and forces the current to take its path nearer the centre. This problem is further discussed later in this paper.

Results after recharging

We apply Eqs (2), (3), (6) and (11), and assume that the furnace temperature was 1500°C before recharging, corresponding to $\bar{p}_\sigma = -49,4 \times 10^4$ W/m^2. Oxidation heat on the electrode in free air will give $\bar{p}_{ox} = 10 \times 10^4$ W/m^2 at 1400°C [9]. We apply physical constants valid at 1400°C. The results are summarized in Table 3.

Table 3. Temperature drop and additional stress at the cylindrical surface after recharging

Time t s	Temperature drop $\Delta\vartheta$ K	Penetration depth δ m	Additional stress σ_φ and σ_z, MPa		
			d=0,284 m	d=0,610 m	d=0,711 m
30	-196	0,024	3,4	3,9	4,0
60	-278	0,034	4,3	5,2	5,4
90	-341	0,042	5,0	6,2	6,4
120	-394	0,048	5,4	6,9	7,1
180	-482	0,059	6,1	8,0	8,3
240	-557	0,068	6,6	8,9	9,3

Comments to Table 3

1. The calculated temperature drop is in good agreement with measurements reported in [7].
2. It is interesting to note that the additional stress is so large, and independent of the current loading. It is not, however, algebraically added to the stress according to Table 2. One reason is that the two part stresses are not at their maximum simultaneously. Another reason is that the modulus of elasticity decreases for higher loads [8].

SLOTS BETWEEN JOINED ELECTRODE UNITS

In the theory about thermal stresses it is found that the axial compressive and tensile stresses balance each other over the main length of the electrode. That is, the tendency to shrinking in colder parts only results in a tensile stress, and the tendency to expansion in the hotter parts only results in a compressive stress. Each cross section being flat before heating remains flat after heating, also when the heating is uneven.

This is true over the main length of an electrode unit. But at the ends of an electrode unit the axial stresses must be zero. The axial stresses are built up over an axial length from the end which is approximately equal to the radius of the electrode. This will result in a bulge of the end surface.

It is possible to derive a formula for the shape of the bulge by applying Saint-Venant's principle and assuming a radial distribution of axial stress on the end surface according to the Bessel function J_0 which leads to [6] :

$$\sigma_z = (\sigma_{zo} - \bar{\sigma}_z) J_0(2{,}405 \times 2r/d)\, e^{-2{,}405 \times 2z/d} \qquad (14)$$

$$w = \frac{1 + \nu}{E} (\sigma_{zo} - \bar{\sigma}_z) \frac{d}{2 \times 2{,}405} J_0(2{,}405 \times 2r/d) \qquad (15)$$

where σ_{zo} and $\bar{\sigma}_z$ = axial stress in the centre and in the cylindrical surface; z = axial distance from the end surface; w = axial displacement of the end surface.

As an example we select the following parameters: $E = 5{,}7 \times 10^9$ Pa; $\nu = 0{,}1$; $\sigma_{zo} - \bar{\sigma}_z = 10{,}7$ MPa; $d = 0{,}427$ m (see Table 2). We get in the centre

$$w_o = \frac{1 + 0{,}1}{5{,}7 \times 10^9} \times 10{,}7 \times 10^6 \times \frac{0{,}427}{2 \times 2{,}405} = 0{,}183 \times 10^{-3} \text{ m}$$

The slot between two joined units can thus be $2 \times 0{,}183 = 0{,}37$ mm near the cylindrical surface. The current will then be forced to take its path nearer the centre of the electrode, and cause a higher overtemperature.

CONCLUSION

The above theory gives an indication on the problems appearing when loading graphite electrodes with high currents. The results are:
- In modern dimensioning of arc furnaces the time for melt-down is so short that steady state temperature distribution radially in the electrode inside the furnace has not time enough to be built up. The internal overtemperature becomes higher near the tip where the diameter is smaller, partly because the current density is increased, partly because the time constant for building up the internal overtemperature is shorter.

- Thermal stresses in the electrode become high towards the end of melt-down because of the internal overtemperature which is due to the current. Additional stresses appear at the cylindrical surface because of the temperature drop after lifting and recharging. The stresses are highest near the tip. Longitudinal cracks appear in the cylindrical surface when the stress exceeds the rupture point.
- The internal overtemperature gives rise to slots between joined electrode units. This will cause a concentration of the current in the centre, and in consequence more overheating.

REFERENCES

1. H.S. Carslaw and J.C. Jaeger: "Conduction of heat in solids", Oxford University Press, New York, Second edition 1959.
2. J. Semmler: "Die Temperaturabhängigkeit verschiedener Eigenschaften von Graphit", Stahl und Eisen 1967, 87(7).364-368.
3. Pamphlet "Make better steel with Tokai Graphite Electrodes", Tokai Carbon Co., Ltd, Japan 1991.
4. Pamphlet "Graphite electrodes. Manufacture, Properties, Forms supplied", Sigri GmbH, Germany 1991.
5. L.A. Dreyfus: "High frequency heating and temperature distribution in surface hardening of steel", Generalstabens Litografiska Anstalts Förlag, Stockholm 1952.
6. S.P. Timoshenko, J.N. Goodier: "Theory of elasticity", Mc Graw-Hill Kogakusha, Ltd, Tokyo 1970.
7. E. Nedopil und H. Storzer: "Messung und Temperaturverteilung im Elektrodenstrang beim Einsatz im Elektrolichtbogenofen", Stahl und Eisen 1967, 87(7).368-373.
8. E.J. Seldin: "Stress-strain properties of polycrystalline graphites in tension and compression at room temperature", Carbon 1966, 4.177-191.
9. H. Jäger et al: "Advanced graphite electrodes - a key to increased arc furnace productivity", Metallurgical Plant and Technology International 1991, (6).24-39.

La modélisation des arcs électriques : des équations théoriques aux applications industrielles.

Clarisse DELALONDRE*, Olivier SIMONIN*, Laurent MINEAU**,
Jean-Luc ROTH***, Colette GEVERS***

* LNH - DER - EDF, 6 Quai Watier, 78400 Chatou, France
** SE - DER - EDF, Les Renardières, Route de Sens, Ecuelles, 77250 Moret sur Loing, France
*** IRSID, Voie Romaine, 57214 Maizières-lès-Metz, France

Depuis les premières expériences de fusion au four à arc, beaucoup de progrès - tant du point de vue de la technologie que du process - ont permis d'améliorer les performances de ces fours en matière notamment de consommation d'énergie ou d'électrodes en graphite. Les fours utilisés sont de plus en plus puissants et pour continuer à progresser dans cette direction, il est nécessaire de mieux comprendre les mécanismes de transfert de chaleur entre l'arc électrique, le laitier et le bain de métal. Pour répondre à cet objectif, le Laboratoire National d'Hydraulique de la Direction des Études et Recherches d'Électricité de France a utilisé et adapté son modèle numérique d'arc électrique développé dans le code différences finies bidimensionnel "Mélodie". On utilise en particulier une modélisation de la turbulence qui permet de représenter la zone laminaire proche de la cathode et les zones turbulentes partout ailleurs. La couche limite se développant sur le bain d'acier a fait également l'objet d'une modélisation qui a permis d'évaluer une partie du transfert thermique entre l'arc et le métal. Ce texte présente les résultats obtenus pour des arcs de 10 et 40 kA dans différentes compositions d'atmosphères (air, CO_2, ...). La configuration choisie d'un arc fonctionnant sur bain plat, correspond à la dernière phase de la fusion des ferrailles au four à arc.

Since the first experiments, considerable technology and process advancements, especially in the field of energy or electrode consumption, have improved the performance of electric arc furnaces. In order to increase furnace power, it is also important to understand better the energy transfer between the electric arc, the slag and the metallic bath. Therefore, the Laboratoire National d'Hydraulique of the Research Division of Électricité De France has used its two-dimensional axisymetric code "Mélodie" adapted for electric arcs modelling in order to simulate some configurations close to electric arcs encountered in industrial furnaces. The calculations take into account the turbulence of the flow by using a low Reynolds number k-epsilon model and try to predict the energy transfer from the arc to the metallic bath by using a one-dimensional model of the electrode boundary layer. The paper presents modelling results obtained for 10 kA and 40 kA arcs burning in different gazes (air, CO_2, etc...). The arc configurations correspond to the modelling of the last step in scrap-iron melting when the arc is burning on a flat metallic bath.

1. INTRODUCTION

Ce texte présente les travaux de modélisation numérique d'arcs électriques réalisés dans le but de mieux comprendre les transferts thermiques entre l'arc et le bain métallique d'un four électrique. Ces simulations ne prennent pas en compte le laitier. Les principales équations utilisées dans le modèle, en particulier celles permettant de prendre en compte la turbulence des écoulements gazeux sont rassemblées dans la première partie de ce document. Les résultats obtenus sont ensuite présentés.

2. LE MODÈLE ET LES PRINCIPALES ÉQUATIONS

Conservation de la masse :

$$\frac{\partial}{\partial t}\bar{\rho} + \text{div}\,\bar{\rho}\,\vec{U} = 0 \qquad (1)$$

$\bar{\rho}$ est la masse volumique moyenne, U est la vitesse moyenne du gaz dont les composantes sont définies par : $U_i = \overline{\rho\,u_i}/\bar{\rho}$

Équation de conservation de la quantité de mouvement :

$$\bar{\rho}\frac{\partial}{\partial t}\vec{U} + \left[\bar{\rho}\,\vec{U}\cdot\overrightarrow{\text{grad}}\right]\vec{U} = -\overrightarrow{\text{grad}}\,p + \text{div}\,\bar{\tau} + \vec{j}\wedge\vec{B} \qquad (2)$$

où p est la pression. $\vec{j}\wedge\vec{B}$ représente la force électromagnétique de Laplace. $\vec{j}$ est la densité de courant. $\vec{B}$ le champ magnétique induit par le passage du courant, est calculé par le théorème d'Ampère. Les fluctuations des termes sources électriques ont été négligées /3, 4/. τ_{ij} est le tenseur des contraintes visqueuses moléculaires et turbulentes.

$$\tau_{ij} = \mu\left\{\left(\frac{\partial U_i}{\partial x_j} + \frac{\partial U_j}{\partial x_i}\right) - \frac{2}{3}\delta_{ij}\,\text{div}\vec{U}\right\} - \overline{\rho\,u''_i u''_j} \qquad (3)$$

Équation de la température moyenne T :

$$\bar{\rho}\,C_p\frac{\partial}{\partial t}T + \left[\bar{\rho}\,C_p\,\vec{U}\cdot\overrightarrow{\text{grad}}\right]T = -\text{div}\,\vec{Q} + \vec{j}\cdot\vec{E} - S_{rad} \qquad (4)$$

où C_p est la chaleur massique à pression constante. Q est le flux de diffusion thermique :

$$\vec{Q} = -\lambda\,\overrightarrow{\text{grad}}\,T + \overline{\rho\,u''_i\,\theta''} \qquad (5)$$

u" et θ" correspondent aux fluctuations de vitesse et de température, et λ à la conductivité thermique moléculaire. S_{rad} représente le terme de rayonnement. Dans les applications présentées ici, il modélise l'émission radiative à partir de données globales fonctions de la température ; ces données sont directement tirées de la littérature /4/. $\vec{j}\cdot\vec{E}$ est l'effet Joule. Le champ électrique $\vec{E}$ et la densité de courant dérivent des équations simplifiées de Maxwell :

$$\text{div}\,\vec{j} = 0 \quad \text{avec} \quad \vec{j} = -\sigma\,\overrightarrow{\text{grad}}\,V \quad \text{(loi d'Ohm)} \qquad (6)$$

où σ est la conductivité électrique et V le potentiel électrique.

La turbulence est représentée par un modèle k-epsilon à bas nombre de Reynolds /5/. En effet, le modèle k-epsilon est maintenant largement utilisé pour la modélisation d'écoulements industriels dans les plasmas et les comparaisons avec les résultats expérimentaux montrent que les simulations numériques réalisées avec ce modèle sont tout à fait réalistes. On constate que dans les plasmas d'arcs électriques, les variations de la viscosité moléculaire et de la masse volumique sont très importantes car elles sont liées aux fortes variations de la température. En conséquence les deux régimes d'écoulement laminaire et turbulent peuvent coexister dans les écoulements plasma et le modèle k-epsilon à bas nombre de Reynolds a été retenu pour modéliser la turbulence /2, 3, 4, 6/. De plus l'opérateur moyenne de Favre est le mieux adapté pour déterminer les variables moyennes du calcul (vitesse, température). Le modèle k-epsilon est basé sur une hypothèse de viscosité turbulente, les corrélations des fluctuations de la vitesse apparaissant dans l'équation de conservation de la quantité de mouvement moyenne sont modélisées par :

$$\overline{\rho u''_i u''_j} = -\mu^t \left[\frac{\partial U_i}{\partial x_j} + \frac{\partial U_j}{\partial x_i} \right] + \frac{2}{3} \delta_{ij} \left[\bar{\rho} k + \mu^t \frac{\partial U_m}{\partial x_m} \right] \quad (7)$$

$$\mu^t = \bar{\rho} C_\mu \frac{k^2}{\varepsilon} \quad (8)$$

Les équations des modèles k-epsilon (valables pour le modèle standard et le bas Reynolds) sont les suivantes :

Équation de l'énergie cinétique de la turbulence :

$$\bar{\rho} \frac{\partial}{\partial t} k + \bar{\rho} U_j \frac{\partial}{\partial x_j} k = \frac{\partial}{\partial x_j} (\bar{\mu} + \frac{\mu^t}{\sigma_k}) \frac{\partial}{\partial x_j} k + \mathcal{P} - \bar{\rho} \varepsilon + \Pi_k \quad (9)$$

où $\mathcal{P}$ est le terme de production due au gradient de la vitesse moyenne.

$$\mathcal{P} = -\overline{\rho u''_i u''_j} \frac{\partial U_i}{\partial x_j} \quad (10)$$

Équation du taux de dissipation de la turbulence :

$$\bar{\rho} \frac{\partial}{\partial t} \varepsilon + \bar{\rho} U_j \frac{\partial}{\partial x_j} \varepsilon = \frac{\partial}{\partial x_j} (\bar{\mu} + \frac{\mu^t}{\sigma_\varepsilon}) \frac{\partial}{\partial x_j} \varepsilon + \frac{\varepsilon}{k} \left[C_{\varepsilon,1} \mathcal{P} - C_{\varepsilon,2} \bar{\rho} \varepsilon \right] + \Pi_\varepsilon \quad (11)$$

Le k-epsilon standard comporte 6 constantes empiriques dont les valeurs sont données dans le tableau suivant.

C_μ	$C_{\varepsilon,1}$	$C_{\varepsilon,2}$	σ_k	σ_ε	Pr^t
0.09	1.44	1.92	1.	1.3	1.

Tableau 1 : Valeurs standards des constantes du modèle k-epsilon (à haut nombre de Reynolds)

Dans l'équation de la température, le terme de diffusion dû aux corrélations vitesse - température est modélisé à l'aide d'une diffusivité thermique de la turbulence en analogie avec la viscosité. On suppose de plus que le nombre de Prandtl (Pr^t) de la turbulence est constant (tableau 1).

$$\overline{\rho\, u''_i\, \theta''} = -\frac{\mu^t}{Pr^t}\frac{\partial T}{\partial x_i} \tag{12}$$

Le modèle à bas nombre de Reynolds proposé par Launder et Sharma 1974 implanté dans le code Mélodie, a été utilisé. Il correspond à une variante du modèle k-epsilon standard dans laquelle on fait varier les constantes empiriques du modèle en fonction du nombre de Reynolds local R^t.

$$R^t = \bar{\rho}\, k^2 / \left[\bar{\mu}\, \varepsilon\right] \tag{13}$$

$$C_\mu = 0.09 \exp\left[-3.4 / (1. + R^t/50.)^2\right] \qquad C_{\varepsilon,2} = 1.92 \left[1. - 0.3 \exp(-R^{t\,2})\right] \tag{14}$$

De plus pour améliorer les prédictions de la turbulence dans la couche limite laminaire en proche paroi, deux termes empiriques ont été ajoutés dans les équations de k et de ε :

$$\Pi_k = -2\bar{\mu}\left[\frac{\partial\sqrt{k}}{\partial x_i}\frac{\partial\sqrt{k}}{\partial x_i}\right] \qquad \Pi_\varepsilon = 2\bar{\mu}\,\frac{\mu^t}{\bar{\rho}}\left[\frac{\partial^2 U_j}{\partial x_i^2}\frac{\partial^2 U_j}{\partial x_i^2}\right] \tag{15}$$

Le système d'équation doit être complété par des fonctions d'état qui relient la masse volumique, le C_p et les coefficients de transport tels que la viscosité dynamique ou les conductivités thermique et électrique, à la pression et la température du gaz. Ces fonctions sont obtenues dans le cadre de l'hypothèse de l'Équilibre Thermodynamique Local, condition généralement satisfaite par les plasmas d'arcs à fortes intensités. Toutefois, l'existence de gradients de température très importants dans la zone d'arc, particulièrement à proximité de l'électrode de graphite implique des déséquilibres thermodynamiques locaux qui ont malgré tout été négligés dans ce modèle /1, 2, 6, 7/.

La représentation de la couche limite se développant sur le bain métallique a fait l'objet d'une attention particulière, car elle nous permet de déterminer le transfert thermique par conduction et convection entre l'arc et le bain d'acier dont la température est fixée à 1550 Celsius. Cette modélisation monodimensionnelle s'inspire des modèles déjà existants de couche limite se développant sur une électrode /1, 7/. On résout une équation de la thermique simplifiée dans laquelle on suppose d'une part que la conductivité électrique est constante, et que d'autre part, il y a équilibre entre le terme source dû à l'effet Joule et le flux de conduction thermique /4/.

3. APPLICATIONS AUX FOURS À ARCS ET RÉSULTATS

Le modèle présenté ci dessus a été utilisé pour le calcul de deux configurations d'arcs fonctionnant sur bain plat et correspondant à la dernière phase de la fusion des ferrailles au four à arc. Le premier cas représente un arc de 10 kA et de 20 cm de long proche des arcs obtenus sur le four pilote de l'IRSID, et le second simule un arc industriel de 40 kA et 40 cm de long. Les calculs ont été menés pour différentes atmosphères gazeuses : de l'air, un mélange air et vapeurs de fer, du CO_2 ou encore un mélange 79% N_2 et 21% CO_2.

Les résultats obtenus dans la configuration du pilote IRSID dans l'air sont d'abord présentés : sur la figure 1, on peut voir que la température calculée à proximité de l'extrémité de l'électrode atteint 64000 Celsius. Cette valeur est probablement trop élevée, des développements récents attribuent ces résultats peu crédibles à une prise en compte incorrecte des phénomènes de transferts radiatifs dans les zones les plus chaudes. En revanche, les autres résultats, tels que la valeur de la différence de potentiel dans la colonne d'arc (250 V) ou le rendement thermique du four estimé à 36% sans laitier (fig. 9), sont en accord avec les mesures réalisées par l'IRSID. Dans le CO_2, le niveau des températures est beaucoup plus faible et semble plus réaliste (fig. 2 et 3) ; il est important de noter que, par manque de données, ces résultats sont obtenus sans aucun terme de rayonnement.

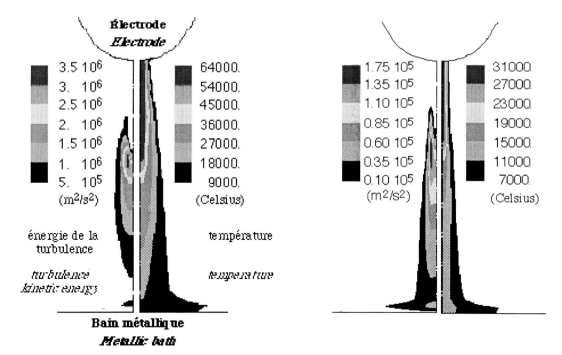

Fig. 1 : Champs de k et de T
Air 10 kA, 20 cm

Fig. 2 : Champ de k et de T
CO_2 40 kA et 40 cm

Les figures 3 à 6 permettent de comparer les températures obtenues dans différents cas. Dans la configuration industrielle, la température est sensiblement égale à celle obtenue pour la configuration IRSID (31000 Celsius dans le CO_2 et 61000 Celsius dans l'air pur). On notera également que les résultats du mélange N_2 - CO_2 sont tout à fait cohérents avec les résultats obtenus dans l'air ou le CO_2. La figure 6 met en évidence que l'augmentation de la conductivité électrique du gaz plasmagène en présence de vapeurs de fer entraîne une diminution de la température de l'arc. En ce qui concerne les tensions d'arc, les valeurs calculées sont peu dépendantes de la nature du gaz : la tension d'arc est voisine de 250 Volts dans la configuration du pilote, et voisine de 450 Volts dans la configuration industrielle.

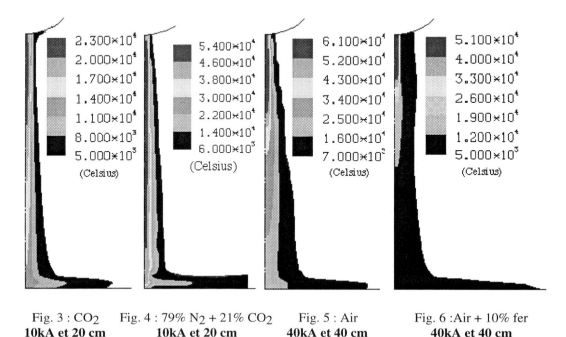

Fig. 3 : CO_2 Fig. 4 : 79% N_2 + 21% CO_2 Fig. 5 : Air   Fig. 6 : Air + 10% fer
10kA et 20 cm   **10kA et 20 cm**   **40kA et 40 cm**   **40kA et 40 cm**

Les figures 1 et 2 illustrent les résultats concernant la turbulence, et montrent que la relaminarisation de l'écoulement dans les zones très chaudes de l'arc est prédite grâce au modèle à bas Reynolds : le maximum de l'énergie de la turbulence est obtenu à proximité de l'axe de l'arc à une distance d'environ 9 cm de la cathode, là où la température commence à varier fortement. De même, la viscosité et la diffusivité moléculaires du plasma sont supérieures à la viscosité et la diffusivité de la turbulence dans la région située juste en dessous de l'électrode. Les prédictions montrent que, dans ce type d'écoulement plasma, la diffusion moléculaire est plus importante pour la thermique que pour la dynamique. Cette remarque tend à démontrer que l'hypothèse d'un nombre de Prandtl turbulent constant et égal à 1 est très perfectible.

On présente ensuite un exemple de résultats de la modélisation de la couche limite anodique se développant sur le bain métallique, qui participe à la détermination le transfert thermique par convection et conduction de l'arc vers le bain. Les figures 7 et 8 ont pour abscisse la distance à la paroi du bain : cette abscisse varie de 0, sur le bain, à 6 10^{-3} m, valeur correspondant à la dimension de la maille du calcul 2D située juste au-dessus du bain.

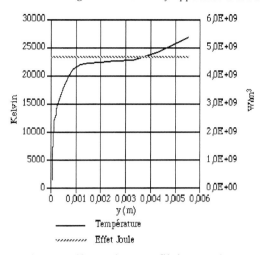
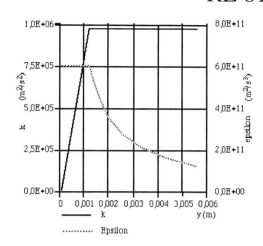

Fig. 7 : Effet Joule et profil de température   Fig. 8 :Profils de k et epsilon

Exemples de modélisation de la couche limite sur le bain (Air 10 kA et 20 cm)

Sur la figure 7 on a porté l'effet Joule supposé constant dans la couche 1D et égal à la valeur du calcul 2D de la colonne d'arc ainsi que le profil de température calculé dans la couche anodique. Les profils de l'énergie et du taux de dissipation de la turbulence qui permettent de calculer localement la diffusivité de la turbulence sont reproduits sur la figure 8. On note en particulier sur cette figure qu'à une distance d'environ 1,3 mm de la surface du bain les profils de k et de epsilon subissent une rupture de pente au passage de la sous-couche visqueuse à la zone turbulente.

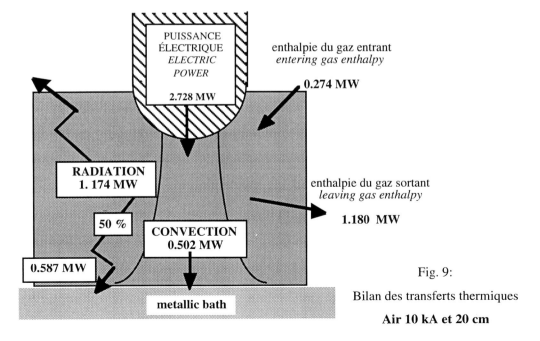

Fig. 9:
Bilan des transferts thermiques
Air 10 kA et 20 cm

Les transferts thermiques (fig. 9), et le rendement peuvent alors être évalués en faisant l'hypothèse que la moitié de l'énergie rayonnée par l'arc est transmise au bain. Le rendement est calculé comme le rapport du flux thermique transféré au bain (= conduction + convection + rayonnement) sur la puissance fournie (= puissance électrique + apport du gaz entrant). Cette valeur calculée valant 36% est tout à fait réaliste et proche des valeurs relevées par l'IRSID sur leur four pilote, comprises entre 30 et 40%, sachant qu'il correspond à un fonctionnement sans laitier donc à rendement relativement mauvais. La sensibilité des transferts thermiques et du rendement à la température de surface du métal a été montrée, en imposant différentes valeurs des températures de surface de bain (1550 ou 3000 Celsius) correspondant à des valeurs extrêmes qui encadrent les valeurs rencontrées dans la réalité. On constate alors que, dans le cadre des hypothèses de calcul du rendement citées plus haut, le transfert par conduction et convection devient négligeable par rapport au transfert radiatif pour la valeur la plus élevée de la température du bain.

4. CONCLUSION

Ces simulations ont permis une meilleure compréhension des phénomènes physiques qui caractérisent l'arc et ont abouti à une première étude de l'influence des paramètres de fonctionnement sur le rendement thermique du four. Les résultats ont été obtenus pour différentes configurations d'arc dont certaines très proches des conditions observées dans les fours industriels. A l'exception des températures obtenues dans l'air, ces résultats sont cohérents avec les données expérimentales disponibles. Nous poursuivons actuellement de nouveaux développements d'abord pour mieux prendre en compte les transferts radiatifs arc - bain d'acier, ensuite pour intégrer l'influence du laitier sur les transferts thermiques.

5. RÉFÉRENCES

1. DELALONDRE and SIMONIN, *Colloque de Physique, Colloque C5*, 1990, 51, supplément au (18), 199-206.
2. DELALONDRE and SIMONIN, *Proceedings of the XIIth IUE congress*, 1992.
3. DELALONDRE, ZAHRAI and SIMONIN, *Seminar on heat and mass transfer under plasma conditions*, Izmir 4-8 July 1994,.
4. DELALONDRE and SIMONIN, *EDF Report HE44/93/026B*, 1995.
5. LAUNDER and SHARMA, *Letters in Heat and Mass Transfer*, 1974, 1, 131-138.
6. SIMONIN, DELALONDRE and VIOLLET, *Pure and Applied Chemistry*, 1992, 64, (5), 625-628.
7. KADDANI, *Thèse de l'Université Paris VI*, 1995

Etude thermique du Chauffage d'Acier liquide par Plasma d'Arc Transféré

P. POILLEAUX [1] - A. BOUVIER [1] - L. TRENTY [1] - L. MINEAU [1] - J. DEBOURCQ [1]
C. DELALONDRE [2] - A. LECLERCQ [3] - J.L. ROTH [4] - C.P. HEANLEY [5]

(1) EDF, Direction des Etudes et Recherches, Centre des Renardières, BP N° 1, 77250 Moret sur Loing

(2) EDF, Direction des Etudes et Recherches, Centre de Chatou, 6 Quai Watier 78401 Chatou Cédex

(3) SOLLAC, 17 avenue des Tilleuls, BP N° 11, 57191 Florange Cédex

(4) IRSID, Voie Romaine, BP N° 320, 57214 Maizières-les-Metz Cédex

(5) TETRONICS, 5 Lechlade Road, Faringdon Oxfordshire SN7 9 AJ, U.K.

RESUME :

Dans ce papier nous présentons les travaux menés à la Direction des Etudes et Recherches d'Electricité de France pour améliorer les performances des systèmes plasma en métallurgie et favoriser leur développement. Dans un premier chapitre on présente la modélisation thermique d'une chambre de chauffage plasma sur un répartiteur de coulée continue. Dans un second chapitre, on présente un modèle numérique bidimensionnel de simulation du chauffage d'un bain métallique par plasma d'arc transféré. Enfin, dans dans un troisième chapitre on présente une installation expérimentale pour l'étude du chauffage de bains métalliques par plasma d'arc transféré.

ABSTRACT:

In this paper we present some works performed at the Research and Development Division of Electricité de France in order to improve the performances of the plasma processes in the field of metallurgy and to facilitate their development. In a first chapter of this paper we present a thermal modelling of a plasma heating chamber on a tundish of a continuous caster. In a second chapter, we show a two-dimensionnal numerical model for the simulation of metal bath heating with transferred arc. At last, in a third chapter, we show an experimental unit assigned to study metal bath heating with transferred arc.

1. INTRODUCTION

La fabrication d'un acier de haute qualité nécessite de contrôler sa composition chimique et sa température tout au long de son élaboration. Ainsi l'on voit se développer des applications pour le plasma d'arc transféré comme chauffage d'appoint au niveau de la métallurgie secondaire (poche d'affinage) ou tertiaire (répartiteur de coulée continue).

Le principe du plasma d'arc transféré est illustré en figure 1 : l'arc électrique est entretenu entre le bain d'acier à chauffer, qui le plus souvent constitue l'anode, et une cathode en tungstène refroidie par eau. Il est stabilisé par un gaz (argon ou azote) permettant d'éviter la pollution des bains que provoquerait une cathode graphite.

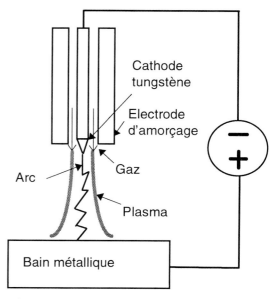

Fig. 1 - Torche à plasma à cathode chaude

Afin de promouvoir le développement des systèmes plasma en France, Electricité de France (EDF) a accompagné la première installation d'une torche à plasma sur un répartiteur de coulée continue, à l'aciérie Sollac-Florange du groupe Usinor-Sacilor (cf. § 2). Par ailleurs, un programme d'étude plus large a été entrepris par la Direction des Etudes et Recherches d'EDF. Il comprend, d'une part, la modélisation couplée des écoulements et des transferts dans un arc et dans un bain métallique (cf. § 3) et, d'autre part, un programme expérimental qui s'appuie sur une installation de laboratoire construite en collaboration avec la société TETRONICS et l'IRSID (cf. § 4).

2. ETUDE THERMIQUE DU CHAUFFAGE PLASMA EN REPARTITEUR DE COULEE CONTINUE

Le chauffage d'acier liquide par plasma en répartiteur de coulée continue s'effectue dans une chambre hémisphérique, constituée d'un capot en béton réfractaire reposant sur le répartiteur (voir figure 2). L'objectif est de réchauffer l'acier d'une dizaine de degrés afin d'éviter sa solidification prématurée.

2.1. POSITION DU PROBLEME

Pour assurer un dimensionnement optimal de la chambre de chambre de chauffage, une étude de son comportement thermique a été réalisée en considérant le rayonnement, largement dominant dans ce type de process, et qui porte le réfractaire à des températures extrêmes. Cette étude a été menée à l'aide du logiciel 3D " TMG " de calcul des transferts radiatifs et conductifs utilisant une résolution par méthode nodale sur maillage de type volumes finis non structuré.

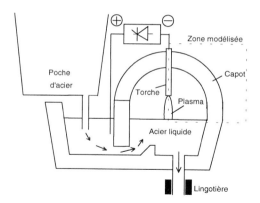

Fig. 2 - Chauffage d'acier par plasma sur un répartiteur de coulée continue

La difficulté première est de connaître à priori le flux thermique rayonné par l'arc. Des travaux antérieurs[1] proposent la répartition suivante des flux : 24 % de la puissance plasma est transférée au bain au niveau de la tache anodique, 5% est perdue dans l'eau de refroidissement du nez de torche, 2% est perdue par les fuites d'argon. Les autres flux (pertes pariétales par le corps de torche et le réfractaire) ainsi que le flux reçu par le bain sont calculés par le modèle.

2.2. RESULTATS

Un premier calcul de référence a été effectué pour une puissance plasma de 700 kW, et une émissivité du bain égale à 0.7. La température maximale du réfractaire atteint 1880 °C en partie basse, face à l'arc. Le rendement de chauffage calculé est de 77 %. Il est supérieur d'environ 5 à 10% à ceux rencontrés expérimentalement,[1] ce qui s'explique par le fait que ce calcul est effectué en régime permanent et sans tenir compte des ponts thermiques dans le réfractaire. D'autre part, la répartition des flux reçus par le bain (figure 3), montre que 50 % de la puissance transmise au bain est rayonnée par le capot et 19% provient du rayonnement direct de l'arc sur le bain. L'émissivité du bain est mal connue et peut éventuellement varier selon l'état de surface du bain. C'est pourquoi, une émissivité beaucoup plus faible de 0.4 (valeur préconisée par [2]) a été également considérée, la température maximale du réfractaire atteint alors 2053 °C. Nous avons aussi regardé l'influence d'une augmentation du rayon intérieur de la chambre de chauffage (R = 0.50 → 0.55 m) qui entraîne une diminution de température du réfractaire (figure 4).

L'influence de la puissance plasma sur la température maximale du réfractaire a également été étudiée (figure 5). On constate que celle-ci croît linéairement en fonction de la puissance plasma. Compte tenu des dimensions de la chambre, une puissance d'environ 500 kW semble être un maximum pour éviter le risque de détérioration rapide du réfractaire, prévu pour des températures limites d'environ 1880 °C. Une campagne de mesures thermiques sur le répartiteur in situ sera réalisée prochaînement.

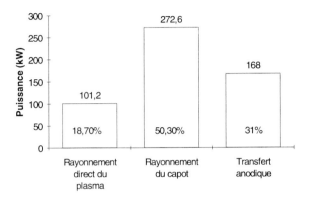

Fig. 3 - Répartition des flux reçus par le bain

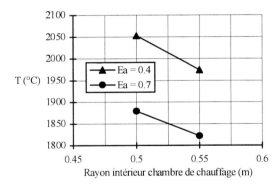

Fig. 4 - Température max. du réfractaire fonction du rayon intérieur de la chambre et de l'émissivité du bain

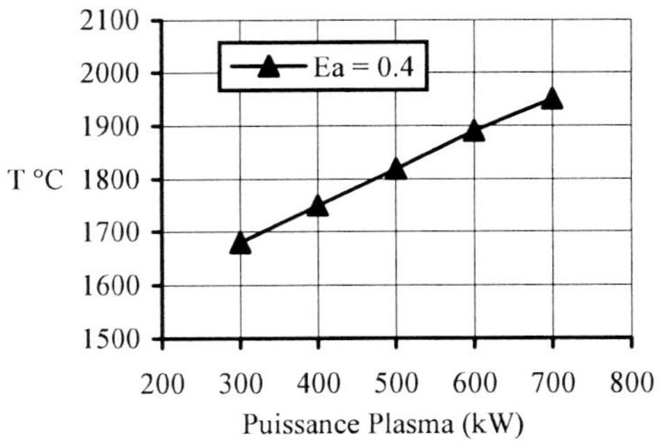

Fig. 5 - Température max. du réfractaire fonction de la puissance plasma

3. MODELISATION NUMERIQUE FINE DU CHAUFFAGE D'UN BAIN METALLIQUE PAR ARC TRANSFERE

Le " Laboratoire National d'Hydraulique " (LNH) de la Direction des Etudes et Recherches d'EDF, a développé à partir de 1990 un modèle de simulation numérique fine de l'arc électrique.[3] De récents développements permettent aujourd'hui d'étendre la simulation à la prise en compte des écoulements et des transferts dans un bain chauffé par arc électrique.[4]

3.1. DESCRIPTION DU MODELE

Le calcul de l'écoulement du plasma est réalisé par la résolution couplée des équations de Navier-Stokes et de l'électromagnétisme. Dans un cas axisymétrique, le champ magnétique est calculé à l'aide du théorème d'Ampère. L'écoulement du bain métallique, lui-même traversé par le courant électrique, est également régi par ces équations.

Bien que la contribution du rayonnement est, comme nous l'avons vu précédemment, très importante, les transferts radiatifs ne sont pas traités dans cette première phase. En l'état actuel on tient seulement compte des pertes radiatives du plasma, supposé non réabsorbant.[5] De même, l'influence des vapeurs métalliques sur les propriétés électriques et radiatives du plasma n'est pas traitée. Ces phénomènes sont en cours d'intégration dans le modèle.

Les calculs de l'arc et du bain sont couplés à la frontière arc-bain en respectant la continuité des densités de flux (quantité de mouvement, courant et chaleur) et de la température à travers la frontière plasma-bain, que l'on suppose plane et indéformable.

3.2. RESULTATS

Les résultats présentés ci-après correspondent à un calcul réalisé sur la configuration d'un four expérimental de 200 kW. <u>Le courant d'arc est de 2000 A, le débit d'argon de 50 Nl/mn et la pression égale à 1 atmosphère.</u> Le bain est constitué de fer pur. Sur les parois du four, constituées de béton réfractaire, on tient compte des pertes thermiques par un coefficient d'échange. Au fond du bain, la température de fusion du fer (1536 °C) est imposée sur une surface circulaire de rayon : 125 mm (électrode de sole). Afin de tenir compte du rôle de la turbulence, un modèle de turbulence k-ε à bas nombre de Reynolds[6] a été utilisé.

Un résultat global de calcul est donné en figure 6 ci-dessous, où sont présentés les champs de température dans le plasma et le bain métallique ainsi que le champ de vitesse dans ce dernier. Les profils de température et de vitesse verticale sur l'axe (figure 7) indiquent respectivement des maxima de l'ordre de 34 500 °C et 1600 m/s, ce qui est réaliste pour ce type d'arc. L'écoulement dans l'arc, non présenté ici, s'apparente à celui d'un jet impactant une surface.

Un bilan thermique du bain est donné en figure 8. On peut noter la contribution non négligeable du flux thermique des électrons entrant à la surface du bain .

L'écoulement dans le bain est caractérisé par deux zones de recirculation et une zone à vitesse quasi nulle. L'une des recirculations provoque une remontée du métal sous la zone d'impact de l'arc, elle est due à l'entraînement de la surface par l'écoulement cisaillé du gaz. L'autre recirculation, qui entraîne une descente du métal sous la zone d'impact de l'arc est due aux forces de Lorentz provoquées par la convergence des lignes de courant électrique vers la tache anodique de l'arc. Les vitesses dans le bain sont de l'ordre de quelques cm/s, ce qui semble réaliste. Le refroidissement à la base du bain provoque une stratification importante (zone à vitesse quasi nulle), d'où une surchauffe du bain de l'ordre de 1000 °C.

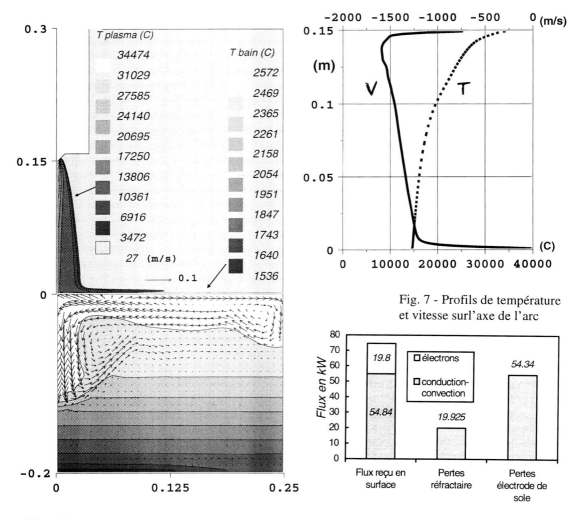

Fig. 6 - Température dans l'arc et le bain et vitesse dans le bain

Fig. 7 - Profils de température et vitesse sur l'axe de l'arc

Fig. 8 - Bilan thermique du bain

4. FOUR EXPERIMENTAL DE FUSION D'ACIER PAR ARC TRANSFERE
(200 kW - 2000 A - 300 kg d'acier)

Les objectifs recherchés par ce programme expérimental sont multiples : valider le modèle de simulation décrit au § 3, diagnostiquer le comportement du réfractaire, caractériser expérimentalement les transferts de chaleur entre l'arc électrique et le bain métallique, acquérir différentes données thermophysiques à hautes températures absentes de la littérature.

4.1. DESCRIPTION DE L'INSTALLATION (figure 9)

L'installation mise en œuvre aux Renardières a été conçue pour faire fondre et maintenir en température une masse de 300 kg d'acier par l'intermédiaire d'une torche plasma à arc transféré.

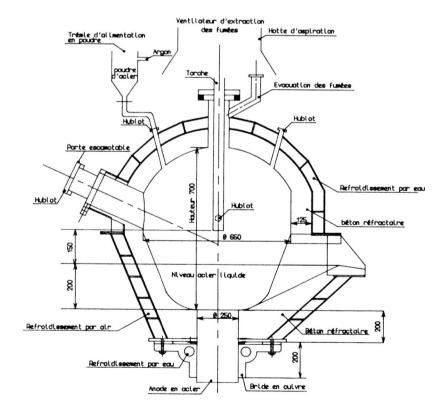

Fig. 9 - Schéma de l'installation expérimentale

LA TORCHE : il s'agit d'une torche TETRONICS à arc transféré alimentée par une unité de puissance à courant continu. Elle fonctionne à l'argon avec un débit de 6 Nm^3/h et sera utilisée avec les caractéristiques électriques suivantes : Iarc = 2000 A - Uarc = 100 V.

LE FOUR : il est basculant et composé de deux parties amovibles (une voûte hémisphérique et un creuset de forme tronconique), garnies de béton réfractaire à haute teneur en alumine. La voûte, refroidie par eau, est équipée de hublots qui permettent, d'une part, l'observation de l'arc et du bain et, d'autre part, la réalisation de mesures fines par voie optique. Un orifice en partie haute permet l'évacuation des fumées. Le creuset, refroidi par air, est équipé d'un bec de coulée. Le fond du creuset est constitué d'une électrode de sole (billette d'acier) dont la base est refroidie par eau.

4.2. INSTRUMENTATION

Les mesures de températures prévues sur ce four sont principalement des mesures par spectrométrie d'émission pour le plasma, par pyrométrie et par thermocouples pour l'acier, par thermocouples pour les réfractaires et pour l'électrode de sole.

La température des réfractaires est mesurée par thermocouples noyés dans le béton réfractaire à différentes épaisseurs. La température de surface n'étant pas accessible, une méthode de résolution inverse du problème de conduction, développée en collaboration l'INSA de Lyon nous permet de déterminer les températures et flux surfaciques sur la voûte du four.

5. CONCLUSION

Pour favoriser la pénétration des systèmes plasma de chauffage de bain d'acier, EDF a apporté son soutien à la première installation de ce type sur une coulée continue du groupe Usinor-Sacilor. Ce soutien s'est concrétisé par une étude thermique de la chambre de chauffage visant à anticiper les problèmes d'usure rapide des réfractaires. Les résultats obtenus ont confirmé l'existence de fortes sollicitations thermiques sur les réfractaires et ont permis d'analyser l'influence des paramètres pouvant les réduire. Il ressort principalement que la puissance du plasma devra être bien maîtrisée pour ne pas dépasser les températures critiques du réfractaire. Une campagne de mesure in situ sera réalisée prochainement afin de valider les résultats de cette étude.

Les travaux d'EDF sur le thème de la modélisation du chauffage de bains métalliques par plasma se poursuivent aujourd'hui par le développement d'un modèle couplant les calculs de l'arc et du bain. Ce modèle, bidmensionnel, prend en compte de nombreux phénomènes : écoulements couplés avec l'électromagnétique, transferts de chaleur et à terme, transferts de masse et rayonnement. Les premiers résultats obtenus ont permis de valider l'approche retenue : couplage itératif des calculs plasma et bain par les conditions aux limites à la surface du bain. Ces résultats montrent également une structure d'écoulement dans le bain caractérisée par une recirculation due à un effet d'entraînement de la surface du bain et une seconde recirculation due aux forces volumiques d'origine électromagnétique (forces de Lorentz). Le refroidissement intense de la base du bain provoque une stratification thermique importante (zone morte) ce qui nuit à l'homogénéisation thermique du bain. Des campagnes de mesure sur un four de laboratoire d'EDF (200 kW - 2000 A - 300 kg d'acier) permettront de valider ce modèle.

BIBLIOGRAPHIE

1. K. MATSUMOTO, Y. HSHIJIMA, K. ISHIKURA, K. UMEZAWA, Y. NURI, Y. OHORI: 'The Implemantation of Tundish Plasma Heater and its Application for Improvement of Steel Qualities', *Proceedings of The Sixth International Iron and Steel Congres*, Nagoya, Japan, 1990.

2. J. HAYASHI, J. AKIMOTO, Y. NAKAMURA: 'Mathematical Modelling of Thermal Plasma Heating', *Proceedings du XXIIe Congrès de l'Union Internationale d'Electrothermie*, Montréal, Canada, Juin 1992.

3. C. DELALONDRE: 'Modélisation Aérothermique d'Arcs Electriques de Forte Intensité avec Prise en Compte du Déséquilibre Thermodynamique Local et du Transfert Thermique à la Cathode', *thèse de Doctorat*, Université de Rouen, 1990.

4. L. TRENTY, A. BOUVIER, C. DELALONDRE, O. SIMONIN, J.B. GUILLOT: 'Numerical modelling of Metallic Bath Heating with Transferred Arc', *Proccedings of ISPC 12*, Minneapolis, USA, 1995.

5. DEVOTO: 'Transport Coefficients of Ionized Argon', *Physics of fluids*, **16** (5). 616.

6. B.E. LAUNDER, B.I. SHARMA: 'Application of the Energy-Dissipation Model of Turbulence to the Calculation of Flow Near a Spinning Disc', *Letters in Heat and Mass Transfer*, 1974, **1**. 131-138.

A Handbook of Electrotechnologies

J. PEÑA

Comité Español de Electrotermia

SUMMARY

The knowledge of Electrotechnologies amongst those who both use and prescribe them takes the form of a series of isolated specialist areas with little intercommunication, and this reduces the opportunities for cross-fertilisation which these technologies permit. The complementary nature to be seen amongst many Electrotechnologies makes it desirable to have a combined base of knowledges in sufficient detail to facilitate initial investigation of the specific application to be implemented.

The developments which have taken place and new applications justify a review of the assumed "squandering" of primary energy when electricity is used. The important contribution of many electrical applications to conserving fossil energy resources should be taken into account. The contribution of Electrotechnologies to business economies in terms of both product quality and company image is an additional factor which furthers interest in their promotion and this is precisely the aim and subject of this Handbook of Electrotechnologies, originally written in Spanish.

RESUME

La connaissance des Electrotechnologies entre prescripteurs et usagers concierne une série de secteurs de spécialisation isolés dont la communication entre eux est difficile, ce qui limite les possibilités de fertilisation croisée offertes par ces technologies. La com-plémentarité existant entre de nombreuses Electrotechnologies justifie le fait qu'il convient de disposer d'une base commune de connaissances suffisamment approfondies permettant une premiére exploration de l'application concrète souhaitée.

Les développements introduits et les nouvelles applications justifient la révision du "gaspillage" d'énergie primaire quand on utilise l'électricité. On doit tenir en compte la contribution importante de nombreuses applications de l'électricité à la conservation des ressources énergétiques fossiles. L'apportation des Electrotechnologies à l'économie de l'enterprise, aussi bien en ce qui concerne la qualité du produit que l'image de l'enterprise, constitue un autre aspect qui renforce l'intérêt de leur promotion, objectif poursuivi dans ce Manuel sur les Electrotechnologies, rédigé initialement en espagnol.

THE SITUATION REGARDING INFORMATION ON ELECTROTECHNOLOGIES

Those connected with the International Union on Electricity Applications (UIE) are in a privileged position regarding availability of the information required for a comprehensive knowledge of Electrotechnologies from the point of view of both theoretic aspects and the major practical applications. Publications from the UIE and National Committees reflect the substantial and effective efforts made to produce technical documentation of a quality and scope which is equal to if not greater than that available in other economic sectors.

Notwithstanding the state of knowledge of Electrotechnologies amongst those who both use and prescribe them, in our opinion this consists of a series of isolated specialist areas with little intercommunication which cuts down opportunities for the cross-fertilisation these technologies permit.

It is very common to find people with a thorough knowledge of a particular Electrotechnology but who lack knowledge of others which could be used in connection with their speciality, and this leads to particularly striking consequences. By way of example, we recall the case of experts on bakery ovens who were unaware of the use of dielectric heating to improve the moisture content in biscuits, or the use of infra-red radiation to extend the life of industrial bakery products which hugely simplifies distribution networks and frequency. A lot of similar examples could be mentioned in connection with metal working and readers will doubtless be able to add many examples from their own experience.

The complementariness in many Electrotechnologies makes it desirable to have a combined data base dealing with them all in sufficient detail to permit initial research into a specific application of interest. Once this first analysis has been made, specialist literature or the appropriate expert can be consulted. The need for this overall approach has formed our motivation for preparing this Handbook.

The paper on "Electroheat, the Motor of Industrial Progress", submitted at the 15th Congress of the World Energy Council in Madrid on 20-25 September, 1992, dealt with a study into ascertaining the situation of Electrotechnology in Spain with respect to:

- information on the decision-making process involved in adopting electrothermal technologies in production processes, specifically in aspects relating to the data sources enabling users to ascertain the existence of technologies, their application to production processes and guarantees that the solution chosen will in fact lead to the desired results;

- the factors which most affect innovative decisions, such as the problems to be overcome, the anticipated repercussions of the innovation and the risks involved in the decision, etc.;

- the results achieved compared with initial expectations and the repercussion on other aspects of the business in terms of both internal operations and market relationships.

INDUCCIÓN ELECTROMAGNÉTICA

- **La inducción electromagnética**
  La corriente inducida
  Potencia transmitida
  El rendimiento
  Ventajas e inconvenientes del calentamiento por inducción
  Principales utilizaciones de la inducción electromagnética

  - **Fusión de metales por inducción**
    Hornos de crisol
    Hornos de canal
    Principales aplicaciones de la fusión por inducción
    Principales ventajas e inconvenientes

- **Calentamiento de metales por inducción para conformado**
  Calentamiento para deformación plástica
  Ventajas e inconvenientes del calentamiento por inducción para el conformado de metales

  - **Tratamientos térmicos de metales por inducción**
    Calentamiento de la pieza a tratar
    Calentamiento para temple superficial
    Enfriamiento de las piezas sometidas a tratamiento térmico
    Principales aplicaciones
    Ventajas e inconvenientes

- **ANEXO: Tratamientos térmicos de los metales**
  Transformaciones estructurales
  Algunos tratamientos térmicos de los aceros
  Tratamientos térmicos en otros metales

Independently of the outcome of the last two points above, which will be the subject of consideration later on to justify objective interest in the use of Electrotechnologies, we shall concentrate on the availability and accessibility of the technological information when considering a handbook with a content similar to the subject matter of this communication.

Trade exhibitions and/or magazines figure prominently as information sources in 54% of the replies given by users who have modified their production processes by adopting electrotechnologies, and thereby learnt of the advisability of their use, whilst the engineering profession itself only accounts for 8.3%. With respect to assessing the information received as input into the decision-making process, primary importance is given to the equipment manufacturer whilst engineers come in last position amongst the six different answers.

The fact that engineers appear in last place and manufacturers in the most decisive position in our view confirms the hypothesis put forward of isolated areas of knowledge revolving around one or a few Electrotechnologies as opposed to the desirable situation of a more general knowledge amongst engineers which would lead to more rational decisions and more effective results.

THE UTILITY OF ELECTROTECHNOLOGIES FROM THE BUSINESS POINT OF VIEW

As is well known, an assessment of the effectiveness of electrical solutions is not solely reflected in the financial aspect of cutting energy bills but, even if the energy balance is negative, there are a lot of circumstances which make the results positive when an overall analysis is carried out, even without taking account of the social advantages which are brought about such as an improvement in working conditions or a contribution to reducing the pollutant emissions involved in all industrial activities.

The aspects which can be financially assessed and which usually reflect an improvement in production costs as a result of adopting Electrotechnologies include savings in raw materials, longer tool life, better process control, better opportunities for automation, fewer rejects and smaller floor space occupied, etc.

In terms of overall business results, an improvement in product quality should be added, along with a better image of the undertaken in terms of technological capability and reliability in supplies, etc., which have a substantial effect on business prospects.

THE UTILITY OF ELECTROTECHNOLOGIES FROM THE ENERGY POINT OF VIEW

Although Electrotechnologies have been known for many years, new developments and applications mean that many concepts relating to their utilisation need to be reconsidered. The most important of these relates to an assumed "squandering" of primary energy with the use of electricity.

Although the improved performance achieved in the production of heat from electricity has helped in part, the most notable progress should be assigned to the Electrotechnologies themselves and their modern applications whose high performance enables the losses originating in the conversion from fossil fuel to electricity to be assimilated.

In order to support this statement, we shall mention some examples of modern Electrotechnology applications which produce a large saving in primary energy.

FUSION DE METALES POR INDUCCION

La posibilidad de transmitir grandes potencias sin utilizar soporte material, y que esta transmisión sea prácticamente independiente de la temperatura del foco a calentar, convierte a la inducción electromagnética en una tecnología de extraordinaria utilidad para la fusión de metales, tanto férreos como no férreos.

Dependiendo de la configuración del campo magnético utilizado como soporte de la transferencia de energía se distinguen dos tipos de hornos: de crisol y de canal. En los hornos de crisol el circuito magnético inductor se cierra a través del material a fundir, mientras que en los de canal lo hace sobre un núcleo de chapa magnética como los utilizados en los transformadores.

Hornos de crisol

(Fours à creuset, coreless crucible Tiegelöfen).

Dentro de los hornos de crisol c: dos tipos principales atendiendo características de la transmisió energía:

- de transmisión directa, er energía se transmite dire material a fundir, con cri material que no permita establecimiento de las Foucault. Pueden ser c conductores refractari conductor (cobre), pe forma que no puedar corrientes menciona son llamados de cri sometidos a refrige

- de transmisión ind que la energía se forma total o par material a fundir realizados en m; electricidad, me que pueden es' de Foucault co elevación de l conducción a

Los hornos de c forma que sean hidráulico, tam! que el inductor móvil. La bobi... por unas culatas de chap... además, sirven de circuito por el qu... cierra el campo mag... así el calentamiento soporte del horno.

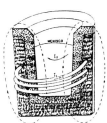

Hornos de crisol

La construcción má...
...ubrir
...rio.
...ap
...lic
...ir

Recocidos subcríticos

Son tratamientos orientados a eliminar tensiones en el material. Una de sus características principales es la menor incidencia de la velocidad de enfriamiento en el resultado.

Los principales son: el recocido de ablandamiento y el contra acritud. Este último similar técnicamente al revenido que suele realizarse sobre las piezas templadas.

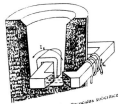

Recocidos con austenización incompleta o globulares

Son tratamientos encaminados a ablandar los aceros mejorando su maquinabilidad por medio de la consecución de una estructura microscópica globular más favorable.

El calentamiento se realiza hasta conseguir una temperatura superior a la crítica inferior, de comienzo de austenización, e inferior a la crítica superior (final de austenización).

El enfriamiento es lento, similar al de recocido de regeneración.

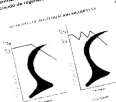

Temple

Se parte de una pieza calentada a temperatura más elevada que la crític superior, y se realiza el enfriamiento forma que la austenita se transforma martensita, componente que propor durezas elevadas. En determinados se precisa la austenización comple que la temperatura de calentamient entre ambas temperaturas crítica

La velocidad de enfriamiento va tipo del acero a tratar. En el ca al carbono, sin aleación y con carbono superior al 0,35%, el debe ser muy rápido. Si se tra aleados puede enfriarse más 1 llegando incluso a ser sufcie enfriamiento al aire en deter...

El martempering es un caso temple con enfriamiento et evita, en parte, la agresivi reduciendo tensiones, se fundamentalmente con p

Consecuencias de la soldadura

Por tratarse de soldadura de fusión, la soldadura con arco puede desencadenar una serie de reacciones que modifiquen sustancialmente las características y propiedades de las zonas de las piezas a soldar localizadas en las proximidades de la unión.

El proceso es tan complejo que nos limitaremos a enunciarlo de forma general para que sirva de motivo de atención cuando se trata de decidir en qué condiciones debe realizarse la soldadura.

La causa de los efectos a considerar radica en una elevación muy importante de la temperatura en una zona limitada de la pieza, elevación de temperatura que ocasiona la fusión del metal, aleación y posterior solidificación y enfriamiento. En las proximidades de la zona de fusión existe otra zona afectada térmicamente en la cual la elevación de temperatura no llega a fundir el metal pero puede llegar a situarlo en condiciones de sufrir modificaciones en su estructura química y cristalina.

En resumen, como consecuencia de la realización de una soldadura de fusión pueden ocasionarse tres tipos de modificaciones: morfológicas, químicas y estructurales.

Las modificaciones morfológicas se deben a la dilatación que origina el calentamiento seguida de la contracción consecuencia del enfriamiento. El resultado de ellas, si no se toman las precauciones adecuadas, pueden ser deformaciones permanentes y aparición de fisuras. A veces es conveniente precalentar las piezas a soldar, para disminuir las diferencias zonales de temperatura, o recurrir a cordones compuestos, depositados en pasadas sucesivas, que evitan en parte los efectos ocasionados por la focalización del calor.

Las transformaciones químicas están ocasionadas por modificaciones en la constitución del material como consecuencia de la mayor reactividad de los productos en contacto al elevar su temperatura. En general, se tratará de oxidación de algunos elementos del metal base en contacto con el oxígeno del aire y de elementos adicionales o reacciones propiciadas por el contenido del electrodo (alma y revestimiento).

Las transformaciones estructurales son las propias de los tratamientos térmicos de los metales, regidos tanto por el incremento de la temperatura como por la velocidad a que se produce este incremento y el enfriamiento posterior (ver fichas "Tratamientos térmicos de los metales por inducción").

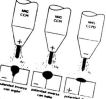

Fig. 5. Incidencia del tipo de corriente y gas inerte empleado en la soldadura MIG

Soldadura en metales

Soldadura por arco en aceros

La soldabilidad es un concepto especialmente importante cuando se trata de soldadura por arco en aceros. Se define así: un acero se considera soldable cuando mediante una técnica adecuada se puede conseguir la continuidad mecánica de la unión.

Las condiciones técnicas adecuadas se refieren a:

- Energía aportada
- Precalentamiento de la pieza previo a la realización de la soldadura
- Velocidad de enfriamiento de la soldadura.

La temperatura de precalentamiento de la pieza se determina mediante la fórmula de Seferian, a través del valor de contenido en carbono equivalente y del espesor de la

Determinación de la potencia transmisible a la carga de un horno

La potencia máxima transmisible por un horno a una carga se calcula como el valor máximo que puede alcanzar la expresión

$$Q = \epsilon \cdot c_1 \cdot S_r (T_h^4 - T_c^4) + h S_c (T_h - T_c) \quad (8)$$

que, para un horno y una carga determinados, es función únicamente de las temperaturas del horno y de la carga.

Los valores de ϵ, S_r y S_c son constantes dependiendo de la configuración de la carga y del horno, y el valor de h, según (17) de T de C, disminuye con la raíz cuadrada de la temperatura del horno.

Evidentemente, para transmitir realmente esta potencia es preciso que la nominal del horno sea superior a la máxima transmisible.

De forma que al estar limitada la potencia transmisible por las temperaturas de operación T_h y T_c, la única forma de reducir el tiempo necesario para efectuar el calentamiento de la pieza será elevar la temperatura del horno. De nada servirá aumentar la potencia nominal del horno si se mantiene el régimen de temperatura.

Al variar las características de la carga ϵ, S_r y S_c, también se modifica el valor de la potencia máxima transmisible, de forma que para hablar de ésta será preciso siempre referirse al horno y a la carga a calentar.

Elevación de la temperatura de la carga

En el caso que nos ocupa, horno calentado por resistencias eléctricas, la transmisión de energía se realiza según el esquema que representa la figura 4.

La resistencia eléctrica transmite calor:

- al gas que constituye la atmósfera (G) del horno Q_{RG}
- a las paredes (P) del horno Q_{RP}

- a la carga (C) a calentar Q_{RC}

Fig. 4 Transmisión de la energía en un horno de resistencias.

La energía recibida por las paredes del horno se divide en dos partes:

- la perdida a través de la pared, que pasa al ambiente exterior del horno Q_{PA}
- la radiada por la pared que, a su vez, se divide en
  - la incidente en la carga a calentar Q_{PC}
  - la absorbida por la atmósfera del horno Q_{PG}

La energía absorbida por la atmósfera del horno se destina a ser

- cedida a la pared por radiación y convección Q_{GP}
- cedida a la carga a calentar por radiación y convección Q_{GC} Q_{GC}

La carga a calentar recibirá la energía:

- radiada por la resistencia (Q_{RC})
- radiada por las paredes (Q_{PC})
- radiada por la atmósfera del horno (Q_{GC})
- transmitida por convección de la atmósfera del horno (Q_{GC})

- The use of electric vehicles in urban goods distribution results in a saving of 38% of the primary energy that an equivalent internal combustion engine vehicle would use.

- The application of ultraviolet radiation to water from treatment plants has an equivalent result to the traditional solution based on chlorine compounds, involving an energy consumption of between 5 and 20 Wh/m^3, substantially less than the primary energy content in the chlorine compounds which are replaced.

- The cryo-concentration of aqueous solutions leads to a 50% saving of the primary energy which would be consumed if twin effect evaporation were used.

A whole host of examples could be mentioned in the use of induction, lasers, reverse osmosis, plasma, electron beams, dielectric hysteresis and heat pumps, etc. which illustrate the contribution of many applications of electricity in conserving fossil fuel resources.

UTILITY OF THE HANDBOOK

Despite the brevity of the above statements, the usefulness of Electrotechnologies can be appreciated as regards promoting industrial concerns, improving product quality and broadening business prospects along with their positive performance in terms of energy conservation.

As a result, it can be seen that the promotion of electrotechnologies assists in the industrial development and energy efficiency which can contribute so much to improving the quality of life. The Handbook is a tool aimed in this direction with a view to placing Electrotechnologies within the grasp of users and prescribing experts as well as those who require a general approach which will enable them to guide users correctly when planning a production process or installation.

The lack of dissemination of the use of Electrotechnologies in developed countries can largely be explained as the result of energy saving policies, often designed around lower electricity consumption, or of administrative barriers to the use of electricity in applications other than lighting and motive power.

As countries develop, the amount of primary energy consumed per product manufactured falls. This trend towards increasing development has historically been accompanied by an increase in the use of electricity in total energy consumption, which illustrates the contribution of electrotechnologies to energy conservation.

CONTENTS OF THE HANDBOOK

The handbook deals with the twelve technologies traditionally considered the basis of electroheat: electromagnetic induction, electrical resistance, direct conduction, electrical arc, plasma, dielectric hysteresis, infrared and ultraviolet radiation, lasers, electron beams, heat pumps, mechanical vapor compression and membranes.

Each technology is dealt with in several subsections where necessary, as indicated in the contents, and is covered firstly from the theoretical, principally physical point of view, followed by a series of applications.

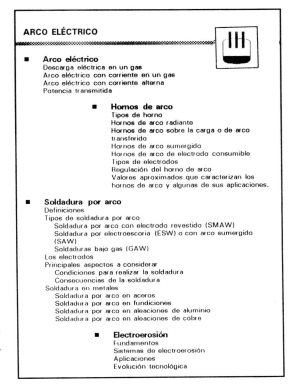

In the basic data we have tried to include a sufficient amount of quantitative facts to enable an initial estimation to be made of the possibilities for the specific application desired. This is the case with reference to the minimum size of load fragments and crucible dimensions in induction furnaces, etc.

In some cases a subsection has been included for the purpose of recalling various basic, non-electrical concepts which will assist in an understanding of the subject matter. This is the case with the subsections on heat transmission, basics of thermodynamics and heat treatment of metals.

Given the scarcity of information available in Spanish, this Handbook places technologies of considerable utility in industrial development within the reach of Spanish speakers.

Contents of the Handbook of Electrotechnologies

- Electromagnetic Induction
  - Electromagnetic Induction
  - Smelting of Metals by Induction
  - Heat Treatment of Metals by Induction
  - Appendix. Heat Treatment of Metals
- Electrical Resistance
  - Resistors
  - Resistance Furnaces
  - Heat Transmission
- Direct Conduction
  - Direct Conduction
- Electric Arc
  - Electric Arc
  - Arc Furnaces
  - Arc Welding
  - Electron Discharge
- Plasma
  - Plasma
  - Applications of Thermal Plasma
  - Applications of Cold Plasma
- Dielectric Hysteresis
  - Dialectric Hysteresis
- Infrared and Ultraviolet Radiation
  - Infrared Radiation
  - Applications of Infrared Radiation
  - Ultraviolet Radiation
- Lasers
  - Lasers
- Electron Beams
  - Electron Beams
- Heat Pumps
  - Heat Pumps
  - Applications for Heat Pumps
  - Basics of Thermodynamics
- Mechanical Vapor Compression
  - Mechanical Vapor Compression
- Membranes
  - Membranes

Education and Research in Electroheat

D. Van Dommelen
K.U.Leuven

The ERE within the UIE

Some thirty years ago, in May 1966, the Education, Research and Laboratories Study Committee began its operation within the Union Internationale d'Electrothermie. It set out as its goal to liaise between educational institutions and research laboratories in order to determine and activate means for developping education and research in Electroheat. It has always kept this course in its activities, but, some ten years ago the Committee's name was changed to become the Education and Research in Education Study Committee. Its responsabilities have recently been defined as encompassing the following projects:

- encouraging technical (basic and applied) research in electroheat via think tanks.
- organizing the dissemination of research findings to economic operators.
- increasing teachers' and students' awareness to electroheat applications.
- promoting education in electroheat at all higher-education institutions.

During these thirty years, electroheat has, indeed, changed its original scope and includes now all those applications of electricity that contribute to its rational use. The ERE committee has kept the pace with the developments in educational methods and contributed by regular exchanges, surveys and seminars to keep its members informed of the relevant trends and developments in the electrotechnologies and their teaching. Stimulated by the good relations among its members not only ideas, but also personnel was exchanged with the support of international or supranational entities.

Presently the ERE committee is meeting at least once a year and preferably at the occasion of a research or educational seminar organized by its members, or a larger convention of international importance.

Committee Meetings

Since the last international UIE Congress in Montreal, the committee met in Magdeburg (Germany) in 1992, where the project was initiated to make texts available at very moderate price for the students in higher-education institutions. Three test subjects were chosen which were induction heating, HF and UHF heating, and automation and control of electroheat processes. For each subject a group of expert was found ready to start the work. At this meeting, future research seminars were proposed and members were found ready to take up the responsability for the organization of seminars on Plasma Processing, Microwave heating and Waste treatment.

In 1993, a first meeting was held in Brussels in January. The terms of reference of the committee were summarized: the committee shall encourage teaching and research in electroheat at all educational institutions and in the continued education programs. The agenda included the immediate preparations of the seminar on plasma processing organized by professor J. Eninger (KTH, Stockholm, Sweden) and supported by the Alfven laboratory, the decision to collaborate with the newly formed microwave power association AMPERE and therefore to cancel the planned seminar on microwave power applications and some further clarifications on the planned seminar on waste treatment. Further on, a tentative content for the HF and UHF notes was presented, meetings were arranged for the team on induction, and the state of matter for the book on automation in electroheat was given. The first impulse was given for the organization of a survey on what teaching aids are presently used and how they electroheat is taught.

The second meeting of 1993 took place at the occasion of the research seminar on Plasma Processing in Stockholm in June. A most interesting presentation of the electroheat education and research in the United Kingdom and France was given by respectively Prof.L.Hobson and Mr.Greboval. Both countries have a well established tradition in the field of electroheat education, which is reflected in the existence of a specific newsletter or journal on educational matters supported by national electric industry, both organize summer courses and publish a number of educational documents. As a consequence of these presentations, Mr.Develey was asked and accepted to organize a formal survey of pedagogical aids available in the different UIE member countries in Paris. This meeting confirmed the planned seminars in Pilzen and Lodz.

In 1994, the committee met for two consecutive days: part of the time was a formal committee meeting and the rest was a closed seminar on the results of Mr.Develey's survey. The most worthwhile multimedia presentations at the latter led to the decision to set up an international event on this topic in 1995 as an open seminar. Profs. M.Machiels and D.Van Dommelen accepted to organize this event. Option was also taken to check on the financial and legal possibilities of having a number of books of the DopÜe collection translated from French into English. The seminar proposed by prof.L.Michalski to take place in Lodz would be set back until after 1996.

In 1995, the meeting took place after the open seminar held in Leuven (Belgium) in April. After updating the list of members, an evaluation of the seminar, and the arrangement of further activities, a proposal for a future seminar was discussed. Prof. Michalski would organize this future seminar in June 1997 on the simulation and identification of EH-processes in Lodz (Poland). As usual, the UIE will support the dissemination of the papers produced for that seminar. Mr. Y. Le Goff was found ready to head a task force on the application of the multimedia in EH education and agreed to present a paper at the UIE Congress in Birmingham on behalf of the ERE Committee. As a further spin-off of the seminar an interuniversity project on multimedia teaching of microwave heating has been started in Belgium with the support of the interuniversity Electrabel chair for Electroheat.

Research and Educational Seminars

Plasma Processing

Prof.J.Eninger organized this seminar at the KTH (Royal Institute of technology) in Stockholm on 21-22 June 1993. 31 experts participated in 45 minutes presentation followed by lively discussions on the following subjects:

- plasma generator performance for various gases and geometries - J. Th÷rnblom
- Finite element induction plasma modeling - Mekideche, Develey, Feliachi
- Non-thermal plasma techniques for gas cleaning - Rutgers, Creyghton, van Veldhuizen
- Removal of gaseous pollutants by corona induces cathalysis - Mattachini
- Evaluation of thermal plasma processes - Harry, Hodge
- Reducing Copper contents of steel by plasma ladle refining - Dembovsky
- Ilserv process for the treatment of EAF and AOD dusts - Bruno, Hunt, Repetto
- Surface modification of polymers "surface photografting" and other methods - Ranby
- Efficient Production of ozone in pulsed volume discharges - Eninger, Nilsson

The seminar was followed by a visit of the plasmafysik, fusion and accelerator facilities of Alfven laboratory. A small brochure with a one-page summary of the presentations is available from the UIE secretariat.

Inventory of teaching tools

Prof.G.Develey (LRTI) and Mr.D.Greboval (CFE-CEE) accepted to organize this seminar in Paris-La-DÚfense (France) on 16-17 May 1994. This was a closed seminar for the committee members and some guests. Educational material from different countries was displayed and could be inspected by the attendants. The program included the following subjects:

- An introduction to the Espace ELEC at the CNIT by Mr. R. Le Goff
- A review of the results of the survey conducted by Prof. Develey and Mr. Greboval on the educational material available for teaching Electroheat in the following countries: Austria, Belgium, Canada, Czech Republic, France, Germany, Italy, Japan, Poland, Russia, The Netherlands, United Kingdom.
- A presentation of the transportable material for inspection by the participants
- Successful demonstrations by Mr. Y. Le Goff of an hypertext computerized teaching aid, and other educational software by Prof. Nemkov and by Prof. Rajagopalan.

This seminar resulted in a brochure made available by the UIE under the title "ERE Seminar - Educational Material - Paris 16-17 May 1994".

Using hypermedia for education in electricity utilisation

As a follow-up of the survey of educational material, an international seminar was organized Profs. M.Machiels and D.Van Dommelen on the premises of the Katholieke Universiteit Leuven (Belgium) on 27 and 28 April 1995. The objective of the seminar "Use of hypermedia for education in electricity utilisation" was to present a number of realisations in this field as it had become clear from the above-mentioned survey that a number of member countries had very valuable material to present.

The potential of this attractive way of transferring knowledge is important not only for regular education, but may indeed offer some new approaches for in-house upgrading of personnel or continued education programs.

The seminar was attended by delegates of the ERE committee and various interested specialists from eight different countries. Demonstrations of practical realisations were presented and emphasized the various acceptances given to the words multimedia and hypermedia. The terminology is often thought to imply the concurrent use of digitally stored information rendering both text, sound, still pictures and animations on a personal computer or workstation. Some speakers have, however, shown attractive and very instructive combinations of non-digital means like videos, simulation tools and experimental set-ups relying indeed on many different media.

The papers presented were accompanied by demonstrations including educational software projected on a large screen, videos and slideshows. The program included the following presentations:

- Introduction, *D. Van Dommelen*
- De l'infographie au multimÚdia le "tout numÚrique" au service de l'enseignement, *J.P.Couwenbergh*
- L'apport des hypermÚdias pour l'enseignement de l'Úlectrothermie, *Y. Le Goff*
- Using Hypermedia for Education in Electricity Utilization, *H.-J. Lessmann*
- Computer Aided Tutor of Power Electronic Systems CATPELS, *M. V. Rajagopalan*
- The Use of Electrical Drive Simulation Packages, *M.G. Jayne & I. Luedtke*
- Using Multimedia in the field of Electrical Machines, *A. Malfait, K. Matthijs, R. Reekmans, K. Hameyer & R. Belmans*
- Teaching of Electroheat by a mixed Computer-Real Process method, *L. Michalski, K. Januszkiewicz*
- Hypermedia Course plus CAD complex: a new Tool to teach Induction Crucible Furnaces, *V. Nemkov, K. Smolnikov, V. Bukanin, D. Kuchmasov, A. Zenkov*
- Enseignement de l'Úlectrothermie en France: une experience en IUT, *G.Develey, A.Teillet*
- The Production of Multimedia, *D. De Grooff*

This seminar was organized with the support of the UIE central office, the Belgian Committee for Electroheat and Electrochemistry, the interuniversity Electrabel chair for electroheat and the electrical energy research group of the K.U.Leuven. The UIE secretariat keeps copies of the presented written material, and contact should be made with the authors for more specific educational supports.

Electric Arc Steel Making

Contents

International recommendation for universal use of UIE/IEC flicker meter 1
Sakulin, M., Renner, H. (Technical University Graz) AUSTRIA, Bergeron, R. (Hydro Quebec) CANADA, Key, T., Nastasi, D. (EPRI-PEAC) USA

The improvement of power quality on the system for arc furnace supply 9
Jozo, P. (University of Montenegro), Budimir, C., Rajko, R. (Steelworks Niksic) JUGOSLAVIA

Energy optimisation practices in electric arc steelmaking 17
Batham, J.K., Wilcox, R. (BOC Gases), Parr, E.A., Twiselton, J.C. (Co-Steel) UK

Future trend of DC arc furnaces in Japan 27
Mori, M., Miyashita, T., Okazaki, K., Kubota, T., Kuba, E. (JEHA) JAPAN

Energy savings using reactor with controlled power regime EAF .. 35
Hradílek, Z. (Technical University Ostrava) CZECH REPUBLIC

A new technology for the control of AC electric arc furnaces 43
Mulcahy, J.A., Kojori, H.A., Scaini, V. (Inverpower Controls Ltd), Burn, D. (Co-Steel LASCO) CANADA

Study of the flicker level in the case where DC and AC arc furnaces exist 53
Ikoma, M. (Kansai Electric Power Co.), Ichihara, H. (Chubu Electric Power Co.), Kondo, O. (Nissin Electric Co.) JAPAN

Development of high efficient melting on DC arc furnace 61
Takashiba, N., Takahashi, K., Yoshida, M., Ueda, A., Ueda, T. (Kawasaki Steel Corporation) JAPAN

Electromagnetic compatibility of EAF and supply power system .. 69
Wasowski, A., Bialek, J., Wilanowicz, R. (Technical Universtiy of Radom), Brociek, W. (Technical University of Warsaw) POLAND, Benghalem, K., Melizi, M.T. (Université de Sétif) ALGERIA

A novel method for the voltage flicker estimation and suppression utilizing the active power filter 77
Ashizaki, Y., Apyama, F. (Toshiba Corporation) JAPAN

Guide to the quality of electrical supply for Industrial Installations 85
Gutierrez Iglesias, J.L. UIE Working Group WG2

International Recommendation for Universal Use of the UIE/IEC Flickermeter

SAKULIN M., RENNER H.: Technical University Graz, Austria, Europe
BERGERON R.: Hydro Quebec, Montreal, Canada
KEY T., NASTASI D.: EPRI-Power Electronics Applications Center, Knoxville, TN USA

ABSTRACT

The UIE flicker measuring method is based on a comprehensive sophisticated simulation of the lamp-eye-brain performance taking into account all technical details of lamp behaviour and all physiological aspects of the human visual sensation system.

However, in its original design it was constructed for the 230 V incandescent lamps only and was therefore limited to countries with 230 V supply voltage. The UIE flickermeter succeeded in replacing the three former flicker measuring methods in Europe - the British gauge point, the French flickerdose and the German FGH Flickermeter, but, it could not succeed in North America and Eastern Asia where 120 V and 100 V lamps are the used.

The paper presented is the result of an international cooperation between Europe, United States of America and Canada. It describes the procedure and the adaptations which had been necessary to include 120 V lamps in the UIE flickermeter design in order to reach worldwide applicability. Furthermore, it presents the first results of field measurements comparing 230 V and 120 V flicker severity.

RESUME

La méthode UIE, à l'égard de l'évaluation du papillotement, s'appuie sur la simulation de la réponse du cerveau, de l'oeil, et de la lampe. Cette simulation reproduit la perception humaine du papillotement influencée par la lampe et par les aspects physiologiques du système visuel.

La conception du flickermètre, selon la méthode originale de l'UIE, se base uniquement sur la lampe incandescente 230 V. Or, seuls les pays avec un système de distribution à 230 V peuvent profiter de cette méthode, laquelle remplace aujourd'hui les méthodes européennes d'autrefois soit; le point étalon anglais, le flickerdose français et le FGH-mètre allemand. Cette limitation, à l'égard de la tension, rend ce flickermètre inutilisable dans les pays nord-américains et dans l'ouest asiatique, lesquelles utilisent un réseau de distribution et des lampes à 100 V et à 120 V.

Ce rapport résulte d'une collaboration internationale entre l'Europe, les États Unis et le Canada. Il décrit la procédure et les mises au point nécessaires pour adapter le flickermètre UIE à la lampe 120 V et pour universaliser la méthode UIE. De plus, il présente les premiers résultats de mesure en site et compare les analyses des Pst's selon l'hypothèse d'une lampe 120 V et 230 V.

INTRODUCTION

As well known, the aim of flicker measurement is to define and assess a criterion which directly expresses the degree of irritation of human beings subjected to luminance variations of lamps caused by fluctuating supply voltages. Flicker measuring therefore means measuring of voltage fluctuations and filtering these fluctuations according to the frequency characteristics of the lamps, the human eye sensitivity and the brain reaction. Lamp frequency characteristics can be different depending on lamp type and wattage. For flicker measurement standardization, therefore, the 60 W / 230 V filament lamp as the most common one was chosen and this lamp is the basis of the original UIE flickermeter. But, lamp construction also depends on the supply voltage level. For the same wattage 120 V lamps need higher currents, this means a construction with thicker filaments leading to higher thermal capacity with higher inertia against changes. Consequently at the same percentage of supply voltage fluctuation 120 V lamps are less flickering than 230 V lamps. These facts are the main reason that the existing UIE flickermeter is not directly applicable for measuring flicker of 120 V lamps. Due to the inherent differences in the frequency characteristics of the lamps a reconstruction of the flickermeter design specifications by adaptation of the filter curve was decided by UIE[8]. PEAC initiated the effort to evaluate flicker characteristics of modern electronic lighting and to establish a criteria for flicker free electronic lamp. TU Graz developed the original idea[4,5] and mathematical basis for translating the IEC flicker limits to other lamp voltages. The new power quality measuring system developed by Hydro Quebec will also include the new 120 V flicker measuring method as described in the paper.

ADAPTATION OF THE UIE FILTER CURVE

For this adaptation the following procedure is chosen. It is assumed that the eye-lamp filter curve for any other 60 Watt incandescent lamp can be derived from the existing UIE filter curve in a theoretical approach by exchanging the lamp frequency characteristics mathematically without performing new lamp flicker experiments with persons.

For incandescent lamps this procedure without practical experiments seems to be justified because frequency characteristics are rather similar. For fluorescent lamps, however, field experiments might be necessary to check the validity of the filter curves constructed in that way.

For the change of the lamp frequency characteristics the transfer functions of the 230 V/60 W-standard lamp and of the 120 V standard lamps must be assessed. It is assumed that the 230 V / 60 W standard lamp means the average of all 230 V / 60 W lamps nowadays on the market, independent of the original lamp which was used in the lamp-flicker experiments for deriving the curve in the past. The 230 V-UIE filter curve for eye + lamp is defined in the specifications[1] as a combined 4th order band pass:

$$H_{230}(s) = \frac{k\omega_1 s}{s^2 + 2\lambda s + \omega_1^2} \frac{1 + s/\omega_2}{(1 + s/\omega_3)(1 + s/\omega_4)} \qquad (1)$$

This eye + lamp transfer function includes the following factors: the lamp fluctuation characteristic which expresses the relative luminance fluctuation due to the relative voltage fluctuation, and the eye sensitivity to luminance fluctuations which itself depends on the average illumination[1]. Assuming a constant average illumination the eye+lamp transfer function can be split into two parts: the eye-transfer-function H_{eye} and the lamp transfer function H_{L230} - (2).

$$H_{230}(s) = H_{eye}(s) \cdot H_{L230}(s) \qquad (2)$$

The new 120 V UIE-filter curve follows from the 230 V-filter by exchanging the lamp filters (3):

$$H_{120}(s) = H_{230}(s) \cdot \frac{H_{L120}(s)}{H_{L230}(s)} \qquad (3)$$

Mathematically this is done in the following way:
- Calculating a new transfer function by multiplying the function $H_{230}(s)$ with the ratio $H_{L120}(s)/H_{L230}(s)$ which leads to a band pass of higher order and
- approximation of the resulting higher order filter curve $H_{120}(s)$ by the same type of 4th order band pass by adapting the filter parameters.

MEASUREMENT OF LAMP CHARACTERISTICS

Experimental and theoretical investigations[1] have shown, that filament lamps can be represented, approximately, as first order low pass filters with the parameters gain factor k_L and time constant T_L:

$$H_L(s) = \frac{\Delta L(s) / L_N}{\Delta V(s) / V_N} = \frac{k_L}{1 + sT_L} \qquad (4)$$

$s = j \cdot 2 \cdot \pi \cdot f_f$ f_f = flicker frequency
V_N = nominal voltage L_N = luminance at nominal voltage

To identify these lamp parameters for the simulation measurements of different lamps from different countries - all on the 1993/94 market - were performed by TU Graz and PEAC. Measurements were done in two ways - using the DC-method at TU Graz and the AC-method at PEAC. In both laboratories the same instrumentation was used and lamps of the same types and of the same manufacturers were measured. To measure the luminance L a photometer (Tektronix) adjusted to the standardized frequency range of visible light radiation was used[1], giving an output proportional to the luminance.

---

[1] It has to be mentioned at this point that in previous publications[4,5] a different photo sensor was used by TU Graz which caused slightly differing results in the measured lamp characteristics - especially in the gain factors. However, this did not affect the ratio (3) for the filter correction.

DC-method:

The correlation between supply voltage and luminance of an incandescent lamp is not linear but can be linearized around the operating point at nominal voltage. For the feeding of the lamp a DC-source instead of AC-source was taken to get a smooth output of the sensor.

In the first test series a measurement of the luminance as a function of the supply voltage $L(V)$ in a range from 80% to 120% of the nominal voltage was done. The gain factor was found by numerical determination of the derivation of this function at **nominal voltage** $V=V_N$.

$$k_L = \frac{d(L/L_N)}{d(V/V_N)}\bigg|_{V=V_N} \quad (5)$$

In a second step the time constant T_L was determined. This was done by analyzing the step response of the luminance after applying a voltage step from 105% to 95% of the nominal voltage. Under the assumption of a first order low-pass characteristic the result should be an exponential function. The time constant T_L was found by an optimal fitting of the theoretical curve to the actual measuring results with T_L as parameter.

AC-method:

In order to characterize a large sample of lamps in a relatively short time, PEAC has developed an automated test system which can produce controlled voltage modulations, measure lamp response, and calculate gain factor k_L. All operations are controlled by a computer, which allows the user to perform a ten-point frequency sweep for gain factor measurements. The automated flicker test setup creates repetitive voltage flicker by amplitude modulating a low voltage 60 Hz signal. Amplitude and frequency of the modulation as well as the wave shape - sinusoidal or square - can be specified by the user. The composite signal is sent to a 1.75 kVA linear amplifier and becomes the driving voltage for the sample under test. An illuminance head is located inside a light chamber at a fixed distance away from the lamp to measure relative light. The light sensor's output is displayed on a digital oscilloscope and automatically sent to the computer for analysis. The algorithm which calculates gain factor ignores the 120 Hz ripple on the light waveform and measures the change in light due only to the modulation.

Table 1 gives the results for the investigated lamps for the DC as well as for the AC approach.

	230 V lamp				120 V lamp			
	k_L		T_L [ms]		k_L		T_L [ms]	
	mean	dev	mean	dev	mean	dev	mean	dev
DC (TU)	3.67	0.06	19.6	0.68	3.61	0.20	28.9	2.37
AC (PEAC)	3.63	0.09	21.9	1.38	3.47	0.09	29.8	1.96

Table 1: Standard lamp parameters (mean and deviation) - measurement results of TU Graz and PEAC

The differences between the AC and DC results can be explained by two facts: Firstly, the test samples did not include physically the same lamps - there are also differences within the same lamp type - and, secondly, the incandescent lamp is not really a first order low pass - it's just a lamp - so, differences in the results are possible. For the filter design the AC results were chosen because these correspond to the natural conditions of lamps.

UIE 120 V LAMP-EYE FILTER:

The transfer functions H_{120} is calculated numerically according to formula (3). Approximation of the resulting curves by 4th order band passes according to formula (1), based on the PEAC lamp characteristics was performed by Hydro Quebec. Fig. 1 shows a comparison of the filter curves for 230 V and 120 V. Table 2 gives the numbers of the 120 V filter coefficients according formula (1).

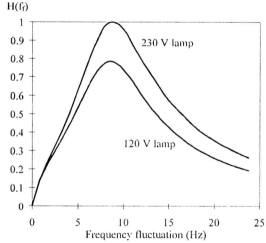

Fig. 1 Comparison of UIE filter curves H(f)

	230 V lamp	120 V lamp
k	1.74802	1.6357
λ	$2\pi\,4.05981$	$2\pi\,4.167375$
$\omega 1$	$2\pi\,9.15494$	$2\pi\,9.077169$
$\omega 2$	$2\pi\,2.27979$	$2\pi\,2.939902$
$\omega 3$	$2\pi\,1.22535$	$2\pi\,1.394468$
$\omega 4$	$2\pi\,21.9$	$2\pi\,17.31512$

Table 2 Comparison of filter coefficients

P_{st} = 1 P.U. CURVES, P_{st} CHECK POINTS

Fig. 2 shows a comparison of P_{st} = 1 p.u. curves calculated for rectangular voltage fluctuations for 230 V and for 120 V lamps keeping all other specifications including statistical evaluation (P_{st}-formula) unchanged. The 120V-UIE flickermeter and the 230 V-UIE flickermeter in that way are physiologically equivalent with $P_{st\,120}$ = 1 p.u. and $P_{st\,230}$ = 1 p.u. meaning the limit curve with the same degree of annoyance.

The 230 V values in Table 3 between 1 and 1620 min^{-1} refer to existing standards[1,2,3] , the values for 2400 and 2875 min^{-1} are based on own simulations (Hydro Quebec and TU Graz) but do not meet the German standard[3] .

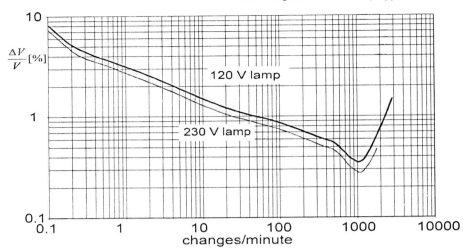

Fig. 2 $P_{st}=1$pu curves for regular rectangular voltage fluctuations

voltage changes [min⁻¹]	ΔV/V [%]		voltage changes [min⁻¹]	ΔV/V [%]		voltage changes [min⁻¹]	ΔV/V [%]	
	230 V	120 V		230 V	120 V		230 V	120 V
0.1	7.400	8.202	22	1.02	1.186	682	0.37	0.445
0.2	4.580	5.232	39	0.906	1.044	796	0.32	0.393
0.4	3.540	4.062	48	0.87	1.000	1020	0.28	0.350
0.6	3.200	3.645	68	0.81	0.939	1055	0.28	0.351
1	2.724	3.166	110	0.725	0.841	1200	0.29	0.371
2	2.211	2.568	176	0.64	0.739	1390	0.34	0.438
3	1.95	2.250	273	0.56	0.650	1620	0.402	0.547
5	1.64	1.899	375	0.50	0.594	2400	0.77	1.051
7	1.459	1.695	480	0.48	0.559	2875	1.04	1.498
10	1.29	1.499	585	0.42	0.501			

Table 3: $P_{st}=1$pu check points for regular rectangular voltage fluctuations

The ratio between the tolerable rectangular voltage changes of 230 V lamps and 120 V lamps varies between 0.89 for low flicker frequencies and 0.69 for high flicker frequencies. This means, that at the same rectangular voltage fluctuation the ratio $P_{st\ 230} / P_{st\ 120}$ varies between 1.11 and 1.44.

Fig. 3 shows a comparison between the proposed UIE 120 V limit curve and the existing US standards IEEE141 and IEEE 519. The frequency axis is scaled in changes per minute, where 2 changes correspond to 1 dip or pulse or fluctuation cycle regarding the US standards[6,7].

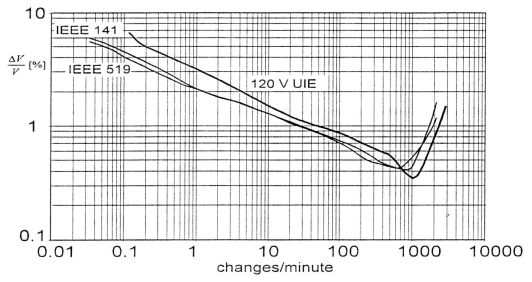

Fig. 3 Comparison between UIE 120 V $P_{st}=1$ pu limit curve and IEEE 141 resp. IEEE 519

FIRST FIELD MEASUREMENTS

As a first example, simultaneous measurement of the flicker in a public supply network with an electric arc furnace nearby was done, using the international flicker analysis system of TU Graz which allows parallel flicker measurement according to different evaluation methods (UIE, $\Delta V10$) and filters for different lamp voltages. Fig. 4 shows the time courses and a correlation analysis of the measured flicker levels. The ratio of the 230 V P_{st}'s versus 120 V P_{st}'s in fig.4 varies between 1.18 and 1.39.

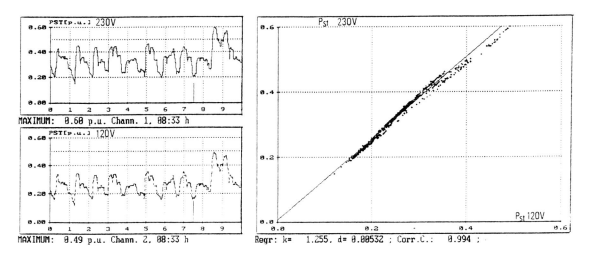

Fig. 4: Time courses and correlation analysis of simultaneous measured UIE 120 V P_{st}'s and UIE 230 V P_{st}'s

CONCLUSIONS

Due to the exact simulation of the lamp-eye-brain performance the UIE flickermeter provides an evaluation of flicker independent of the shape of the voltage fluctuations. It can be used, therefore, for all kinds of disturbers, like electric arc furnaces, welding machines, motor startups etc. The investigations presented in this paper shall be the basis for a guide how to adapt the existing 230 V UIE flickermeter for measuring flicker of 120 V lamps giving the same information about flicker severity for all types of disturbers. Design specifications and P_{st} limit curves for a lamp corrected 120 V UIE flickermeter are derived in the paper and compared.

In future work PEAC will continue to characterize the flicker response of fluorescent lamps, - together with Hydro Quebec a standard test protocol for measuring human response to lamp flicker shall be developed in order to get a better understanding of the flicker of fluorescent lamps and to establish appropriate flicker limits.

In a further step, a UIE flickermeter version for Japanese 100 V lamps could be derived in the same way.

REFERENCES

1. UIE Disturbances Working Group: 'Flicker measurement and evaluation', 1992
2. IEC 868 ' Flickermeter - functional and design specifications', 1986
3. VDE 0846 'Flickermeter - Funktionsbeschreibung und Auslegungsspezifikation', 1994, Germany
4. M. SAKULIN, H. RENNER: 'Strategy for worldwide applicability of the UIE/IEC flickermeter', *3rd international conference on power quality PQA94, Amsterdam,* 1994
5. M. SAKULIN, H. RENNER, R. BERGERON: 'UIE/IEC flickermeter for 120 V incandescent lamps', *4th international conference on power quality PQA95, New York,* 1995
6. ANSI/IEEE Std 141-1993: 'IEEE recommended practice for electric power distribution for industrial plants'
7. IEEE Std 519-1992: 'IEEE recommended practices and requirements for harmonic control in electrical power systems'
8. UIE 'Flicker measurement and evaluation', amendment 1996 (in preparation)

PRALAS JOZO, *University of Montenegro, EED, Podgorica*
ĆETKOVIĆ BUDIMIR, *Steelworks Nikšić*
RADULOVIĆ RAJKO, *Steelworks Nikšić*

The Improvement of Power Quality on the System for Arc Furnace Supply

ABSTRACT

The electric arc furnace (EAF) represents accidental variable non-linear load that unfavourably influences the network supply. A research has been carried out at the Nikšić Steelworks to improve a power quality on PCC. According to the measurement performed on the power supply system (35 kV and 110 kV) and the evaluation of results it has been concluded that the existing reactive power compensation system almost completely compensates reactive power, but effects the flicker level and doesn't suppress higher harmonics. It has also been concluded that differences exist between ideal and real resonant frequency, which are different per individual phases and contribute to the increase of klir factor and flicker effect in relation to the case of system operation without SVC.

RÉSUMÉ

Le four à arc électrique présentent la plus instable, accidenntalle variable non-lineaire charge, qui est tres defavorable pour les reseaux d'alimentation. Dans l'usine des acieries Nikšić ont exerce les recherches en raison de perfectionner la qualité de puissance au point de raccordement (pcc).

À la base des mesures, realiseé sur les reseaux d'allimentation (35KV et 110KV) et du traité des results, on a constaté que le systèm pour compensation de l'énergie réactive, mais défavorable influent au niveau des flickers et n'éloignent pas les harmoniques supérieurs.

Par analyse des résultats mesuré, on a etabli que existent considérable la diférence parmi theorique et réelle raisonente fréquence, qui sont encore différentes pour individuelles phases. Ce contribuent au agrandissement de klir factor et l'effet de flicker, en cas que le réseaux d'allimenation travaillent sans le systèm pour compensation.

1. INTRODUCTION

Electric arc furnace (EAF) is a very unstable consumer which causes different disturbances at the point of common coupling (PCC), such as higher harmonics voltage and current, non-symmetrical load of network supply, fast fluctuations of voltage, flicker effect and transient over voltages.

In order to find out, which characteristics must have supply system providing optimal energy conditions for the furnaces and other consumers connected to the same PCC, it's necessary to make a detailed analysis of the system of network supply-arc furnace.

For that purpose, a definite experimental and numerical research has been done in the Steelworks Nikšić, Montenegro - Yugoslavia.

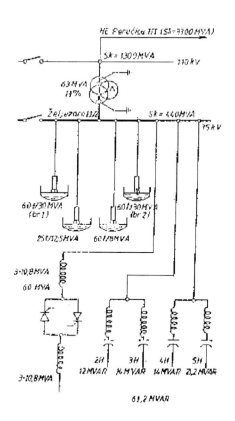

Fig 1 - Elementary diagram of Nikšić steel plant power supply

Two EAF 60t, 30MVA; one EAF 30t, 12MVA i SKF plant 8MVA, are connected to common point 35kV through step-down transformer 63MVA, 110/35kV, Uk=11%.

As shown on the fig.1 on the same bus-bars (35 Kv) the SVC (Static VAR Control) for reactive power compensation is connected. Capacitor banks of total power 61.2 MVAR are arranged in four filters for second, third, forth and fifth harmonics for these are the most expressive ones. Thyrstor converter together with air chokes reacts as a variable inductance because the current of the choke can be changed continuously operates as a continuously variable condenser which follows the changes of load.

2. THE BASIC PARAMETERS OF THE QUALITY OF ENERGY

2.1 THE PHENOMENON OF HIGHER HARMONICS

higher harmonics generated by the EAF operation are random values due to random arc variations during EAF melting process, especially at the beginning of bore-in periods of heat. Arc current is non-periodic and the analysis reveals a continuous spectrum of harmonics components including not only the whole number values. The harmonious analysis has shown that the whole number harmonics, especially the third, the fifth and the seventh dominate, and the harmonics amplitude decreases approximately with the factor $f^{-1/2}$.

The main source of higher harmonics is electrical arc, Then magnetizing current of transformer. The second and the fourth harmonics can be especially high which must be considered at the filters designing. With delta connection the harmonics multiple 3 in spectrum are eliminated.

By the summing up all simultaneously existing harmonics of voltage in one phase an instantaneous value of voltage u is obtained.

$$u = \mathrm{Re} \sum_{n=1}^{\infty} e^{j(n\omega t + \varphi_n)}$$

Where : n - rank of harmonic,

ω -the circular frequency,

U_n - effective value of the n-th harmonic,

φ_n -phase angle of the n-th harmonic.

Usual measure of distortion sinusoidal wave is distortion factor (klir factor, THD) which is defined as:

$$D = \frac{\sqrt{\sum_{n=2}^{\infty} U_n^2}}{U_1}$$

Distortion sinusoidal wave form has random character and depends on many factors such as current value, period of furnace operation, number of furnaces in operation at the same time connected on the PCC.

Capacitor's banks for reactive power compensation, with one or more frequencies, are coming in parallel resonance with the inductivity of the SVC system. Harmonics generated in the power supply system at resonant frequencies are amplified, what is unwanted phenomenon, and because of that, more filters which are adjusted on one of the frequencies of the second, the third, the fourth and the fifth harmonic have been used.

2.2. ASYMMETRICAL NETWORK LOADING

Asymmetrical network loading is provoked by single-phase or three-phase loads of the EAF with unbalanced arc impendances. Unequal voltage drops are caused by asymmetrical phase loading resulting in different line voltage in the inverse current component. The tolerable limit of voltage asymmetry is

$$\alpha_{U_{dozv}} = 0.202$$

2.3. VOLTAGE VARIATIONS AND LIGHT FLICKER

Variable current drawn by EAF passes through the network impedance causing voltage fluctuation at the PCC.

Due to short circuits between electrodes and the scrap, load variations have an amplitude up to 50% around a mean value and the frequency of 0,5 Hz. Simultaneously with these slower changes, some faster current changes with an amplitude of approximately 20% and a frequency of (2-30)Hz appear as a consequence of the internal arc changes.

The voltage variations are mostly the consequence of the changes of the reactive power. At the normal operation mode, the reactive power of EAF changes considerably faster then the active power.

Human eye is the most sensitive at the flicker disturbances at a certain amplitude and frequency of 10 Hz. The irritating action of flicker increases with amplitude square "a10" and linearly with duration of disturbance. Term "flicker dosage"

$$G = \int_0^T a_{10}^2 dt$$

represents "the quantity" of irritation. Decrease of flicker disturbance can be attained by the increase of short circuit power at PCC and by the decrease of the variations of reactive power.

2.4. TRANSIENT OVERVOLTAGES

Interruptions of the circuit of EAF cause overvoltages, which are consequences of fast changes of high currents. These overvoltages are approx. (1,2 - 3,5) Un.

3. SOME OF THE INVESTIGATION AND MEASUREMENT RESULTS

To achieve access into the above mentioned occurrences following power supply system and the corresponding parameters of power quality at PCC, certain measurements on the bus bars 35 and 110kV, for different operating conditions, different configurations of network, with and without SVC have been performed.

Voltages, currents, active and reactive powers have been measured at 35 kV bus bars.

At the data processing of the above described measurements, the values for harmonic distortion, flickers, voltage variations at switching the furnaces' breakers, variations active and reactive power at the different stages of the operation EAF, with and without SVC have been obtained. A numerical simulation and the analysis of the discussed system have also been performed.

Here are some of the typical results:

- When SVC in operation the reactive power is almost completely compensated.

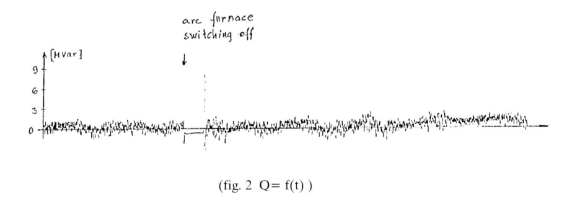

(fig. 2 Q = f(t))

- The total distortion factor THD has been determined by the reliable measuring instrument RMS and represented in the form of time function (fig. 3.a) and in the form of histogram (fig. 3.b).

By analysis of these diagrams, it can be concluded that the SVC affects the increasing of THD factor. Extremely high value, about 11%, has been registered during the switching the EAF off the network (fig. 3.a), but in the melt down periods of heat the value of 8% is often registered.

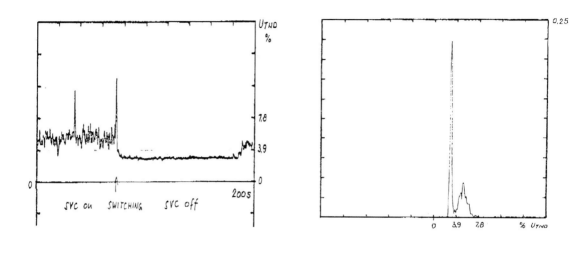

Fig.3.a Fig. 3.b

-Voltage specter shows that even and odd harmonics have an approximately same value. Current histograms with and without SVC also show significant differences.

- Flicker level is higher during the SVC operation, but is too high in both cases: (1,37 - 4,67)% at 35 kV and (0,21 - 0,7)% at 110 kV bus bars.

- When the SVC is not in operation the ratio of current harmonics is:

$$\frac{I_2}{I_1}=0.04; \quad \frac{I_3}{I_1}=0.05; \quad \frac{I_4}{I_1}=0.03; \quad \frac{I_5}{I_1}=0.05; \quad \frac{I_7}{I_1}=0.02$$

When the SVC is in operation, this ratio of higher current harmonics in relation to basic is increasing.

- Transient overvoltages have been registered at any of switching off the furnace breaker.

- Dynamic performance of filters is not sufficiently efficient for the fast changing harmonics.

All this leads to the conclusion that filtering effects (of built in filters) are limited. Control of filters' characteristics and their adjustment in operating conditions is necessary, for the deviation of adjustable values of only a few percents can affect the resonant curve of filters and so cause unwanted consequences.

According to the performed measurements the characteristics of filters have been derived: Quality factor Q_0 representing the inverse value of relative frequency band

$$Q_0 = \frac{1}{\frac{\Delta f}{f}}$$

resonant frequency and resonant impedance. An example of resonant curve for the second harmonic is shown on the figure 4. The frequency band is determined by these curves.

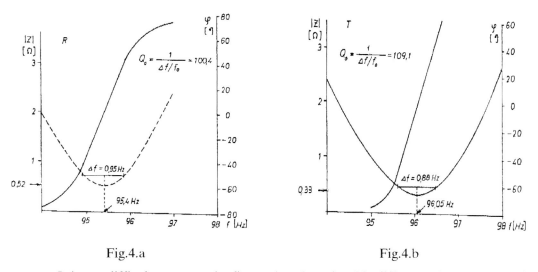

Fig.4.a   Fig.4.b

It is not difficult to see at the figure that there is a big difference between the ideal resonant frequency (100 Hz), and the actual resonant frequencies still different per individual phases. The measurements have shown that, in our case, actual resonant frequencies the biggest observed deviation is 8,2%. The problem that the deviations of measured and required values in many cases are smaller than those achievable by the minimal degree for adjusting the capacity, should be mentioned, too.

Since the quality factors are different per phases, the filtering effects per phases are different, too, and that affects the differences between harmonics distortion voltages and currents at network supply. The highest level of asymmetry has been registered with filter of the 5-th harmonic. The filters are placed outside one above other, and the level of Asymmetry depends on the geometry of the reactors.

In order to overcome the above problems, the solution could be found in the selection of another filter type with wider frequency band such as damped filters (L.2). These filters can be designed as C-type filter (fig.5a) or as third-order filter (fig. 5.b)

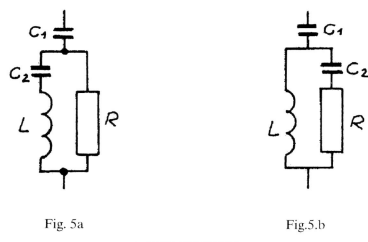

Fig. 5a   Fig.5.b

4. CONCLUSION

The electric arc furnaces, in the discussed case of the Nikšić Steelworks system cause severe disturbances in the network supply.

The reactive power compensation system fully compensates the reactive power, but increases the levels of voltage distortion and flicker effect. The cause of the increased distortion is in the asymmetry and difference of quality factor filters of higher harmonics per individual phases. The relatively small deviations of resonant frequency of the set value cause considerable difficulties. In the discussed case it was not possible to set resonant frequency due to relatively high minimal degree for adjusting the capacity.

The system analysis of thypistors control in the SVC plant and use of other types of filters with wider frequency range is recommended.

REFERENCES

1. A. DAN, M. KOVACCS, Z. CZIRA: 'Power system study on the onsite measurements', *Steelworks, Nikšić, 1989*

2. B. ĆETKOVIĆ: 'Metode mjerenja naponskih prilika u sistemu napajanja elektrolučne peći', Postgraduational thesis, *Zagreb, 1987*

3. N. GOTHELT, P. LUNDIN, M. RUBINSTEIN: 'X-filter - a computer program for optimizing filter circuits', *ABB REVIEW, 5/91*

Energy Optimisation Practices in Electric Arc Steelmaking

J.K. BATHAM *BOC Gases*
E.A. PARR *Co-Steel Sheerness*
J.C. TWISELTON *Co-Steel Sheerness*
R. WILCOX *BOC Gases*

Summary

The world's first Fuchs Shaft Furnace at Co-Steel Sheerness has demonstrated significantly lower electrical power consumption than in conventional electric arc furnace steel production. Scrap pre-heating procedures and BOC's modern oxy-fuel burner system have been implemented to provide a high degree of chemical energy input. Following the very succesful installation of the furnace in 1992, joint development programmes between Co-Steel and BOC have investigated energy optimisation technology using novel oxygen injection devices. Large amounts of energy in the form of chemical energy and heat are lost during conventional electric arc steelmaking. Oxidation of CO to CO_2 in the furnace by oxygen injection and the optimal transfer of the released heat to steel scrap or the molten bath (so-called post combustion) provide more efficient energy utilisation. A fast response gas analysis system suitable for use under extreme conditions has been developed to provide continuous monitoring of CO, CO_2 and O_2 concentrations in the waste gases of the shaft furnace and in the conventional arc furnace at Sheerness.

Résumé

Le premier four à arbre conçu par Fuchs et installé chez Co-Steel Sheerness a démontré une consommation moins élevée d'énergie électrique que la production d'acier par four électrique traditionnel. La préchauffage de la feraille et les brûleurs d'oxygène et d'hydrocarbures modernes de BOC [la Compagnie Britannique d'Oxygène] ont été installés afin de fournir un niveau de puissance chimique important. Des programmes de développement commun entre Co-Steel Sheerness et BOC ont étudiés la technologie d'optimisation d'énergie utlisant des dispositifs d'injection d'oxygène nouveaux ont suivis l'installation avec succès du four en 1992. Des grandes quantités d'énergie chimque et thermale sont perdues pendant l'élaboration d'acier à four électrique traditionel. L'oxyde de carbone s'oxyde en dioxyde de carbone dans le four par l'injection d'oxygène et par le transfert optimal de la chaleur à la feraille ou dans le bain (la dite poste-combustion), ce qui abouti à une utilisation d'énergie plus efficace. Un système d'analyse des gaz à vite réponse convenable à l'usage dans des conditions difficiles a été développé afin de fournir un contrôle continu des concentrations de CO, CO_2, et O_2 dans les gaz perdus du four à arbre et du four électrique traditionnel à Sheerness.

1. INTRODUCTION

Co-Steel Sheerness is a mini mill steel works comprising two electric arc furnaces, two ladle furnaces, two casting machines and separate bar and rod rolling mills. The plant was originally built to produce reinforcing bars, but over the years the emphasis has shifted to higher quality grades. During the 1980s, production in the melt shop continually improved to an annual output of 750,000 tonnes/annum, but it became apparent that the design of the original A furnace at Co-Steel Sheerness did not meet the needs of modern UHP furnace operation.

Because of the proximity of the plant to the town of Sheerness, Co-Steel Sheerness embarked on a project for a replacement furnace with design requirements of zero melt shop fume emissions and reduced noise levels. Additional requirements were increased productivity, low costs and low energy consumption. The two environmental issues required the entire furnace to be contained within a "doghouse" enclosure. The resulting C Furnace went into production in March 1992.

2. FURNACE EQUIPMENT AND CONTROL STRUCTURE

The energy balance for a typical arc furnace in Fig 1 shows much of the energy applied to an arc furnace is simply lost in the fume extraction. Some of this energy can be recovered by using the fume to pre-heat the scrap. Co-Steel Sheerness examined systems installed at other plants, but the requirement of a furnace "doghouse" enclosure precluded the adoption of these ideas. Traditional scrap pre-heating has many limitations so the solution implemented for the C Furnace was based on the Fuchs Shaft principle utilising a water cooled shaft mounted as an integral part of the roof. Hot exhaust gases from the furnace, assisted by six BOC oxyfuel burners at the base of the shaft, preheat the scrap in the shaft and it subsequently falls into the semi-molten bath.

There are several advantages to this approach, most notably that energy is recovered in the shaft from the two exothermic reactions:

$$2C + O_2 \rightarrow 2CO \quad +2.85 \text{ KWHr/KgC}$$
$$2CO + O_2 \rightarrow 2CO_2 \quad +6.55 \text{ KWHr/KgC} \text{ ("post combustion")}$$

Each burner has a fully adjustable fire rate, fully adjustable oxygen/gas ratio and can also be used to inject an adjustable 100% Oxygen flow into the furnace. The burners run permanently on a pilot flame to prevent blockage, and follow a fire programme for each charged basket. The principle is summarised in Fig. 2.

Plantwide control systems are traditionally based on a multi-layered structure, and Co-Steel Sheerness follows this well established procedure. At the highest level is an IBM AS400 mainframe. Below this is a process control level which was originally based on DEC VAXen computers, but is currently being changed to a distributed network of PCs. Below the process control computers, the actual plant control is usually performed by Programmable Controllers (PLCs), the majority of which are of Allen Bradley manufacture and designed in house by the Co-Steel Sheerness Engineering Department [1].

In the furnace pulpit, as far as possible, all controls are positioned to aid operator ergonomics and reduce manning requirements. Touchscreens are widely used, and the HCI aspects of these studied in some detail. The net result of these design decisions was to greatly simplify the pulpit controls[2]. Burner programmes can be created and modified in the pulpit to optimise energy inputs according to charge and process parameters.

The furnace operated initially for nine months without the "doghouse" enclosure, which allowed the benefits to be closely studied.[3] In particular there was no fume emission into the shop during melting and noise level on the floor around the furnace decreased by over 22dba (from 116dba to 94dba). Fume removal from around the furnace allowed it to be operated at a slight positive pressure to encourage post combustion of carbon monoxide in the shaft. This improved energy usage and reduced the carbon monoxide levels passed to the Baghouse.

3. OPTIMISATION OF ENERGY INPUT

Following the commissioning period, advances in melting and refining technology progressively lowered the electrical consumption and power on time of the shaft furnace. The furnace, at the end of 1993, showed an electrical consumption of 360 kWh/tonne billet with overall savings of £8.77/tonne billet (Table 1). Using the versatility of the BOC rocket burners, extended shaft preheat established that electrical consumption could be lowered much further (Table 2). This procedure involved prolonged periods, immediately after charging the furnace, with energy input only from the burners. Electrical power was not utilised until later in the furnace cycle and this procedure can be valuable in periods of high electricity price tariffs.

Although the shaft furnace technology ensures that high levels of waste gas energy are recuperated, CO emissions were still high at certain times during the furnace cycle. Consequently, potential energy was available to recover in the molten bath, or solid scrap, if this CO were burnt to CO_2. These high levels of CO also presented environmental concerns and the potential for explosions and physical damage in the offtake hot gas ducting. Consequently, a collaborative project was initiated between Co-Steel Sheerness and BOC in order to develop an effective system for *post combustion* in the shaft furnace.

3.1 POST COMBUSTION

The basic objective of post combustion is simple: convert the chemical energy present in CO (from decarburisation) and possibly hydrocarbons or H_2 (associated with scrap and carbon additions) into sensible heat in the charge by combustion to CO_2 and H_2O. The difficulties facing this objective include (i) the changing furnace conditions, as the transition occurs between heating solid scrap and heating liquid steel covered by an insulating slag layer, (ii) air ingress which inefficiently provides combustion of CO, (iii) high electrode losses from a strongly oxidising atmosphere, (iv) short residence time of the gaseous species in the furnace and (v) reversion of product CO_2 by contact with the carbon present in Fe droplets and slag. Factors such as these indicate that success in achieving the full benefits available from post combustion will come only by developing an effective strategy for oxygen lancing.[4]

In reported industrial studies of post combustion, electrical power input has been reduced to 290 kWh/tonne billet with a 100% scrap charge.[5] However, it appears likely that such figures are achieved with sacrifice of steel yield.

3.2 FURNACE WASTE GAS ANALYSIS

The effectiveness of post combustion in the electric arc furnace is largely determined by the degree to which the injected oxygen can be mixed with CO to give burning to CO_2, within the limited residence time of the gases in the furnace. Consequently, it was essential that a reliable offgas analysis system was developed in order to monitor continuously gas composition during the furnace heats. A purpose built analytical system was developed jointly by BOC and Orbital Gas Systems, Market Drayton, UK. Incorporation of N_2 purging systems and fail safe alarms has allowed reliable and continuous gas measurement by Siemens analysers, following an intensive period of system integrity improvement. A fast response time, less than 10 secs from furnace elbow to VDU display in the furnace control cabin, is obtained.

3.3 CARBON-OXYGEN IMBALANCE

Waste gas analysis revealed that high levels of CO were present in the offtake gases for most of the furnace cycle. The oxygen content mostly showed less than 10% by volume, associated with CO often in excess of 20%. At this time the sources of carbon within the shaft furnace were coke breeze in the charge, roof carbon injection, water cooled door lance injection, and electrode wear. The sources of oxygen in the furnace were provided by the water cooled door lance, the oxy-gas burners and from air ingress. A simple furnace mass balance showed that the minimum carbon addition of approximately 1400 kg/heat could not be fully combusted to CO_2 by the oxygen supplied by the door lance and the oxy-gas burners. Comparison with the K-ES process at Acc. Venete, as shown in Table 3, revealed a greater "oxygen deficit" at Sheerness when deliberate oxygen injection volumes were considered. Clearly, there was potential at Sheerness for the realisation of post combustion energy within the existing steelmaking practice.

3.4 POST COMBUSTION TRIALS - DEVELOPMENT OF THE TECHNOLOGY

Waste gas analysis revealed that CO contents in excess of 35% were produced when water cooled lance carbon was introduced and that low hot gas extraction rates promoted gas reactions in the furnace. These latter reactions could give serious gas blow-outs from the door.

Initially, burners 1 and 6, the first to be clear of scrap in the cycle, were utilised at super-stoichiometric ratios or with pure oxygen. Without the restriction of charge scrap, with a gas stream angled down at 32° and positioned at the base of the shaft, burner gas flow is across the surface of the molten bath allowing hot, turbulent, oxygen rich gas to react with CO rising from the foaming slag. These changes produced a reduction in CO levels, generally to less than 10%. Although electrical consumption stayed constant at 360 kWh/te billet through these furnace changes, reductions in burner gas and water cooled lance oxygen produced immediate cost savings of approximately £0.25/tonne steel.

The shaft furnace offers clear advantages when attempting to realise post combustion benefits. Firstly, the residence time of the gases is longer than in a conventional furnace and, secondly, the scrap in the shaft can absorb energy from post combustion performed early in the heat cycle. When the furnace approaches flat bath conditions, energy recovery from post combustion is more difficult and, as for conventional arc furnaces, entails combustion of CO rising from the foamy slag. Consequently, the development of post combustion could be considered as a division into the three distinct topics: (i) generation and combustion of CO early in the heat cycle, (ii) combustion of CO under flat bath conditions and, of direct relevance to both of these, (iii) effective methods of oxygen enrichment or oxygen injection within the furnace vessel.

Oxygen Injection

There were four basic methods available for oxygen enrichment: (i) utilising the burners at super-stoichiometric ratios (ii) utilising the burners on 100% oxygen (iii) increasing O_2 flow through the water cooled door lance and (iv) installing an additional lance or lances.

Although the burners are seemingly installed at an advantageous angle and position, operation at super-stoichiometric ratios has practical limitations because the flame becomes piercing, with reduced side flame radiation. Super-stoichiometric flames are more applicable during flat bath conditions and can provide benefits in CO reduction between approximately 18 MWH to tap (30 to 33 MWH).

Power savings are increased by burner operation at 100% oxygen, essentially through the flat bath period. BOC has established and patented flat jet burner and flat jet O_2 lance technology, which is commercially applied in glass and secondary aluminium melting.[6] The configuration of the flames and jets in this technology were eminently applicable to efficient combustion of CO in the electric arc furnace. Development of O_2 lance technology was initiated as described below.

The Generation and Combustion of CO Early in the Heat Cycle

The realisation of post combustion energy early in the heat cycle was approached by utilising roof carbon injection. A water cooled oxygen lance with flat jet nozzle was positioned in the fixed roof of the shaft furnace, local to the pre-heater shaft, so that a broad sheet of O_2 gas could complete combustion of the carbon introduced by the roof lance and the heat generated could be absorbed in the scrap in the shaft. Injection of oxygen through the roof also promoted post combustion during flat bath and CO content in the waste gases could be contained below 10%.

Roof carbon injection rate is restricted to a maximum of 40 kg/min. With reference temperatures of 25°C at inlet and 1500°C at outlet, combustion with pure oxygen can deliver approximately 7 kWh/kg carbon, with post combustion ratio = 1.0. Assuming a 20 min roof carbon injection period, whilst scrap is in the preheater shaft, the total attainable energy release is approximately 60 kWh/tonne charge. A heat transfer efficiency of 40%-50% can produce valuable energy savings.

The Combustion of CO under Flat Bath Conditions

Supplemental additions of oxygen were introduced to the slag/metal interface by means of a lance mounted on the water cooled Fuchs door lance or by flat jet nozzle inserts in the burners. In each case, a broad stream of oxygen, generated by flat jet lance technology, combusted the major source of CO evolution during the refining period, namely that localised source immediate to the

Fuchs lance. As shown by the analytical trace in Fig.3, CO content was significantly reduced after flat bath by oxygen contributed by the supplemental lance (the "piggy-back" lance). CO peaks are limited to the 2-3% level whilst CO_2 contents are 25 - 30%.

Post Combustion using the BOC Rocket Burners

The two outer burners at the periphery of the base of the preheater shaft were operated with 100% O_2 (at 6MW power equivalent: 1200 m3 O_2/hr) from early in the heat cycle until furnace tap. By this procedure, and with optimisation of oxygen input through the door lance, the electrical power input was reduced by approximately 40 kWh/te billet and the power on time reduced by 4 mins (Table 4) with no loss of scrap to billet yield (92%). Fig.4 illustrates comparative furnace heats produced with and without pure oxygen through the burners. With the instigation of this post combustion technology, the shaft furnace has operated at an electrical input of approximately 300 kWh/tonne charge.

Since the inception of post combustion processes at Co-Steel Sheerness, the average monthly electrical input on the Shaft furnace has decreased progressively, as shown in Fig. 5. The waste gas analytical system has proved invaluable in the realisation of on-line regulation of oxygen additions to the furnace. O_2 injection, through burners and/or lances, is now controlled automatically from the CO, CO_2 and O_2 content in the offtake gases.

Work is continuing to quantify the power savings to be gained from the supplemental oxygen additions. Experimentation on the conventional furnace at Co-Steel Sheerness has also revealed high CO levels which offer potential for post combustion.

REFERENCES

1   E.A. Parr "Control Aspects of the Fuchs Shaft Furnace at Sheerness Steel",
       Proc. of the 4th European Electric Steel Congress, Madrid, 1992

2   E.A. Parr "Safely Interfacing Men and Machines", Control & Instrumentation Nov 93

3   J. Twiselton "Shaft Furnace Environmental Implications",
       51st Electric Furnace Conference, Washington, November 1993

4   J.C. Twiselton, J.K Batham & C. Felderman,
       European Electric Steel Congress, Paris, June 1995

5   H Mueller et al Singapore Seminar, November 1995

6   For further details contact BOC Group Technical Centre, Murray Hill, New Jersey

Table 1 Cost Savings, Shaft vs Replaced Furnace

	£/tonne
Metallic Yield	£ 3.58
Refractory	£ 0.21
Electrodes	£ 2.04
Dust Disposal	£ 0.96
Electrical KWhr	£ 3.48
Oxygen Usage	-£ 0.90
Natural Gas	-£ 0.81
Overall Saving	£ 8.77

Table 2 Shaft Furnace Compared with Conventional Furnace

	Conventional Arc Furnace	Sheerness Shaft Furnace
Electrical Energy KWhr/tonne	514	275
Burner Oxygen cu m/tonne	3.3	35.5
Burner Gas cu m/tonne	1.7	17.7
Lance Oxygen cu m/tonne	12	5.5

Table 3   THE SHAFT FURNACE

Could do better with *post combustion?*

Mass Balance

		Sheerness	Acc. Venete (K-ES)
Carbon:	basket charge	8.9 kg/te bl	7.0 kg/te bl
	lance inject	4.4 kg/te bl	10.0 kg/te bl
	electrode wear	2 kg/te bl *:	2.3 kg/te bl
Oxygen:	door lance	11 m³/te bl	12 m³/te bl
	burners	21 m³/te bl	12 m³/te bl
	(gas)	(10 m³/te bl)	(6 m³/te bl)
	post combust lances	—	18 m³/te bl
	kWh/tonne billet	360	345
For $C + O_2 = CO_2$:			
	"O_2 deficit"	18 m³/te bl	6 m³/te bl

If 1 m³ O_2 ≡ 4 kWh/te, 12 m³ O_2 deficit could provide ~ 50 kWh/te billet.

* Now achieving 1.8 kg/te bl.

Table 4   RECENT FIGURES ON C FURNACE PERFORMANCE

90 TONNE BILLET WEIGHT PER CAST
ALL FIGURES ARE TONNES BILLET

	Power On Min	Furnace Electrical Power kWh/te billet	Ladle Furnace Electrical Power kWh/te billet	Door Lance Oxygen m³/te	Burner Oxygen m³/te	Burner Natural Gas m³/te	Average Ratio Oxygen/Natural Gas	Total Oxygen m³/te
STANDARD SHAFT CHARGING	34	365	30	11.1	21.1	10	2.11	32.2
WITH POST COMBUSTION	30	326	30	9	27.5	9.8	2.81	36.5

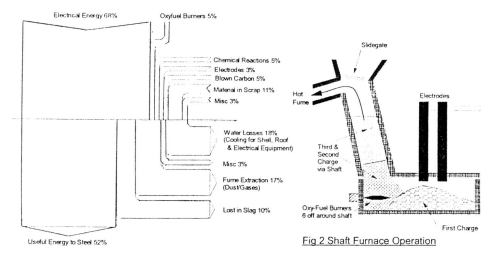

Fig 1 Energy balance for a typical electric arc furnace

Fig 2 Shaft Furnace Operation

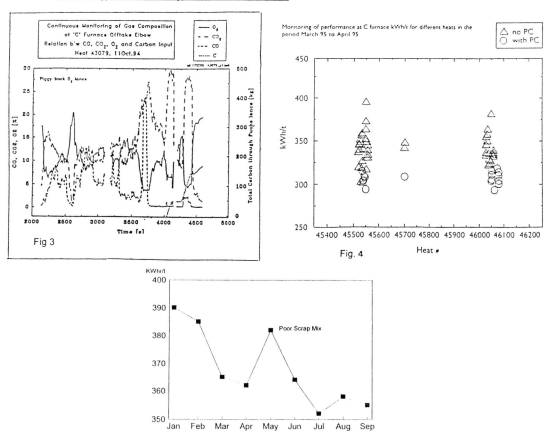

Fig 5 C Furnace KWhr/tonne billet

Future Trend of DC Arc Furnaces in Japan

MASAYUKI MORI
TETSUROU MIYASHITA
KINZOU OKAZAKI
TSUTOMU KUBOTA
EIJI KUBA

Member of the Arc Heat Working Group of the Japan Electroheat Association

Abstract

In July 1993, the Arc Heat Working group of the Japan Electroheat Association launched a survey of DC arc furnaces in Japan, focussing on past performance and expected future directions. This paper analyzes the results of questionnaires sent to Japanese steelmakers operating DC and/or AC arc furnaces. Our basic conclusion was that while Japanese steelmakers rate the DC arc furnace highly and expect to employ such furnaces in the future, there are still technical, economic, and construction-related issues that remain to be solved before the DC arc furnace can replace currently operating AC arc furnaces.

RÉSUMÉ

En juillet 1993, le groupe de travail de chauffage à l'arc de l'Association Japonaise de Chauffage Electrique a commencé l'ètude des fours à l'arc au courant dtrect au Japon faisant la mise au point sur la réalisation du passé et l'évolution dans l'avenir. Cet article tient á analyser les résultats de la questionnaire envoyée aux aciéries japonaises mettant en oeuvre des fours à l'arc au courant direct/alternatif. L'analyse des résultats montre principalement que les aciéries japonaises apprécient bien la performance des fours á l'arc au courant direct et pensent á les adopter dans l'avenir; toutefois, diverses problèmes techniques, économiques ou d'intervention restent á résoudre avant qu'ils ne remplacent les fours á l'arc au courant alternatif qui sont actuellement en service.

INTRODUCTION

This study surveyed Japanese steelmakers by questionnaire regarding six points: (1) desired improvements in arc furnace technology, (2) current state of furnace operations, (3) actual state of furnace facilities (including auxiliary equipment), (4) automation of furnace facilities, (5) environmental control, and (6) their evaluation of DC arc furnace technology. The questionnaire was sent to 49 facilities belonging to domestic Japanese arc furnace steelmakers, of which 36, corresponding to 47 AC furnaces and 5 DC furnaces, responded. This represents a response ratio of 73.5%.

Since the actual operating experience with DC arc furnaces described in the responses to this questionnaire was limited, it can be said that it is still too early to use this technique to evaluate the DC arc furnace.

Accordingly, in this survey we focussed on how steelmakers view the DC arc furnace from the standpoint of a firm operating AC arc furnaces. We asked steelmakers who had already introduced DC arc furnaces to base their responses on their actual experiences, and we asked for opinions from steelmakers who had not yet employed DC arc furnaces.

The responses indicated that although the DC arc furnace can be effective in reducing flicker, reducing unit electrode consumption, easing automation, and labor reduction, there are problems with this technology, including high initial costs, the space required for the renovations not being available at some sites, the period required for those renovations being excessive, and problems related to the electrode at the base of the furnace.

While 38% of these facilities are considering adopting DC arc furnaces in new furnace facilities, a mere 6% have decided against the DC arc furnace. However, 56% of the respondents felt that actual experience with DC arc furnaces is still inadequate and that it was not possible to decide.

CURRENT STATUS AND CONSTRUCTION PLANS FOR DC ARC FURNACES IN JAPAN

Remarkable progress has been made in DC arc furnace research. Early research began in the latter half of the 1970s, mainly in Europe, and in 1985 NUCOR, Inc. in the US put the world's first DC arc furnace, a unit with a 35-ton capacity, into operation. In Japan, the first DC arc furnace, also a 35-ton unit, was put into operation in 1988 by Topy Industries Co., Ltd. at their Toyohashi Works.

Based on the results obtained from these plants, the trend towards larger scale DC arc furnaces progressed rapidly.

Table 1 lists the currently operating DC arc furnaces in Japan and plans for future construction.

Table 1 DC Arc Furnaces in Japan: Operating and Planned Sites

	Company		Nominal capacity	Date placed in operation
1	Topy Industries Co., Ltd.	(Toyohashi)	35t	January, 88
2	Tokyo Steel Mfg. Co., Ltd.	(Kitakyushu)	130t	September, 89
3	Daido Steel Co., Ltd.	(Hoshizaki)	20t	December, 89
4	Kyoei Iron and Steel	(Wakayama)	60t	May, 90
5	Daiwa Steel Co., Ltd.	(Mizushima)	100t	December, 90
6	Nakayama Steel Mfg. Co., Ltd.	(Osaka)	40t	April, 91
7	Kobe Steel Ltd.	(Takasago)	30t	September, 91
8	Kansai Billet Center	(Sakai)	120t	January, 92
9	Tokyo Steel Mfg. Co., Ltd.	(Okayama)	150t	April, 92
10	Kyoei Steel	(Nagoya)	110t	October, 92
11	Nakayama Steel Works, Ltd.	(Osaka)	70t	December, 92
12	Daiwa Steel Co., Ltd.	(Mizushima)	100t	December, 93
13	Mitsubishi Steel Mfg. Co., Ltd.	(Muroran)	100t	April, 94
14	Mitsuboshi Metal Industry Ltd.	(Tsubame)	60t	April, 94
15	Takunan Seitetu K.K.	(Okinawa)	40t	November, 94
16	Toa Steel Co., Ltd.	(Kashima)	150t	May, 95
17	Tokyo Steel Mfg. Co., Ltd.	(Utsunomiya)	250t	95
18	Sumitomo Metal Industry Ltd.	(Osaka)	40t	95
19	Shimizu Iron and Steel	(Muroran)	60t	95
20	Yamato Kogyo Co., Ltd.	(Himeji)	130t	96
21	Tokyo Steel Mfg. Co., Ltd.	(Takamatsu)	60t	96

DESIRED IMPROVEMENTS IN ARC FURNACE TECHNOLOGY

When asked what current arc furnace technologies most needed to be improved, Japanese steelmakers responded as follows based on their continued experience with arc furnace steelmaking.

1st Automation and labor savings	7th Flicker control
2nd Energy savings	8th Continuous scrap charging
3rd Dust control	9th Improvement of preventive maintenance
4th Noise control	10th Malodor control
5th Scrap selection	11th Continuous production
6th DC arc furnace	12th Reduced upper harmonic components in voltage flicker

The main result here was that automation and labor reduction are seen as the most important issues for steelmakers when considering arc furnace technologies, and that other important issues include improvement of the working environment, reduction of power costs, small-lot production of multiple products, and high productivity.

Saving energy was weighted nearly as highly as automation and labor savings. This is because arc furnaces use electric power, which is a high-cost form of energy, and how effectively this energy can be used directly relates to reducing the overall cost of the steelmaking operation. This ranking reflects the fact that steelmakers are working on this issue actively. The DC arc furnace received a surprisingly low ranking here. This is probably because steelmakers who are currently using arc furnaces need to solve the technical problems with their current AC arc furnaces before they introduce DC arc furnaces in their operations.

CURRENT STATE OF FURNACE OPERATIONS

The practice of furnace operation is closely related to power rates, and steelmakers make serious efforts to reduce power costs by using the cheaper off-peak (Sunday, holiday, and weekday night) power.

As a result, the operation schedule changes radically between weekdays, Saturdays, and Sundays, as listed in table 2. The percentages of sites operating continuously (24-hour operation) on Saturdays and Sundays are a high 68.8% and 77.1%, respectively.

Table 2 Operating Formats   Units:%

Operating format	Weekday	Saturday	Sunday
Nighttime only	56.3	2.1	6.3
Daytime only		8.3	4.2
Daytime + nighttime	20.8	14.6	8.3
24-hour operation	16.7	68.8	77.1
No operations	2.1	2.1	
No response	4.2	4.2	4.2

While there are large variations in the characteristics of power, electrode, auxiliary energy, and other items consumed in arc furnace operation, these factors have the following values. Power consumption (average values) for AC furnaces centers in the 376 to 400 kWh per ton range while that for DC furnaces is 375 kWh per ton or lower; in general DC furnaces consume less power than AC furnaces.

Electrode consumption for most AC furnaces is somewhat under 3.0 kg per ton, but there is a wide distribution extending to the under 2.0 kg per ton range. Since 80% of DC furnaces have an electrode consumption of under 1.5 kg per ton, we can say that DC furnaces tend to have lower electrode consumption than AC furnaces.

Although oxygen consumption figures sometimes include the oxygen used for decorbonization, most furnaces consume about 20 to 35 Nm^3 of oxygen per ton of product.

Note that the rate of oxygen consumption increases with increasing use of auxiliary firing agents.

There is a negative correlation between oxygen consumption and power consumption, i.e., as oxygen use increases, power use decreases.

While only 25% of AC furnaces employ EBT for steel outlet, 100% of DC furnace use EBT. This is probably because DC furnaces are introduced with the intention of adopting automated production.

CURRENT FURNACE FACILITY CONDITIONS (INCLUDING INCIDENTAL FACILITIES)

The relationship between arc furnace capacity (in tons) and furnace transformer capacity (in kVA) is usually in the range 400 to 1000 kVA per ton.

The size of the electrodes used in AC furnace is between 350 and 600 mm, whereas that for DC furnace electrodes is between 600 and 700 mm.

The number of heats a water-cooled small ceiling furnace can withstand is relatively small, usually under 2000 heats, although there have been two furnaces in the range 4000 to 5000. The life of water-cooled panels varies widely depending on the materials and usage conditions, ranging from 1000 to over 20,000 heats. The frequency of repairs is once every 2 to 5 heats, and that per production unit is one for every 2 to 5 kg, for more than half the respondents. Scrap is preheated to between 200 and 400℃, and this reduces the power consumption by 20 to 40 kWh per ton. Scrap preheating is adopted by about 30% of furnaces. While about 70% of furnaces employ an auxiliary firing unit, only one of the five reported DC furnaces used such a unit.

The amount of coke usage was between 5 and 15 kg per ton for 70% of the respondents. While slag is received in pots and transported by crane or carts, the use of carts is increasing. Molten steel is stirred by gas stirring in 30% of furnaces. However, a stirrer was only adopted in one furnace. There was no observable correlation between furnace capacity and air capacity in dust collectors. While local hoods are installed on 30% of all furnaces, they are installed on four of five recently constructed DC arc furnaces, and only eight furnaces have a furnace enclosing chamber. About 78% of respondents expressed dissatisfaction with the level of cleanliness of their operations. The percentage of furnaces employing spray cooling for the electrodes was 76%.

FURNACE FACILITY AUTOMATION

We asked Japanese steelmakers what was, for them, the most important reason for introducing automation. Power input control during melting was the most highly ranked, being specified by 79% of respondents. Closely following at 76% was power input control during refining. Hot repair, demand control, and data logging were specified by 70% of respondents.

These automation techniques have all been demonstrated to be effective.

On the other hand, 58% of respondents had already installed control units for electrode connecting and dust collectors, but these systems have not proved as effective as the previously mentioned systems.

Figure 1 shows the actual performance and the hoped-for effect of control equipment that is actually installed and that had the highest response ratios.

This figure shows that there is great interest in control of melting power, control of refining power, and hot repair in that order.

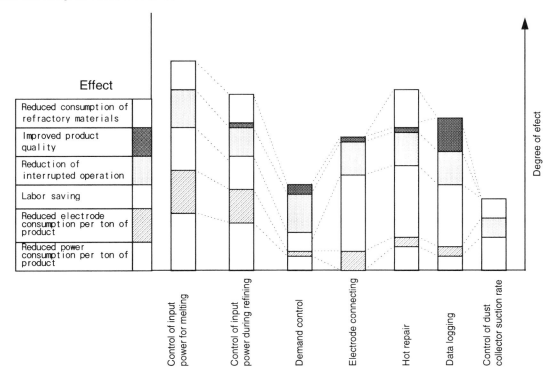

Fig. 1 Degree of Effect (Actual or Expected) for Different Control Functions

ENVIRONMENTAL MEASURES

FLICKER CONTROL

Of all sites, 35.3% have serious flicker problems. Thus flicker is a major problem for arc furnace operations, and must be resolved.

Flicker suppression units have been installed in 88% of plants included in this questionnaire. Of these 52.7% are static-type reactive power control units of the reactor-control type using thyristors.

It should be noted here that only three sites have installed active filter (self-commutated inverter) units.

UPPER HARMONIC COMPONENT CONTROL

Functions to suppress upper harmonics are included in flicker suppression units. Of the respondents, Two plants are using active filters.

ENVIRONMENTAL PROTECTION

While 59% of plants surveyed in 1986 were using some form of acoustic shielding, of the 34 plants responding to this survey, 30 (88%) employ some form of noise reduction. This reflects the fact that noise pollution is a major social problem that has received much attention recently.

All plants responding employ various forms of soot and dust removal, and these are further reduced by water sprays and plants (greenery) inside the factory.

Currently adopted techniques for malodor control include preheating waste gas to high temperatures, elimination of turning preheating, not preheating oily scrap, cooling exhaust gas to under 100℃, limitation or elimination of preheating, and selection of raw materials.

EVALUATION OF THE DC ARC FURNACE

Since actual experience operating DC arc furnaces is still quite limited, it is to early to evaluate DC arc furnace performance using questionnaire-based methods.

Accordingly, in this survey we focussed on how steelmakers view the DC arc furnace from the standpoint of a firm operating AC arc furnaces. We asked steelmakers who had already introduced DC arc furnaces to base their responses on their actual experiences, and we asked for opinions from steelmakers who had not yet employed DC arc furnaces.

ADVANTAGES OF THE DC ARC FURNACE

The result of this survey is that the DC arc furnace is seen as extremely effective at reducing flicker and reducing electrode consumption per ton of product, and that it is somewhat effective at promoting automation and labor savings.

Overall, the DC arc furnace is not regarded as "extremely effective" but rather most steelmakers regard it as "effective."

We adopted a scoring technique in which weights were assigned to answers and a point score was calculated. Figure 2 shows the results of this evaluation.

The contributions of the DC arc furnace to flicker reduction and reduction of electrode consumption per unit product were rated the most highly, with contributions to automation and labor savings being the next most highly rated item. Other contributions were seen as being of roughly the same magnitude. Thus steelmakers currently rate the DC arc furnace highly.

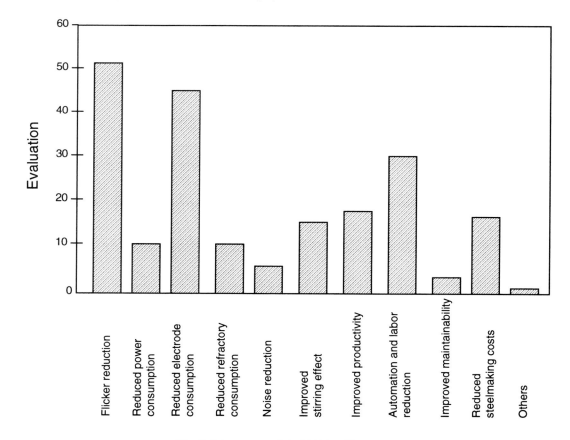

Fig. 2 Advantages of the DC Arc Furnace

TECHNICAL PROBLEMS WITH THE DC ARC FURNACE

Technical difficulties with the DC arc furnace were evaluated using the same scoring technique as in the previous section. Figure 3 shows the results.

Of the technical issues facing the DC arc furnace, the one that was mentioned most often was the life of the furnace bottom electrode. Electrode reliability and lead control were the second and third most often mentioned issues. Although preventing arc deflection and increasing capacity were also mentioned, some steelmakers do not see these as problems. Thus there was a difference of opinion on these issues.

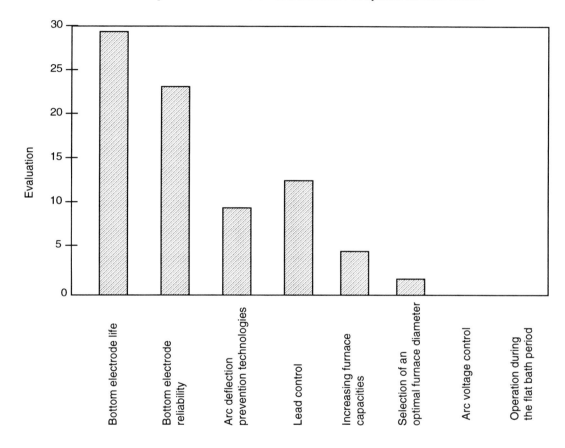

Fig. 3 Technical Problems Facing the DC Arc Furnace

PLANS FOR ADOPTION OF DC FURNACES IN NEW FURNACE FACILITIES

While the number of plants planning to adopt DC furnaces when introducing new furnaces is 38%, the number that have decided against the DC furnace is a low 6%.

Since actual experience with DC arc furnaces is still limited, 56% of the respondents felt that it was not possible to decide whether or not to employ DC furnaces.

OBSTACLES TO THE ADOPTION OF THE DC ARC FURNACE

High initial costs, lack of space for renovation, extended down time for renovation, and problems with the furnace bottom electrode are seen as obstacles to the introduction of DC furnaces, as mentioned above.

SUMMARY OF DC ARC FURNACE EVALUATIONS

The results of this survey make it clear that the DC arc furnace is rated highly. The survey also pointed out that enhanced reliability and longer life remain as technical issues, and that steelmakers would like to see eased maintenance management of the furnace floor. While there is an ongoing trend towards increased adoption of the DC arc furnace, it appears that there are still technical, economic, and construction problems that need to be solved to allow currently operating AC arc furnaces to be replaced with DC arc furnaces.

CONCLUSION

This was the second survey of the current state of arc furnace operations by the Japan Electroheat Association, following the first survey in 1986. Unlike the previous survey, this survey follows the major technical innovation known as the DC arc furnace. Furthermore, since 7 years have passed since the first survey, we are confident that this survey serves to grasp the current state of arc furnace operations and collects important views on future directions.

We will be pleased if this report can contribute to progress and development in the arc furnace industry.

ENERGY SAVINGS USING REACTOR WITH CONTROLLED POWER REGIME EAF

Z. HRADÍLEK
Technical University Ostrava
Czech Republic

ABSTRACT

Final economic results of melting in electric arc furnaces /EAF/ are substantially influenced by control of their power regime. A reactor must be introduced whenever we utilize the maximum output of the furnace transformer on its highest voltage levels. Regime of introducing the particular stages of the reactor is specific for each of the arc furnaces. This makes necessary both previous measuring of basic electric parameters and variables, and also energetic analysis of the current melting regime. Then only we can determine the proper time mode of the reactor usage. Given examples are taken from the 50-tons arc furnace in the Steelwork Vítkovice a.s., Czech Republic.

RÉSUMÉ

Les résultats économiques finals de fusion dans les EAFs sont essentiellement influancés par la réglage du régime énergétique des fournaux arcs. Si on demande le puissance maximal de transformateur de fournau sur les voltages les plus hautes il faut utiliser le réacteur. Le régime de rangement des particulaire degrés du réacteur est spécifique pour chaque fournau arc. Pour ca il est nécessaire de mésurer des fondamentaux parametres et variables électriques et analyser le régime contemporain de fusion. Puis on peut déterminer le mode de temps d´utilisation du réacteur. Les résultats sont obtenus en utilisant le réacteur sur le 50-tonnes fournau a Vítkovice aciérie.

UTILIZATION OF A REACTOR WITH AN EAF

There was no need to use a reactor for most of the time if the short-arc melting regime was used. The regime requiring only the short-arc technology had no need to introduce the reactor for most of the melting time. A reactor, if there was any, was used only for reduction of the big currents amplitude, which arise mostly during the first phases of melting. These currents mostly originate from short-circuits between the electrodes and the charge. It is only the power engineering view that makes the reactor necessary at this stage. Undamped power drops disturb other appliances feeded from the same supply system. However, the power regime using the short-arc technology with a reactor suffers from decrease of the total output, which is caused by drop of the instantaneous voltage.

The long-arc technology, which is being introduced in growing numbers nowadays, brings quite different situation. The furnace works with the highest voltage and a long arc as long as possible. The aim is to reduce the charge-melting time and to use the full power of a transformer for the longest possible time. The long-arc operation mode requires introducing the reactor. Reactors, if put in series with the transformer and arcs in a furnace, help to stabilize the long arc. They enable increasing the current-voltage phase shift, with the voltage phase advanced, in a phasor diagram. This phase shift leads to the global effect of arcing stabilization.

CHANGES OF THE TIME BEHAVIOR OF THE WATTLESS POWER AS A PARAMETER OF STABILITY OF THE ARC

The arc is quite labile during the melting-down stage. Reasons of this lability are:

- arc flashing-over on spikes and edges of the charge
- elongation of the arc caused by the electromagnetic forces of neighbor phases
- different degree of ionisation of the burning arc column

The arc is often extincted during short circuits in the melting-down stage. These short circuits together with big currents bring instantaneous rise of the wattless power demand. The critical situation comes at the moment when short circuits appear and a great deal of the whole power is made by the wattless component. The time behavior of the wattless power is therefore the parameter characterizing the arc stability.

50 TONS EAF IN THE STEELWORK II, VÍTKOVICE A.S.

This EAF with the charge of 50 tons of scrap is fitted with a 10 MVA transformer. Primary voltage of the transformer is 22 kV, secondary voltage is uncontinuously controlled in 19 stages ranging from 120 to 350 V. The transformer can be overloaded up to 12 MVA for 2 hours during the melting-down stage. The three highest voltage taps can work with the whole rated power, which however drop proportionally for lower voltage taps. All voltage taps can be regulated under load. The transformer core is made of oriented lamination, oil cooling is forced with a water radiator. The reactor, which core is made of oriented lamination, shares the same tank with the transformer. Its two stages of reactance, 1 MVA and 2 MVA, can be selected during work under load.

FINDING THE OPTIMAL REGIME OF WORK WITH THE REACTOR

One of the alternative methods of the arcing stabilization, as already has been said above, is introduction of the reactor. measurings with the the PSCOPE device have been performed on the 50 tons EAF in the Steelwork II, Vítkovice a.s.to establish the optimal reactor-using working regime. This apparatus allows instantaneous exact readings of voltages, currents and powers in each phase.

Amplitudes of the wattless power during melting without reactor are much bigger and more numerous than for melting with it. 'Fig 1'. A big difference, especially evident during the first 35 minutes of melting, shows that if the reactor is not used, substantial lability of the arc as well as much more frequent short circuits between electrodes and the charge appear at this stage.

If considered from the angle of stability of arcing, the above mentioned time behavior of the wattless power is the most advantageous method for determination of the reactor-using working regime. The wattless power time behavior for melting with a reactor is shown in 'Fig 2'. Big amplitudes and frequent changes of the wattless power are evident especially during the first 35 minutes after the start of melting and after introducing the second basket.

The time regime of introducing the reactor for the 50 tons EAF was designed using the 'Fig 2'.

ECONOMIC CONTRIBUTION OF THE REACTOR

Following principles were kept to evaluate the contribution of the proposed melting regime:
- heats were similar in amount and composition of the charge
- heats were charged in the same number of baskets
- heats with continuous work

Economic contribution of the proposed regime.

Electric power savings.

Reduced power demand was proved in all cases when the proposed regime of introducing the reactor was kept. The average energy saving of the observed EAF was 32 kWh t^{-1}.

Shortened time of melting

If the proposed regime is followed, not only energy saving is expected, but also shortening of the aggregate time of the melting-down stage and thus increasing productivity of the furnace. An average shortening of the melting time was 22 minutes.

Linig wear reduction

Shortened melting time is closely connected with reduction of the thermal stress of the lining and thus also with prolongation of its service life. One of the factors favorably influencing the service life is starting the melting with lower voltage and thus with shorter arc. This short-arch technology prevents the furnace roof from direct irradiation of the arc. After some 5 minutes, when the arc is safe protected by the surrounding charge, it is advantageous to apply the long-arc technology, this means the highest possible voltage.

Saving of electrodes

The expected saving of electrodes, resulting from lower power demand and shorter melting time, should not be neglected. Consumption of the material of electrodes is proportional to the second power of current. Saving of electrodes is thus considerably accented by the long-arc technology with lower current.

SUMMARY

Determination of the optimal working regime is the ideal solution from the angle of costs, and ensure essential rise of the steel production efficiency. Analysis of measuring the 50 tons EAF in the Steelwork II in Vítkovice was used for designing the regime of utilization of the reactor. This power regime shows time limits for introducing of the higher reactance of the reactor, with positive effect on stabilization and calming down the arc, The reactor must be put out of service when it only uselessly reduce efficiency of the furnace.

Bibliography

1. Z.HRADÍLEK: 'Electro-metallurgical Problems of Steelmaking Arc Furnaces *Thesis ČVUT Praha*, Prague,1981,116-135.
2. Z.HRADÍLEK and J.GAVLAS: 'Concept on Improving of Power and Economic Results of Melting in the 50 tons EAF in the Steelwork II, Steelwork Vítkovice a.s.'*Report nr.1072 TU Ostrava,*' Ostrava, 1994,5-17.

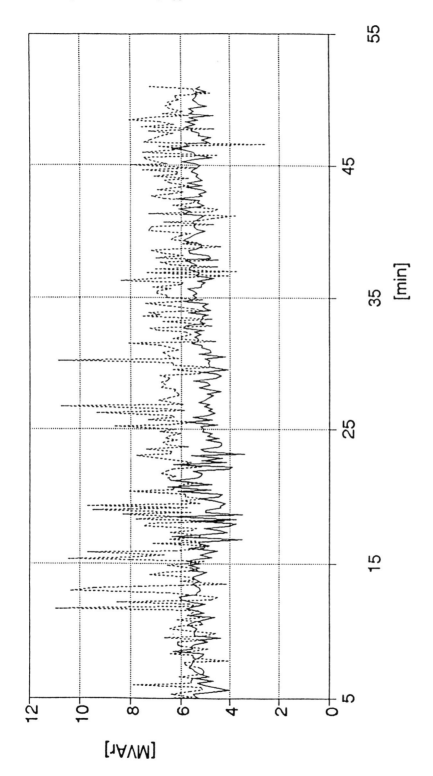

Fig.1: Amplitudes of the wattless power during melting without reactor /------/ and with reactor /———/ for the 50 tons EAF

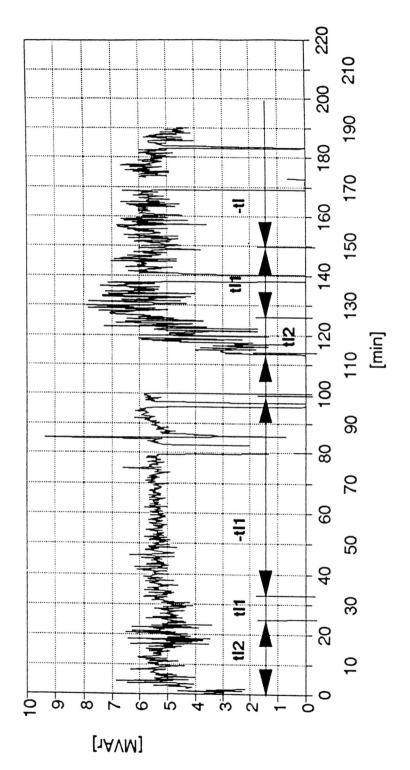

Fig.2: The wattless power time behavier for melting 50 tons EAF. The time regime of introducing the reactor

A NEW TECHNOLOGY FOR THE CONTROL OF AC ELECTRIC ARC FURNACES

JOSEPH A. MULCAHY	J. Mulcahy Enterprises, A Division of Inverpower Controls Ltd.
DR. HASSAN A. KOJORI	Inverpower Controls Ltd.
DAVID BURN	Lasco
VINCENT SCAINI	Inverpower Controls Ltd.

SUMMARY

Design, installation and operating results of a Smart Predictive Line Controller (SPLC) for reducing arc furnace flicker problem at Co-Steel LASCO in Whitby is presented.

The installed SPLC is comprised of a 3f, 46kV thyristor-based AC switch rated at 100MVA, a real time central controller for on line control of the AC switch and a state of the art data acquisition system for the control/monitoring the performance of the arc furnace system.

Real time adaption to changes in furnace conditions, allows the SPLC to predict the correct parameters of the arc furnace system and utilize a pre-calculated look-up table for adjusting the delay angles of the AC switch to reduce flicker.

The installation and operating results of the SPLC at LASCO along with the limitations of the design and requirements are discussed in detail.

Conception, installation et fonctionnement résultent d'un Predictive Intelligent Règle Contrôleur (SPLC) tremblote pour chaudière de l'arc réductrice problème à Co-Steel LASCO dans Whitby est présenté.

L'a installé SPLC est compris d'un 3f, 46kV AC thyristor-installé à change a estimé à 100MVA, un vrai temps contrôleur central pour change sur contrôle de la ligne de l'AC et un état du système de l'acquisition des données de l'art pour le contrôle/ écoute la performance du système de la chaudière de l'arc.

Vrai [adaption] du temps à changements dans chaudière conditionne, permet le SPLC prédire les paramètres corrects du système de la chaudière de l'arc et utilise un pre-calculé apparence-en haut table pour mise au point le délai oriente de l'AC change réduire le vacillement.

L'installation et fonctionnement résultent du SPLC à LASCO avec les limitations de la conception et exigences sont discutées dans détail.

INTRODUCTION

Flicker on the power system due to disturbances caused by arc furnaces has been a problem since the inception of the arc furnace. Generally utility growth kept pace ahead of demand except in some minor installations.

With the advent of large U.H.P. furnaces the utility requirements increased but could generally be met by operating the furnaces at low voltage and high arc currents in very stable arc regions. If necessary on larger installations synchronous condensers were added to stiffen the network.

With the advent of numerous large furnaces, i.e., with over 100T capacity, operating with low current arcs to reduce consumption of electrodes, Flicker became a widespread problem.

This was met with techniques to reduce flicker, e.g., S.V.C., Saturable Reactors in shunt or in series, and of course a change to the highly complex D.C. Electric furnace. None of these solutions really stood a rigorous test on Flicker Reduction to acceptable levels.

The D.C. furnace has direct Thyristor Control over the current to the electrode and exhibits some desirable features, such as quieter and less violent operation with some reduction in Flicker. Most of these installations have been shunted by S.V.C.'s. Yet even the most rigid bus exhibits some unacceptable Flicker levels.

At Co-Steel Lasco the solution chosen was

a) Current control by Thyristor to an A.C. Furnace.
b) A predictive control scheme based on derived models of the arc calculated through a LOOK-UP TABLE.

2.0 BASIC DESCRIPTION OF ARC FURNACE SYSTEM

Co-Steel Lasco was founded in 1964 and was the first plant in North America to depend solely on continuous casting to supply its mill. There were no Ingot Molds.

2.1 LASCO ARC FURNACE SYSTEM

In 1979 an expansion was started comprising a 110-t tap weight furnace with a low voltage high current design. This plant produces approximately 800,000 tonnes of finished product per annum with a work force of 650 including the executive staff.

The plant lies between two large Nuclear Generating stations on the North Shore of Lake Ontario. The utility supplying power is the Ontario Hydro. Due to system configurations and their desire for reliability a maximum fault level of 3000MVA is available on the 230KV bus. With the present operation of the furnace at 65MW demand with arc currents of 55,000 Amps some Flicker is produced and complaints have been received.

Co-Steel Lasco with Ontario Hydro have approved the installation of the SPLC to achieve a reduction of Flicker based on the Schwabe Meter to 0.28% (see Figure 1), or zero complaints.

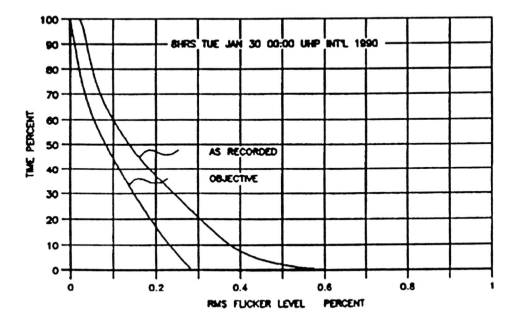

Fig. 1. Cumulative Flicker Level

2.2 A.C. SWITCH

Single line diagram, Figure 2 shows the furnace in its final mode of evolution.

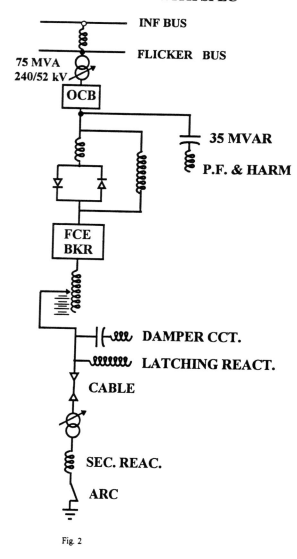

Fig. 2. Furnace Single Line Diagrams

2.3 AUXILIARY H.V. COMPONENTS

The high voltage components comprise an engineered system of energy delivery covering the present and some future modifications. The principle components of this system are as follows:

i.   Incoming Power Transformer

The 230KV/46KV Step Down Transformer which has ON-load secondary tap changing down to 20KV. This low voltage is to be used during SPLC commissioning and as a future Power Regulator if the SPLC has a breakdown.

ii.   Protection

One 46KV O.C.B. provides Short Circuit Protection for the complete circuit. All trips and emergency isolation is accomplished in this device.

iii.   Fault Limiter

A line filter choke of approximately 1Ω was inserted ahead of the SPLC to limit bus faults in the 46KV system to acceptable levels.

iv.   P.F. Correction

An existing power Factor/Harmonic Shunt is provided tuned to the 3RD Harmonic. This is rated 35MVAr to achieve .90 P.F.

    v.   *Underground Cable Protection*

A damper circuit has been installed ahead of the underground 46KV cables to the E.A.F. This protection was felt necessary because of the age of the cables and a possibility of voltage ringing in the cable inductive capacitive loop to ground when the SPLC imposed a continuous series of high transient voltages e.g., during the energization of the Furnace Transformer and prior to arc initiation.

    vi.   *Latching Reactors to ensure Switch Turn On*

2.4. COOLING

The SPLC comprises an Air-Cooled arrangement. A closed air circulation system cooled by a Water/Glycol Heat Exchanger is provided. The heat is then dissipated over non-evaporative cooling tower, so that both air and water are recirculated and not contaminated.

2.5 CENTRAL CONTROLLER

The central controller is the brain of the SPLC and is comprised of a 386 VME based PC, custom designed H/W for data acquisition, a VME-based DSP56001 for digital signal processing and a MC68332 micro-controller for the generation of gating signals for the control of the thyristors of the AC switch and system diagnostics. The VME card cage, a monitor for operator interface, fibre optic transmitters/receivers for the gating signals/diagnostics along with Instrumentation isolation transformers and terminal blocks are installed in the central controller cabinet. This cabinet is sealed and pressurized with air for reliable system operation.

Figure 3 shows the control block diagram for the SPLC central controller

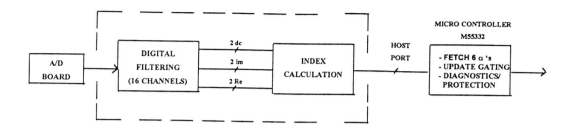

Fig. 3. A basic Control Diagram for the SPLC Control Controller

The DSP56001 is used as a co-processor for the MR8332 micro-controller for digital signal processing. The A/D conversion and discrete Fourier analysis for all the 16 analog signals by the DSP at every sampling point takes around 256TS. [DSP56001 operates at a clock rate of 20MHz and can execute 10.24 million instructions per second (MIPS) including up to 30.72 million concurrent arithmetic and dual data move operations per second (MOPS)]. The most important aspect of the design is the use of a Digital Signal Processor DSP56001 as a high speed math co-processor for the micro-controller, and the use of a look-up table where the values of alpha are stored to be called up in a few nanoseconds. A very fast VME-based memory board with 128MB of DRAM is available for the look-up table.

This look-up table is precalculated for the arc furnace system based on all possible furnace operation points and can be accessed on line quickly to effectively reduce the flicker problem.

2.6 DATA LOGGING SYSTEM

A state of the art data logging system has been installed at LASCO. This includes a 486 PC with 8MB RAM and a hard disk of 520MB. The data acquisition board allows simultaneous sample/hold for 16 analog channels. All the system information is stored on the hard disk during the furnace operation 32 times every 60Hz cycle. This allows for monitoring the performance of the SPLC independent of the central controller, either on-line or remotely from another PC. A removable hard disk with 270MB storage capacity is provided for very fast back-up of a complete heat cycle with all the associated history files (5-10 minutes available during furnace reloading).

2.7 OPERATOR INTERFACE

A major effort has been spent on the operator compatibility to realize a man-machine interface which is user friendly and fault tolerant. The implemented operator interface provides an easy way of operating the SPLC with the arc furnace. All the relevant system information can be monitored, modified or stored. The software comes complete with on-line help. The system is password protected and maintains a list of history files. This makes it possible to isolate any real-time problems and provides the option of duplicating these real time problems in an event driven simulation.

3.0 DESCRIPTION OF SYSTEM OPERATION

3.1 REAL TIME ADAPTIVE ARC FURNACE MODEL

An arc furnace is a non-linear, time varying load. Quick variations of arc voltages and line currents result in a flicker problem which is mainly associated with low frequency harmonics. Time variations of electrical quantities are due to arc length fluctuations which can be caused by electrode movements, collapse of metal scrap and/or electromagnetic forces. As a consequence, the arc furnace constitutes a considerably unbalanced load, sometimes quickly varying between short circuit and open circuit conditions.

The arc furnace models described in the literature are mainly linear and single phase, where electric arc is described as voltage regulation with time-modulated amplitude [2,3]. Non-linear 3f models have also been proposed [4] based on white-noise time variation laws. However, such models can not be used reliably for on-line modelling of the arc furnace system.

The key for successful reduction of flicker by SPLC is the prediction of an accurate on-line model for the arc furnace system.

Figure 4 shows an on-line arc furnace model which is used for predictive control of the arc furnace. The predicted current very accurately follows the actual current, to the point that they virtually coincide. By this model, the line currents can accurately be predicted at any time during the arc furnace operation cycle. This predictive model is then used to prevent excessive peak line currents which cause the flicker.

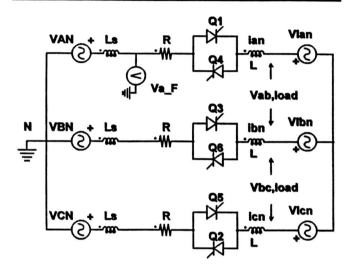

Fig. 4. Arc Furnace System On-Line Model with the Compensator

4.0 CONCLUSION

The installation at Co-Steel Lasco represented a real challenge - to reduce Flicker on a long arc furnace operation. This has been met. Our next step is to convert the furnace to a high reactance low current furnace. Nominal open circuit secondary voltages up to 1650V are planned.

With this final modification we feel confident that our Electrode Consumption will approach the best numbers achieved by the D.C. Furnaces with greater elegance and simplicity.

The first complete heat under SPLC control is scheduled for December 1st, 1995.

5.0 ACKNOWLEDGEMENTS

The project has been actively supported by Co-Steel Lasco and funding has been received from Energy Mines and Resources Canada, Industry Energy Resource and Development Corporation, Research and Development Department, a Federal Government Body.

Technical support has been provided by Ontario Hydro and U.H.P. Corporation.

The results are attributable to the engineers and technicians who have lived with this project for the past three years.

REFERENCES

[1] J.A. Mulcahy, S. Dewan, "Modern Switching Techniques for A.C. & D.C. Electric Arc Furnaces", AISE Cleveland 1994.

[2] M. Loggini, G.C. Montanari, L. Pitti, E. Tironi, D. Zaninolli, "The effects of series inductors for flicker reduction in electric power systems supplying arc furnaces" IEEE/IAS Ann. Meeting, pp. 1496-1503, Toronto, Canada, October 1993.

[3] W.S. Vilcheck, D.A. Gonzalez, "Measurements and simulation-combined for state-of-the-art harmonic analysis" IEEE/IAS Ann. Meeting, pp. 1530-1534, Pittsburgh, USA, October 1988.

[4] G.C. Montanari, M. Loggini, A. Cavallini, L. Pitti, D. Zaninelli, "Arc-furnace model for the study of flicker compensation in electrical networks", IEEE PES Winter Meeting, New York, January 1994.

Study of the flicker level in the case where DC and AC arc furnaces exist

M. Ikoma, Kansai Electric Power Co., Inc.

H. Ichihara, Chubu Electric Power Co., Inc.

O. Kondo, Nissin Electric Co., Ltd.

ABSTRACT

In recent years, there has been a trend towards the sudden popularity of DC furnaces, which have the advantages of low unit consumption of electrode, electric power and refractory as well as low noise and flicker. Since the first 30-ton DC arc furnace in Japan was installed in 1988, there have been a succession of them and already there is a 130-ton furnace in operation.

We presented various data about flicker estimation at the UIE-12 Congress. Subsequently, we have collected and analysed data from another three DC furnaces. This report is concerned with flicker values in AC and DC furnaces and when several furnaces are superimposed.

RESUME

Ces dernières années, l'utilisation des fours à arc CC s'est développée de façon sensible. Outre les économies qu'ils permettent de réaliser grâce à leur faible consommation en électrode et en énergie réfractaire et électrique, ceux-ci présentent l'avantage de générer peu de bruit et de scintillation. Depuis la mise en exploitation au Japon, en 1988, du premier four à arc de 30 tonnes, beaucoup d'autres installations de ce type ont été réalisées, l'une d'elles atteignant une capacité de 130 tonnes. Les auteurs ont présenté de nombreuses données sur l'estimation de la scintillation lors du Congrès UIE-12. Ils ont également regroupé et analysé les données de trois autres fours CC et présentent dans ce document les valeurs de scintillation dans les fours CA et CC et dans le cas où plusieurs fours sont superposés.

1. INTRODUCTION

Using a data recorder, we collected current data from three DC furnaces with different numbers of rectifier phases (12, 18 and 24 phases). In the UIE-12 report we presented DC furnace P and Q characteristics, harmonics, flicker estimates and flicker values but this time we have concentrated on analysing flicker values due to the superimposing of several DC furnaces and the superimposing of AC furnaces and DC furnaces. The main points are summarised below.

(1) Generally, a DC furnace is said to produce less flicker than an AC furnace of the same capacity. We have studied the extent of this difference.

(2) Concerning the relation between $\Delta V10$ and UIE Pst, a correlation of 1:3 is reported for AC furnaces but we have confirmed a correlation for DC furnaces and for superimposed DC furnaces, based on our current data.

(3) In the past, there have been various discussions in Japan about AC furnace flicker superposition but now root mean square method is used. The changes are completely random in the case of AC furnaces but with the DC furnaces which we measured this time, since the rectifiers would keep the current constant, we checked whether the same root-mean-square method could be applied.

Now, with regard to the relationship between flicker and the number of rectification phases in a DC furnace, if flicker is produced by variations in the reactive and active power, the difference due to the number of rectification phases will affect the higher harmonics. So, there is nowhere for the fundamental harmonic to participate and we believe, therefore, that flicker is unrelated to the number of rectification phases. Even in predicting flicker in an AC furnace, which produces more low order higher harmonics than a DC furnace, no particular harmonic factor is introduced. From this sort of viewpoint, we have omitted from here any detailed study of the relationship between flicker and the number of rectification phases.

2. METHOD OF ANALYSIS

(1) TYPES OF ELECTRIC FURNACES

Analyses were carried out on the following four types of arc furnaces:

Short Name	Type	Circuit Voltage (Furnace transformer primary)	Frequency	Rectification phases	Furnace transformer capacity
1) AC :	AC furnace	33 kV	60 Hz		40 MVA
2) DC1 :	DC furnace	22 kV	60 Hz	12 phases	60 MVA
3) DC2 :	DC furnace	33 kV	60 Hz	18 phases	85 MVA
4) DC3 :	DC furnace	22 kV	60 Hz	24 phases	78 MVA

(2) METHOD OF FLICKER ANALYSIS

From the voltage and the current data for each furnace recorded on site, we imitated the power supply system shown in Figure 1 and simulated flicker with the same value as when four furnaces are connected to the same bus.

Figure 2 shows the flicker reproduction block diagram.

To reproduce the voltage containing the flicker component, the voltage between R and S was synchronized by PLL to make a perfect sine wave voltage, that is, an infinite bus voltage Vo(R-S) was made, then we added the voltage fluctuation component ΔV_{RS} due to the power supply inpedance, which was obtained by differentiating the R-phase – S-phase current of the arc furnace current. The superposition of several arc furnaces was handled by adding the currents of the furnaces to be superimposed. Flicker reproduction was carried out on a single phase circuit but Scc and Pmax are 3-phase converted quantities for easier handling on an actual system.

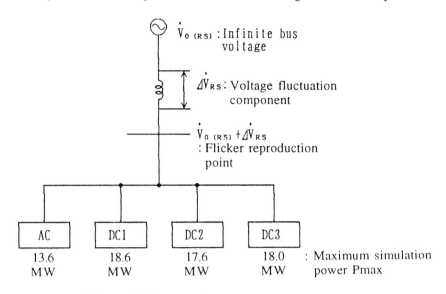

Figure 1 : Diagram of power system under simulation

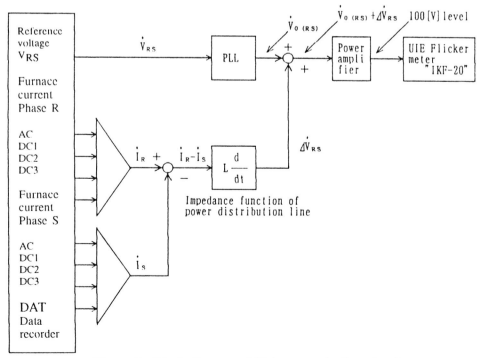

Figure 2 : Block diagram of flicker reproduction circuit

In analysing flicker superposition, by using the same data repeatedly in order to avoid different flicker values each time they were reproduced, the following points were considered carefully:
1. All arc furnace currents were synchronized by standardizing the voltage signals for a particular arc furnace and recorded on one digital audio tape (DAT). This enabled the currents of all furnaces to be the same as if they had been recorded simultaneously. Figure 3 shows an example of voltage and current waveforms.
2. Flicker superposition was handled by combining the furnace currents with an adding machine.
3. The start of the flicker meter operation was based on the seconds display at the time of DAT replay.
4. The flicker levels for each single furnace were adjusted carefully.

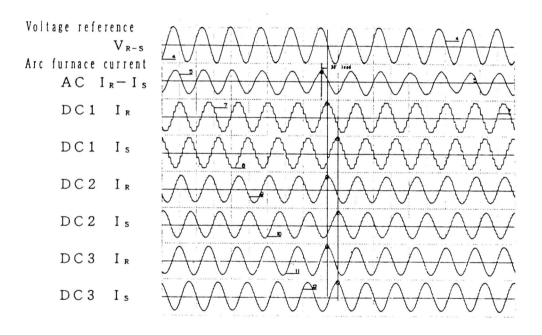

Figure 3 : Waveforms for voltage and each furnace current

(3) FLICKER ANALYSIS FACTORS

Pst and $\Delta V10$ were analysed in respect of flicker in individual furnaces and also superimposed on several furnaces. The flicker meter which was used is Model IKF-20 made by Fukuoka Denki Keiki Co., Ltd., which has functions for analysing both Pst and $\Delta V10$. A single case of flicker analysis took 60 minutes (about one charge) and generated 6 lots of Pst data at 10-minute intervals and 60 lots of $\Delta V10$ data at 1-minute intervals.

3. ANALYSIS RESULTS

(1) SINGLE ARC FURNACE FLICKER ASSESSMENT

Up to now, the method of predicting and determining relative flicker levels has been to take the ratio of the short circuit capacity of the critical bus and the electrode short circuit capacity of the furnace has been used as the parameter. But, as in this case with DC furnaces, suppressing and controlling the current passing through the reactor and thyristor in the DC furnace circuit makes it difficult to stipulate the extreme short circuit capacity of the furnace. Therefore, the maximum power consumption of the furnace would be a suitable parameter which expresses the scale of the furnace. Here, by choosing $K = Pmax/Scc \times 100$ (%) from Pmax, the maximum power when flicker is reproduced and Scc, the short circuit capacity of the critical bus, we have obtained the relationship between the flicker value and K.

Table 1 : Assessment of flicker in single AC and DC furnaces

	Simulation Results			K (%) for Pst=1	Pst for K=1 (%) : Ratio of AC to DC furnace flicker
	Pmax (MW)	K(%) at Scc = 2000 MVA	Pst		
AC	13.6	0.68	0.827	0.82	1.216 (1.0)
DC1	18.6	0.93	0.816	1.14	0.877 (1/1.39)
DC2	17.6	0.88	0.784	1.12	0.891 (1/1.36)
DC3	18.0	0.90	0.685	1.31	0.761 (1/1.60)

From the Table 1, DC furnace flicker of the same scale of power ranges between 1/1.4 and 1/1.6 of that of the AC furnace. In other words, it shows that the flicker level in the AC furnace is about 1.5 times that of the DC furnaces. Also, if Pst = 1.0, the corresponding values of K are:

AC furnace K = 0.82 (%)   (Scc/Pmax = 120)

DC furnaces K = 1.12 to 1.31 (%)   (Scc/Pmax = 80)

In other words, to ensure an output of 10 MW requires a power source with a short circuit capacity of at least 1200 MVA for an AC furnace or 800 MVA for a DC furnace.

(2) FLICKER SUPERPOSITION RESULTS

From the results of reproducing flicker by combining actual currents, and the results of analysing the composite relationships of the simultaneous flicker from flicker values obtained from the results of reproducing the flicker in each single furnace, we found that they are composed of root-mean-squares. Tables 2 and 3 show examples of flicker values composed of actual currents and results obtained by calculating the root-mean-squares of single furnace flicker values. From the tables, we see that the errors between the root-mean-squares agree within a few percent. Superpositions which are not shown here have similar results.

Table 2 : Example 1 - Flicker superposition results

	Single furnace Pst		Result of superimposing A and B	Calculated root-mean-square	Error (%)
	AC A	DC1 B	Pst C	$D=\sqrt{A^2+B^2}$	D-C/C×100
10-min	0.827	0.666	1.032	1.062	+2.9
20-min	0.369	0.816	0.920	0.896	2.6
30-min	0.526	0.440	0.691	0.686	-0.7
40-min	0.379	0.695	0.810	0.792	2.2
50-min	0.163	0.449	0.478	0.478	0.0
60-min	0.567	0.429	0.686	0.711	+3.6

Table 3 : Example 2 - Flicker superposition results

	Single furnace Pst			Result of superimposing A to C	Calculated root-mean-square	Error (%)
	DC1 A	DC2 B	DC3 C	Pst C	$E=\sqrt{A^2+B^2+C^2}$	E-D/D×100
10-min	0.666	0.784	0.625	1.208	1.204	-0.3
20-min	0.816	0.538	0.661	1.190	1.180	-0.8
30-min	0.440	0.387	0.449	0.732	0.738	+0.8
40-min	0.695	0.652	0.685	1.121	1.154	+2.9
50-min	0.449	0.429	0.403	0.710	0.740	+4.2
60-min	0.429	0.394	0.431	0.736	0.725	1.5

(3) STUDY OF METHODS OF COMBINING FLICKER VALUES

From the discussion above, we found that simultaneous flicker is a combination of root-mean-squares but in estimating actual flicker, we have the estimated or measured values of flicker (the maximum value of Pst) in each furnace and, when they are superimposed, they are not necessarily root-mean-squares.

In the case of AC furnace flicker combinations, as a result of comparing various methods of combination, we are now using root-mean-square method. In the case of DC furnace combinations and AC and DC furnace combinations, we studied which method of combination should be used. The results are shown below. Taking the maximum Pst value for flicker in a single AC or DC furnace as the flicker value for that furnace, the results of comparing the root-mean-square, 2.37th root-mean-2.37th power, cube root-mean-cube and 4th root-mean-4th power derived from those values with the maximum Pst value after supervision are shown in Table 4. As can be seen from the table, in this simulation, 2.37th root-mean-2.37th power gives the smallest mean error.

Table 4 : Study of flicker combination methods

	Maximum Pst values for single furnaces				Maximum Pst value after combination	Root-mean-square		2.37th root-mean-2.37th power		Cube root-mean-cube		4th root-mean-4th power	
	AC	DC1	DC2	DC3		Calculated value	Error (%)	Calculated value	Error (%)	Calculated value	Error (%)	Calculated value	Error (%)
AC + DC1	0.827	0.816			1.032	1.162	12.6	1.101	6.7	1.035	0.3	0.977	-5.3
AC + DC2	0.827		0.784		1.064	1.140	7.1	1.080	1.5	1.016	-4.5	0.959	-9.9
AC + DC3	0.827			0.685	1.044	1.074	2.9	1.019	-2.4	0.961	-8.0	0.911	-12.7
DC1 + DC2		0.816	0.784		1.014	1.132	11.6	1.072	5.7	1.008	-0.6	0.952	-6.1
DC1 + DC3		0.816		0.685	1.050	1.065	1.4	1.011	-3.7	0.953	-9.2	0.903	-14.0
DC2 + DC3			0.784	0.685	1.041	1.041	0.0	0.987	-5.2	0.930	-10.7	0.879	-15.6
DC1 + DC2 + DC3		0.816	0.784	0.685	1.208	1.323	9.5	1.215	0.6	1.008	-16.6	0.952	-21.2
AC + DC1 + DC2 + DC3	0.827	0.816	0.784	0.685	1.422	1.560	9.7	1.401	-1.5	1.035	-27.2	0.977	-31.3
Mean error							6.9		0.0		-9.6		-14.5

(4) CORRELATION OF PST AND ΔV10

The relationship between Pst and ΔV10 in respect of one DC furnace was assessed and reported to be 1 : 3 at UIE-12. Here, we show the relationship between Pst and ΔV10 in respect of four types of furnaces as well as the superposition of those furnaces in Figure 3. Here, Pst means the maximum of six Pst values taken over a 60-minute period. It should be noted that ΔV10 is the 4th highest value of 60 items taken in 60 minutes. From Figure 4, we could see the relationships between Pst and ΔV10 with regard to flicker for the AC furnace alone, the DC furnaces alone and after various superpositions of these furnaces, and we found that these ratios are about 1:3, which is the same as that previously reported.

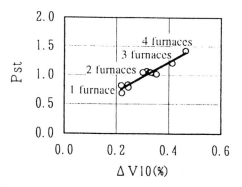

Figure 4 : Correlation between Pst and ΔV10

4. CONCLUSION

This report can be summarized as follows:

(1) In comparing flicker in AC and DC furnaces of the same scale of maximum power, flicker in the AC furnace is about 1.5 times higher than in the DC furnace.

(2) Simultaneous superimposed flickers can be combined with root-mean-squares irrespective of the type of furnace.

(3) For the data which we obtained, combining flicker from several furnaces by taking the 2.37th root of the mean of the 2.37th power of the individual flicker values gave the smallest mean error.

(4) As with AC furnaces up to now, the relationship of about 1 : 3 between Pst (max) and ΔV10 (4th max) DC furnace flicker applies, even when DC furnaces are superimposed or when DC and AC furnaces are superimposed.

REFERENCES

1) Electrical Institute Technical Reports No. 26, 1974
2) Electrical Institute Technical Reports No. 69, 1965
3) Japanese UIE-DSC Shadow Committee, 1985
4) UIE-12 Congress, 1992

Development of High Efficient Melting on DC Arc Furnace

NOBUYOSHI TAKASHIBA, KIYOSHI TAKAHASHI, MASAHIRO YOSHIDA
HIROJI NUMATA, SHUNJI OHTSUBO
Mizushima Works, Kawasaki Steel Corporation
Kawasakidori 1-chome, Mizushima, Kurashiki 712, Japan

ABSTRACT

Kawasaki Steel Corporation developed a uniform melting technology comprising a technique for preventing arc deflection and a technique for free control of the arc direction. The technology was applied to the DC EAF with a capacity of 100 tons, realizing uniform melting. A reduction in electric power consumption of approximately 30kWh per ton of molten steel was achieved.

To reduce the operating power consumption of the dust collecting fan and other equipment, a common power supply for the simultaneous operation of two furnaces was developed and applied, reducing operating power consumption by approximately 30%. Finally a technique for using of molten pig iron in the DC EAF was developed, markedly reducing melting power consumption.

RESUME

Kawasaki Steel Corporation a mise au point une technologie de coulée homogène, combinaison d'une technique préventive de déflexion d'arc avec une autre technique permettant d'exempter de contrôle sur la direction d'arc. Cette technologie, appliquée à DC EAF avec une capacité de 100 tonnes, réalise une coulée homogène. Il en été ainsi obtenu dans la consommation en énergie électrique une diminution approximative de 30 kWh par tonne d'acier coulé.

Par ailleurs, dans le but de réduire la consommation en énergie au niveau de la soufflante séparatrice de poussière et d'autres installations, un système commun d'alimentation en énergie a été développé et mis en place pour le fonctionnement simultané de deux fours en permettant un gain d'environ 30%. En outre, il a été mis au point une technique pour employer la fonte brute coulée dans DC EAF, donnant une économie remarquable dans la consommation en énergie de la coulée.

INTRODUCTION

The DC arc furnace offers excellent performance in the unit consumption of melting power and electrodes, as this furnace type basically uses a cathodic top electrode and requires only one electrode. However, uniform melting is not realized in large-scale DC EAFs which have simply been scaled-up without considering the need for techniques to prevent arc deflection and the following problems can be expected. [1,2,3,4,5]

(1) Increased melting power consumption due to areas of delayed scrap melting

(2) Furnace wall water cooling panel damage and other equipment problems

These problems were solved by developing a uniform melting technology comprising a technique for preventing arc deflection and a technique for free control of the direction of the arc.

A common power supply system for simultaneous operation of two EAFs was therefore developed. This technology reduced the investment cost by enabling the common use of equipment and made it possible to increase productivity to approximately 1.5 times that when a one-power-source one-furnace system is used, achieving a reduction of about 30% in the unit consumption of operating power for dust collection fans and other equipment. [6]

Further, a technique for using molten pig iron in the DC EAF was developed as a standard operating procedure, markedly reducing melting power consumption, while also enabling economical production of steel with a low level of impurities. [7]

These technologies were applied to the two commercial 100-ton EAFs at Daiwa Steel's Mizushima Works, achieving uniform melting, the targeted production capacity, reduction in power consumption, and energy savings.

The following presents an outline of this newly developed technology for high-energy efficiency melting with the DC arc furnace.

DEVELOPMENT OF UNIFORM MELTING TECHNOLOGY

In DC arc furnaces, a DC magnetic field forms around the power supply conductors and the arc is also a direct current; therefore, a horizontal electromagnetic force will normally affect the arc if the arrangement of the conducts is not appropriate, causing a "deflection" phenomenon in which the arc is oriented in a fixed direction which does not coincide with the furnace center.

Because arc deviation causes higher melting power consumption and other cost increases, prolongs the melting time, and reduces the working ratio (i.e. productivity), a uniform melting technology which prevents arc deflection is an essential element technology for high-current DC furnaces where the current reaches 100kA.

TECHNIQUE FOR PREVENTING ARC DEFLECTION

Arc deflection is prevented by an appropriate arrangement of the power conductors and bottom electrodes, which should minimize the sum of the horizontal vectors of the electromagnetic force acting on the arc based on an electromagnetic analysis. Accordingly, arc deflection is prevented by analyzing the strength of the magnetic field at the point of arc generation due to the power conductors in a three-dimensional orthogonal coordinate system in the design stage using the Biot-Savart equation (1), and the strength and direction of the electromagnetic force acting on the arc using equation (2), and minimizing the sum of the vectors obtained, considering also the magnetic shield effect of the steel furnace shell and other influencing factors. [3,4]

Strength of magnetic field ΔH attributable to short conductor

$$\Delta H = \alpha \cdot I \cdot \Delta l \cdot \sin \theta \cdot (4 \pi r^2)^{-1} \quad (A \cdot m^{-1}) \quad \ldots \ldots (1)$$

Electromagnetic force ΔF acting on arc

$$\Delta F = Ia \cdot \mu_0 \cdot \Delta H \quad (N \cdot m^{-1}) \quad \ldots \ldots (2)$$

α	:	Coefficient considering attenuation due to steel shell (-)
I	:	Current in conductor (A)
Δl	:	Minimum length of conductor (m)
θ	:	Angle between conductor and arc (rad)
r	:	Distance from conductor to arc (m)
Ia	:	Arc current (A)
μ_0	:	Magnetic permeability (H·m^{-1})

TECHNIQUE FOR CONTROLLING ARC DIRECTION

In addition to the technique for preventing arc deflection discussed above, a technique for positively controlling the direction of the arc was also developed to ensure uniform melting under all conditions, including uneven distribution of large lumps of scrap in the furnace.

In this directional control technique, multiple power supply systems are adopted for the bottom electrodes, as shown in Fig. 1, and transformers and rectifiers capable of individually controlling the current in each system are configured to allow monitoring and control of the current and voltage. The effectiveness of this arrangement is supported by a high-accuracy electromagnetic analysis in the design stage. [1,3,4,5]

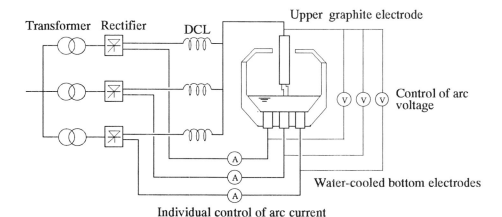

Fig. 1 Composition of arc direction control system by current control

CONFIRMATION OF UNIFORM MELTING

Fig. 2 shows the results of the maximum load thermal flux value which occurs in the furnace wall water cooling panels directly above the surface of the molten steel bath during one heat, as obtained from the cooling water flow rate and the difference in the temperature of the cooling water at the water inlet and outlet, and indicates that maximum thermal flux value was substantially uniform in the circumferential direction. A region of somewhat higher thermal load can be seen from the exhaust side to the tapping side, but the higher load here is attributable to the gas flow in the furnace, in which the post-combustion gas resulting from combustion of CO gas evolved from the bath and oxygen injected from the deslagging hole flows toward the exhaust, and is not the result of arc deflection. [1,3,4]

If arc deflection occurs, the top graphite electrode will show unbalanced wear similar to that seen at the tip of the electrodes used in AC arc furnaces, with some differences depending on conditions. However, when the uniform melting technology was applied to the DC arc furnace, wear was uniform around the circumference of the electrode, as shown in Photo 1, and no cracking or falling off of material could be seen at the tip. These observations confirmed that electrode wear was the result of surface oxidation and sublimation of the tip, as predicted theoretically. [1,3,4]

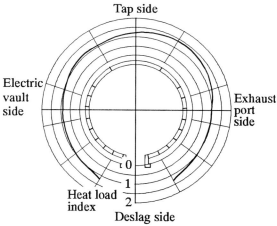

Fig. 2 Circumferential load thermal flux distribution at furnace wall

Photo 1 Wear of top graphite electrode

DEVELOPMENT OF COMMON POWER SUPPLY FOR SIMULTANEOUS OPERATION OF TWO EAFs

Generally, EAF shops use a one-power source one-furnace configuration with one set of furnace electrical equipment and one power supply for each furnace. However, the maximum equipment capacity is necessary only during the melting period. Under these circumstances, it cannot be said that the equipment capacity is being fully utilized. In fact, if the equipment utilization ratio is defined as (operating time ratio x capacity utilization ratio), there are many instances in which the power supply and dust collecting equipment, among others, are functioning at below the 50% level. A technology for common use of the power supply system in the simultaneous operation of two furnaces (referred to below as the "common power system") was therefore developed in which two arc furnaces make common use of this equipment to hold down investment costs by maximizing the use of equipment unit capacity and to strengthen the production capacity of the shop. At the same time, this technology improves energy efficiency by making it possible to reduce the operating power consumption of dust collectors and the cooling water supply and treatment equipment.[6]

Fig. 3 and 4 show the composition of the main electrical equipment and the power input pattern for each furnace. Even when the two furnaces are operated simultaneously, the total input for two furnaces does not exceed the allowable capacity of the power supply equipment for one furnace.

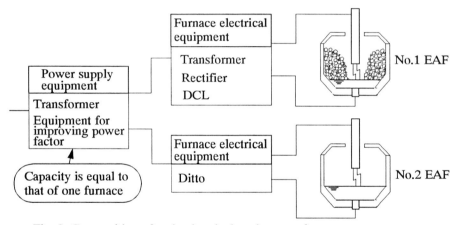

Fig. 3 Composition of main electrical equipment of common power system

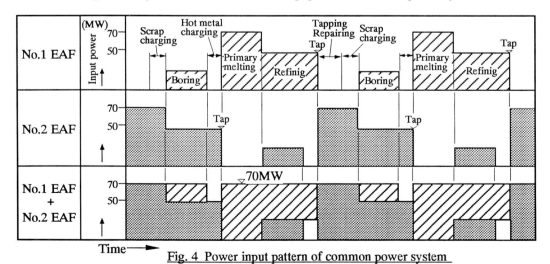

Fig. 4 Power input pattern of common power system

DEVELOPMENT OF TECHNOLOGY FOR USE OF MOLTEN PIG IRON

Hot metal operation is a technique for EAF operation which takes maximum advantage of the superior location of the arc furnace in an integrated steelworks. This technique reduces melting power consumption and improves productivity by using blast furnace hot metal as an iron source.

When using hot metal in an arc furnace, it is important to prevent the phenomenon known as "sudden rimming," an explosive phenomenon caused by the sudden generation of CO gas by a reaction of the non-deoxidized molten steel and scrap in the furnace with high-carbon hot metal. In other words, the technique of hot-metal operation is essentially a technique for preventing sudden rimming. Table 1 shows the hot metal charging technique and melting and refining techniques developed in the present work to prevent sudden rimming. Fig. 5 shows the hot metal operation pattern. The establishment of this new technique has made it possible to use hot metal stably and in large quantity even in the DC furnace, which requires hot heel operation (leaving of some metal in the furnace after tapping) to ensure good conductivity between the bottom electrodes and scrap during the primary melting period. [7]

Table 1 Hot metal charging technique and melting/refining techniques for preventing sudden rimming

	Purpose and content		Necessity
Hot metal charging technique	Optimization of timing of charging	If early	The charge of hot metal solidifies, and sudden rimming tends to occur more easily in secondary melting period.
		If late	Heat efficiency is reduced during the secondary melting period because the scrap in the furnace collapses. The melting time may become excessive because some amount of time is also required to decarburize the hot metal, which invites decreased productivity.
Melting and refining (decarburization) techniques	Optimization of carbon content of steel during melt-down		Level of carbon content during meltdown relative to prescribed content:
		If low	Because the molten steel is in an over-oxidized condition, an increased amount of an added carbon agent is required in order to balance the amount of deoxidizing agent and target carbon content.
		If high	Because the temperature of the bath is sufficiently high, sudden rimming occurs easily due to scrap collapse.

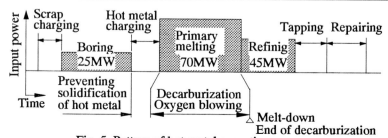

Fig. 5 Pattern of hot metal operation

BENEFITS OF NEW MELTING TECHNOLOGY

The technologies described above for uniform melting, common power system, and the use of hot metal have been realized and applied in the two 100-ton DC arc furnaces at Daiwa Steel's Mizushima Works. Table 2 shows the basic specifications of the DC arc furnaces.

Table 3 compares the unit consumption of electric power, operability, and productivity of an DC arc furnace using the technologies for uniform melting and hot metal operation, the conventional AC furnace, and a DC arc furnace without the uniform melting technology. These newly developed methods can therefore be called effective energy saving technologies.

As shown in Table 4, not only is it possible to reduce the operating cost from that with the general one-power supply one-furnace system, but it is also possible to reduce the investment cost and improve productivity. Thus, this technology is effective both in saving energy and in reducing iron and steelmaking costs.

Table 2 Basic specifications of DC arc furnaces using new melting technology

		Specification
Furnace capacity		100 tons
Transformer capacity		100 MVA (33.4MVA × 3)
Upper graphite electrode	Number	1
	Diameter	28 inches
Bottom electrode	Number	3
	Cooling system	Water-cooled type

Table 3 Benefits of uniform melting technology and hot metal operation of EAF

	AC arc furnace	DC arc furnace		
		Without uniform melting	With uniform melting technology	
	Without hot metal	Without hot metal	Without hot metal	With 40% hot metal
Unit consumption of electric power for melting	400 kWh/t	390	370	250
Operability Productivity	Standard	Somewhat improved But requires measures for skull onthe furnace wall	Superior With little skull on furnace wall	Excellent Requires decarburization technology

Table 4 Benefits of common power system technology for simultaneous operation

	One-power source one-furnace system	Common power system for simultaneous operation
Investment costs	1.0 (Standard)	0.75
Operating power consumption	1.0	0.7 (-14 kWh/t)
Productivity	1.0	≥1.6

CONCLUSION

The development and establishment of a uniform melting technology, comprising a technique for preventing arc deflection in large-scale DC arc furnaces and a technique for controlling the direction of the arc by controlling the current, a technology for the simultaneously operation of two furnaces with a common power supply, and a technology for the use of hot metal has made it possible to realize a DC arc furnace shop with dramatic reductions in both melting power consumption and capital investment costs.

The practical application and establishment of this high-energy efficiency melting technology has not only promoted energy saving, but has also made it possible to produce and supply iron and steel with greater economy, and is expected to have an important ripple effect in this field of steelmaking in the future.

REFERENCE

1. N. TAKASHIBA: 'Development of DC Arc Furnace by CLECIM-KSC,' KAWASAKI STEEL TECHNICAL REPORT, 1993, No.29, 45-53
2. J. DAVENE: Steel Technology International, 1988, 159-164
3. K. KOIDE: 'Control of Arc Direction and Uniform Melting on DC Arc Furnace,' CAMP-ISIJ, 1992, 5 (1), 298
4. N. TAKASHIBA: 'Uniform Melting of DC Arc Furnace by Controlling Arc Direction,' 51st ELECTRIC FURNACE CONFERENCE PROCEEDINGS, 1993, 209-214
5. A. UEDA: 'Development of Melting Control Technique for DC Arc Furnace,' CAMP-ISIJ, 1991, 4 (4), 1275
6. M. YOKOYAMA: 'Development of Arc Furnace Shared One Power Source with Another Furnace for Common Use,' CAMP-ISIJ, 1995, 8 (1), 199
7. T. YAHATA: 'Melting Technique for DC Arc Furnace,' CAMP-ISIJ, 1993, 6 (4), 976-979

ELECTROMAGNETIC COMPATIBILITY OF EAF AND SUPPLY POWER SYSTEM

A. WĄSOWSKI, J.BIAŁEK, R.WILANOWICZ

Technical University of Radom, POLAND

W. BROCIEK

Technical University of Warsaw, POLAND

K. BENGHALEM, M.T. MELIZI

Universite' de Se'tif, ALGERIA

SUMMARY:

The paper presents the analysis of the real working conditions of the arc furnace and its influence on the quality of the delivered energy (variability unbalance and distortion of current waveform) and the working characteristics of the arc furnace. The considerations have been illustrated by the results of the analysis of 50 tons furnace.The investigations have been done at different working conditions , including the case of compensation of the reactive power , as well as inclusion of filters of higher harmonics.

RESUME:

Dans ce rapport on présente l'analyse de l'influence des conditions reelles de fonctionnement d'un four à arc (variabilité déséquilibre et défomation des courants d'arc) sur la qualité de tension dans le réseau d'alimentation et sur les caractéristiques operationnelles de four. Les considérations sont illustrées par les résultats de l'analyse effectuée sur l'exemple de four de 50 tonnes capacité. Les recherches sont effectuées pour divers conditions y compris dispositifs pour compensation de la puissance réactive , et des filtre d'harmoniques.

1. INTRODUCTION

One can mention , as the potential threat for EMC of three-phase EAF and supply power system , two factors:

* random and dynamic changes of length and conductivity of arcs , appearing independently in each phase,
* non-linearity of arc current-voltage characteristics.

In result phase currents delivered by supply network are characterized by dynamic variability , unbalance and distortion from sinusoidal transients , deteriorating operational characteristics of EAF on the one hand , and on the other hand causing disturbances in supply system - voltage flicker, voltage unbalance and voltage harmonics, deteriorating service conditions of other receivers supplied by the same HV system.

In this paper we present an analysis of influence of these factors on the most essential operational characteritics of EAF (useful power and overall efficiency) and on disturbances in supply system, as well as satisfying of EMC criterions, and advantegous influence of appliances for capacitive compensation of reactive power.

2. VOLTAGE FLICKER AND UNBALANCE

Very dynamic changes of currents during scrap melting create voltage flicker and unbalance in supply network, deteriorating operation of other receivers supplied by the same network, provoking threat for EMC of furnace and supply system.

Especially sensitive on voltage changes are the lighting receivers - in the effect of light flicker. During the passed decades one has made, in this field, many researches, reported among others on UIE Congress. To facilitate international cooperation UIE has created Study Committee "Disturbances", later transformed to Working Group "Power Quality". These researches have been concentrated on initiation, comparison and generalization of theoretical and experimental works led in different countries ; as result of them one have elaborated special measurement appliance - flickermeter, and one have compared different means used for decrease voltage flicker [1] at present investigations are concentrated on coordination of international recomendations concerning the EMC conditions.

Simplifying one can estimate random and irregular changes of arc current and arising as effect voltage flicker by standard deviations $D(I)$ and $D(U)$, connected by the formula :

$$D(U) = D(I^a)R_s + D(I^r)X_s \approx D(I^r)X_s \approx 2D(I)E(I)X_s/I_t \qquad (1)$$

R_s, X_s - resistance and reactance of supply network in PCC, $D(I^a)$, $D(I^r)$ - standard deviations of active and reactive current components; $E(I)$ - mediocre value of arc current, I_t - theoretical current of short-circuit of electrodes with charge.

Decomposing the envelope of irregular voltage flicker on Fourier's series, one can calculate their efficace amplitude $a_{eff} = 2\sqrt{2} D(U)$ [2], and taking into account the spectrum of flicker sensitivity in dependence of its frequency one can estimate their equivalent amplitude of 10 Hz frequency - v_{10} (or measure it with aid of flickermeter).

In approximate it can use the dependence $v_{10} \approx 0,7\ a_{eff} \approx 2D(U)$ [3].

In order to decrease excessive voltage variations, one has used many means [1], in the report we will pay attention to one of them - the bank of parallel condensers controlled by fast- operating thyristor systems. But current changes provoke not only voltage flicker, they have also unfavourably influence on working characteristics of EAF. We will make the comparison on example of 50 tons capacity furnace, with transformer rating 25/30 MV·A and transformation ratio 110 kV/501..167 V [4], mean values of arc supply circuit resistance and reactance are equal $R= 0.62$ mΩ, $X= 3.67$mΩ, heat losses of furnace $P_{th} = 3.1$ MW. For supply system 110 kV we admit $S_{sc} = 1200$ MV·A ($X_s = 0.21$ mΩ).

We will calculate active power delivered from supply system P, useful power transferred to the charge P_{us} and overall efficiency η_0, from the following formulas

$$P = \sqrt{3} I \sqrt{U^2 - 3I^2 X_f^2} \qquad (2)$$
$$P_{us} = P - 3I^2 R - P_{th} \qquad (3)$$
$$\eta_0 = P_{us}/P \qquad (4)$$

($X_f = X_s + X$ - full phase reactance). The most important for consumers it is working characteristic of useful power - presented on Fig.1, and characteristic of overall efficiency presented on Fig.2 - curves 1; we have admitted operation on transformer tap II - 483.5 V and range of current mean value variation from 20 to 40 kA (nominal current $I_n = 29$ kA).

Current range recommended for rational operation is limited between current I_e ensuring maximum of overall efficiency, and current I_g ensuring maximum of useful power transferred to the charge. These currents are determined by formulas;

$$I_e = U \sqrt{P_{th}/(P_{th} + 0,5 U^2 R/X_f^2)} /(\sqrt{6} X_f) \qquad (5)$$
$$I_g = U \sqrt{1 - R/\sqrt{R^2 + X_f^2}} /(\sqrt{6} X_f) \qquad (6)$$

For analysed furnace we obtain $I_e = 31,8$ kA ($\eta_0 = 0.792$) and $I_g = 46.7$ kA ($P_{us} = 22,6$ MW).

But the current I_g is 60 % greater than nominal current and menaces to overload of furnace transformer, so as the upper limit we admit 125% of I_n, and recomended current range 32..36 kA.

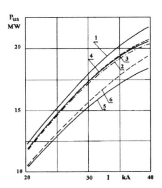

Fig. 1. Characteristics of useful power

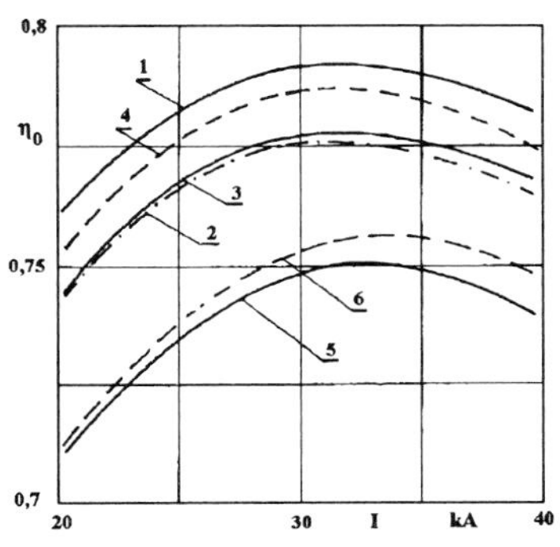

Fig. 2. Characteristics of overall efficiency

It can estimate variations of phase currents by standard deviation of currents assessed as 20...40 % of I_n, we will admit $D(I) = 0.3 I_n \approx 10 \text{kA}$. So for mediocre current varying in limits $E(I) = 20....40$ kA we obtain the equivalent amplitude of flicker $v_{10} = 0.8...1.6 \%$, depassing recommended value 0,3 % of nominal voltage [1].

On operation of EAF, besides its construction asymetry [6,7] influence also current variability so it is necessary correct formulas (2) and (3) to the form:

$$P = \sqrt{3} I \sqrt{U^2 - 3I^2(1 + 3v_i^2)X_f^2} \qquad (7)$$

$$P_{us} = P - 3I^2(1+v_i^2)R - P_{th} \qquad (8)$$

(where $I = E(I)$ $v_i = D(I)/E(I)$).

The corrected characteristics are presented on Fig.1 and 2 - curves 2.

In order to analyse the influence of current unbalance we will admit unbalance coefficient k equal to the ratio of negative sequence component I_- to the positive sequence component I_+ ($k_i = I_-/I_+$). The current unbalance occurring during melting of scrap can be assessed on $k_i = (0.2...0.4)I_n$, so for calculation we admit the negative sequence component $I_- = 0.3 I_n = 10$ kA, and the unbalance sequence component of voltage in supply network will be assessed by formula

$$k_v = U_-/U_+ = \sqrt{3} I_- X_s / U_n \qquad (9)$$

assuming $U_+ = U_n$ we have $k_v = 0.75 \% U_n$. This value is lower than proposed admisssible value 2 % [5].

For correct operational characteristics we admit symmetry of phase resistances and reactances (triangular secondary circuit). Then one must correct the formulas (2) and (3) to the form:

$$P = \sqrt{3} I \sqrt{U^2 - 3I^2(1 + k_i^2)^2 X_f^2} \qquad (10)$$

$$P_{us} = P - 3I^2(1 + k_i^2)R - P_{th} \qquad (11)$$

($I = I_+$) The influence of current unbalance on operational characteristics is also presented on Fig.1.and 2- curves 3.

3. ARC FURNACE AS THE SOURCE OF HIGHER HARMONICS

The paper presents the model of arc furnace taking into account the parameters following from the voltage-current characteristic of the arc.

Taking into account that there are no possibilities separate the characteristics of the arc and the high current circuit, both elements are treated together at the formation of the mathematical model of the arc furnace.

Fig.3a illustrates the exemplary characteristics of the 50 tons furnace, obtained on the basis of the measurement for the period of melting. In this period the furnace is treated as the source of higher harmonics

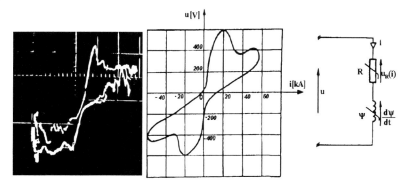

a)   b)

Fig.3a. Current - voltage characteristics of arc   Fig 3b. Circuit model of arc.

In numerical experiments the characteristic of Fig.3a. has been substituted by its equivalent in the form of series connection of nonlinear element R and Ψ (Fig.3b).

The circuit equation of Fig.3b has the form:

$$u = u_R + \frac{d\Psi}{dt} = u_R + \frac{d\Psi}{di}\frac{di}{dt} \tag{12}$$

The current-voltage characteristic of the nonlinear resistor has been approximated by the polynominal of the form:

$$u_R = ai - bi^3 \tag{13}$$

Similary, the characteristic of the linkage flux of the coil versus current has been approximated by the polynominal of the second degree

$$\Psi = ki - li^2 \tag{14}$$

Where a,b,k,l are the constant coefficients.

As a result of the numerical analysis of the problem we have stated that the assumed approximation leads to the correct results with simulated currents and voltages equal to real measured values in the system.

Taking into considerations expressions (13) and (14) we get the equation (12) in the form

$$u = ai - bi^3 + (k - 2li)\frac{di}{dt} \tag{15}$$

At sinusoidal current we get

$$i = I_m \sin \omega t \tag{16}$$

$$\frac{di}{dt} = \omega I_m \cos \omega t = \omega I_m \left[\pm \sqrt{1 - \left(\frac{i}{I_m}\right)^2} \right] \tag{17}$$

$$u = ai - bi^3 + (k - 2li)\omega I_m \left[\pm \sqrt{1 - (\tfrac{i}{I_m})^2} \right] \quad (18)$$

To obtain the values of coefficients a,b,k,l we have to solve four equations. The sign (+) in equation (7) corresponds to this part of the characteristic, where $\frac{di}{dt} > 0$, and sign (-) to $\frac{di}{dt} < 0$. Equation (15) describes the circuit model of Fig 3b. for the case defined by (16). In real circuit the currents are distorted and equation (18) only approximates the general equation (12).

Transformation of equation (12) to the general equation for nonsinusoidal currents is impossible in an explicit way, because the currents are unknow at the phase of formation of these equations. For the characteristic of Fig.3a. equation (2) and (3) are of the form:

$$u_r = 8.6 \cdot 10^{-3} i - 1.7 \cdot 10^{-12} i^3 \quad (19)$$
$$\Psi = 2 \cdot 10^{-5} i - 4.6 \cdot 10^{-10} i^2$$

Assuming the approximating functions in the form of (19), the higher harmonics of currents and voltages, generated by the furnace have been evaluated. The distortion coefficient (THD) of the voltage are described in the form:

$$v_{U\%} = \sqrt{\sum_{n=2}^{n=25} U_{n\%}^2} \qquad v_i = \sqrt{\sum_{n=2}^{n=25} \frac{I_n}{I_1}} \quad (20)$$

where
$U_{n\%} = \frac{U_n}{U_1} \cdot 100\%$ - the rms value of the n-th harmonic expressed as the percentage of the basis one U_1.

During the numerical analysis we are able to obtain useful power value according to the formula

$$P_{us} = \sum_{n=1}^{n=25} U_n I_n \cos\varphi_n - P_{th} \quad (21)$$

where: n - order of harmonic

In analysis circuit total harmonics distortion in PCC are $v_{U\%} = (0,7...0,48)$, this value is lower than proposed admissible value 3% [1]. The influence of harmonics current on working characteristic is presented on Fig. 2 and 3 - curve 4.

4. TAKING INTO ACCOUNT REAL CONDITIONS OF EAF OPERATION ON ITS WORKING CHARACTERISTICS.

Each factor analysed previously threats to EMC of furnace and supply system, and causes decreasing of useful power and overall efficiency too. In order to estimate the joint effect of these factors on furnace working characteristics we have used the idea proposed by Kohle [10], introducing to the supply circuit additional reactances dependent on current variations, unbalance and nonlinearity of arc. Assuming as current $I = E(I_+)$ its mediocre value of first harmonic of positive-sequence component, we have estimate active power delivered from supply system by formula

$$P = \sqrt{3}\, I \sqrt{U^2 - 3I^2 X_{tot}^2} \quad (22)$$

where:
$$X_{tot}^2 = X_f^2[(1+k_i^2)^2 + 3v_i^2] + X_h^2 \quad (23)$$

(reactance X_h can be calculated from formula (2) assuming $X_f^2 + X_h^2$ instead of X_f^2, and power calculated with taking into account higher harmonics of currents and voltages)

and useful power by formula
$$P_{us} = P - 3I^2 R(1 + v_i^2 + k_i^2 + v_i^2) - P_{th} \quad (24)$$

The values corrected according to these formulas and previous coditions are presented on Fig. 1 and 2 - curves 5. It can state, that for considered simplifying assumptions, useful power, in comparison with ideal conditions of EAF operation, is decreased by 2...3 MW, and overall efficiency by 0,04..0,05, but for these decreased values recommended current range is also 32... 36 kA.

5. PROFITABLE INFLUENCE OF ARRANGEMENTS FOR REACTIVE POWER COMPENSATION.

System disturbances produced by EAF operation, have caused elaboration of many solutions [1] ensuring their decreasing, in order to satisfy the EMC criterions. We will consider the profitable influence of static condensers, controlled by fast-operating thyristor systems enabling to decrease variations of reactive power delivered by network, and in effect voltage flicker in PCC too. Control of condenser powers individually in each phase enables not only improvement of power factor, but also decreasing current and voltage unbalance in supply system. Condenser bank often is equiped in filters, eliminating selected harmonics in system voltage.

These solutions are used in many cases and their effects are very good, considerably decreasing network disturbances and satisfying all EMC conditions. But on the other hand these arrangements influence also on operation of EAF itself (in the effect of improvement the quality of supply voltage). In order to estimate roughly this advantageous influence we have assumed ideal conditions, i. e. total elimination of loading of supply system by current reactive and negative-sequence components and by current harmonics, as the effect we have perfect voltage quality in PCC, and improved working characteristics - curves 6 on Fig. 1 and 2.

CONCLUSION

Proposed methods enable, in simple way, to estimate the influence, on EAF working characteristics, of real operational conditions and of compensating arrangement used for improve its EMC with supply system. Influence of these last arrangements is, saying true, less than it results from comparison of curves 5 and 6 on Fig. 1 and 2, because of inevitable time-lag of control system, and its discontinuous action, but is quite clear.

REFERENCES

1. UIE STUDY COMMITTEE " Disturbances"; Arc Furnace Disturbances - *State of Art.*, *EWT* 1982, No . B 3/4

2. A. WĄSOWSKI : Fluctuations de tension (flicker) provoquées par plusieurs fours a' arc fonctionnant simultanément. *RGE* 1979 Nr 3

3. B.BOWMAN:Flicker Variation de tension sur le résean d' un four a' arc.*JFE* 1978 Nr.8

4. A.WĄSOWSKI: Current and voltage unbalance balancing operation of arc - resitance furnace, *ETEP* (proposed to publish)

5. UIE WORKING GROUP " Power Quality " WG 2 Guide pour la qualité de l'alimentation électrique des installations industrielles

6. A.WĄSOWSKI , K.BENGHALEM , T. MELIZI :Equilibrage des puissances d'arcs d'un four à arc dissymétrique . *JFE*, 1989, No. 40

7. A.WĄSOWSKI, J. BIAŁEK, A. NOWOCIEŃ : Begründung eines Anderungsvorschlags für das Kriterium zur Beueurteilung der Asymmetrie von Lichtbogenöfen in den internationalen Normempfehlungen, *EWT* ,1988 , No. B3

8. W. BROCIEK : Circuital representation of cooperation of three-phase UHP arc arrangement with an electric power system, *RE*, 1987, No. 3-4

9. W. BROCIEK, R.WILANOWICZ : Digital modeling of nonlinear elements of the nonuniqe driving point characteristic. *XVII Seminar of fundamentals of electrotechnics and circuit theory , 1994 Poland Conf. pap. 219 - 224.*

10. S.KOHLE, M.KNOOP, R. LICHTERBECK: Lichtbogenreaktanzen von Drehstrom-Lichtbogenöfen, *EWT*, 1993, No. B4

UIE XIII Congress on Electricity Applications 1996 MI 77

A Novel Method for The Voltage Flicker Estimation and Suppression Utilizing The Active Power Filter

Y. ASHIZAKI
TOSHIBA Corporation   1-1-1, Shibaura, Minato-ku, Tokyo, JAPAN
F. AOYAMA
TOSHIBA Corporation   1, Toshiba-cho, Fuchu-shi, Tokyo, JAPAN

ABSTRACT

Voltage flicker suppression devices are used to suppress voltage flicker caused by the electric arc furnaces used in steelmaking. Although only the reactive power was previously considered as a factor that causes voltage flicker, factor analysis has shown that there are components due to unbalanced current(negative phase sequence component) and distorted wave current(harmonic component) and that these components have magnitudes that cannot be ignored.
Since the voltage flicker suppression device based on the thyristor phase controlled reactor(TCR) that were used previously cannot, in principle, suppress distorted wave carrent, they did not provide adequate performance. Given this problem, we designed an active-filter type flicker suppression using a PMW inverter that uses a large-capacity GTO thyristor. This device achieved an excellent suppression effect by separating out the voltage flicker factors to detect and suppress voltage flicker.

Résumé

Dispositif pour suppression de papillotement de tension sont utilisés pour contrôler le papillotemnet de tension provoqué par le four à l'arc électrique pour la sidérurgie. Conventionellement, seule la puissance réactive a été considérée comme facteur causant le papillotement de tension, mais l'analyse de facteur de ce papillotement de tensiona monté qu'il existent des composants dus aux courants déséquilibrée(antiphase) et aux courant de distorsion(harmonique), dont la magnitude ne peut être négligée.
Comme les dispositifs pour suppression de papillotement de tension basés sur le réacteur de contrôle de phase de thyristor(RCT) qui ont été utilisés conventionellement ne peuvent supprimer les ondes de distorsion, ile ne peuvent apporter des resultats escomptés. Vuceproblèm, nous avons conçu un dispositif auto-excité pour suppression de papillotement de tension pourvu d'un inverseur PWM utilisant un thyristor GTO de plus grande capacité. Ce dispositif permet d'obtenir un excellent effet de suppression de papillotement de tension en isolant des facteurs de papillotement de tension pour les détecter et supprimer.

1. INTRODUCTION

The power fluctuations associated with load fluctuation in steelmaking arc

furnaces and similar equipment cause voltage fluctuation, voltage flicker in the power system. While voltage flicker suppression devices are generally available to suppress these phenomena.
TOSHIBA has a proven track record in this area, including TCR-type flicker suppression. Furthermore, TOSHIBA was the first in industry to produce, in 1987, a practical active filter type voltage flicker suppression device that applied a voltage-type PWM inverter.

Factor analysis of voltage flicker has shown that unbalanced current(negative phase sequence component) and distorted wave current(harmonics component) have a large influence to voltage flicker. this paper presents a factor analysis procedure, control technique for active filter type flicker suppression devices that apply those analysis techniques, an overview of the circuits in that device, and examples of the application to steelmaling plants.

2. SIMULATOR-BASED VOLTAGE FLICKER FACTOR ANALYSIS AND EVALUATION OF VOLTAGE FLICKER SUPRRESSION

2.1 FACTORS THAT CAUSE VOLTAGE FLICKER

We have maintained that voltage flicker in steelmaking arc furnace is induced not only by reactive power fluctuation but also by several other factors as well. We have analyzed the voltage flicker that occurs in several arc furnaces, and have concluded that it is appropriate to consider the following four factors that induce voltage flicker.

(1) Real power fluctuation (ΔP)
It is normally thought that reactive component(X) is much larger than resistance component(R) in transimission line or transformer, so voltage flicker due to real power is ignored. However, in case where R is large, this flicker cannot be ignored.

(2) Reactive power fluctuation (ΔQ)
This has been widely used in voltage flicker calculations, in fact, is the major factor determining the amount of voltage flicker.

(3) Fluctuation of the unbalanced current(negative phase sequence current); iN
In AC arc furnace, normally, each of three phases is controlled independently and thus the behavior is irregular. As a result, current is always thought to be in an unbalanced state. In extreme cases, loads only exist on two phases. Therefore it is conceivable that the fluctuating unbalanced current, as a fluctuating negative phase sequence voltage component, could become a factor that causes voltage flicker.

(4) Fluctuation of distorted wave current(harmonic distortion current); iH
Harmonic components at integer multiples of the fundamental frequency will not cause voltage flicker unless they fluctuate over time. However, it has been demonstrated that fluctuations in harmonics over time appear as voltage flicker. Significant levels of fluctuating distorted wave current that includes

non-integral multiples in the lower harmonic frequencies occur in AC arc furnace, and this is one of the factors that causes voltage flicker.

AC steelmaking arc furnaces differ from the generally–use load (ex. motor, inverter). In arc furnace, current distortion is large, and current waveform is not fixed but is constantly changing, and lower harmonic components($\leq$5th) are large. Therefore, we can infer that the influence of the unbalanced current(iN) and the distorted wave current(iH) is large.

2.2 FACTOR ANALYSIS TECHNIQUE USING SIMULATION

This section describes an analysis technique that divides the voltage flicker into the four factors described above. The purpose of this analysis is to grasp the ratio of the voltage flicker that occurs due to each of these factors. Fig.1 shows the conceptual block diagram of the analysis circuit used. The block in this circuit have the following functions:

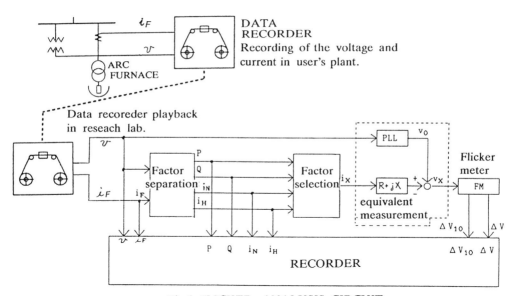

Fig.1 FLICKER ANALYSIS CIRCUIT

(1)Data recorder: The input current of the furnace iF and the bus voltage V at user's substation as analog signal. This data is reproduced at TOSHIBA's reseach lab and input to the analysis circuit.

(2)Factor separation: This circuit consists of P and Q calculation circuits and filters. This circuit accepts the arc furnace current and bus–line voltage and separates the four factor P, Q, iN, iH. The magnitudes of each of these factors is recorded on a recoreder.

(3)Factor selection : Selects one of the four factors and outputs the current signal(=ix) for the selected component.

(4) Equivalence measurement : This circuit determines the equivalent voltagefluctuation due to the selected factor. This circuit uses an infinitely large bus-line voltage Vo simulated with a PLL (Phase Locked Loop) circuit and a simulated system impedance(R+jX) to calculate the voltage Vx due to the corresponding factor with the following formula.

$$Vx = Vo - ix \times (R + jX)$$

(5) Flicker meter: The output Vx is input to the flicker meter and the voltage flicker $\Delta V10$ and the voltage variation ΔV. The $\Delta V10$ value is a voltage flicker evaluation method widely used in Japan.

2.3 RESULTS OF THE VOLTAGE FLICKER EVALUATION

Fig.2 shows sample results from a voltage flicker factor analysis.
As results of performing this analysis for several arc furnaces, we have verified that the ratios of these four factors as follows.

Flicker due to real power fluctuation
$\Delta V10-P$ 10 %
Flicker due to reactive power fluctuation
$\Delta V10-Q$ 70 %
Flicker due to unbalanced current
$\Delta V10-iN$ 50 %
Flicker due to distorted wave current
$\Delta V10-iH$ 50 %

From the results of these analysis, it is clear that the voltage flicker components have the following relationships.

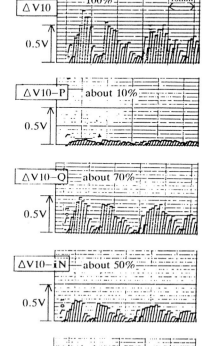

Fig.2 EXAMPLE OF FLICKER ANALYSIS RESULTS

$$\Delta V10 = \sqrt{(\Delta V10-P)^2 + (\Delta V10-Q)^2 + (\Delta V10-iN)^2 + (\Delta V10-iH)^2}$$

We arrived at the following conclusions from the factor analysis and the results of the evaluation described above;

(1) The voltage flicker due to real power fluctuations(ΔP) occupies a relatively small part of the total voltage flicker, about 10 %.
(2) The voltage flicker due to reactive power fluctuations(ΔQ) occupies a relatively large part of the total voltage flicker, about 70%.
(3) The voltage flicker due to unbalanced current(iN) and to distorted wave current(iH) are both about 50% and thus both have magnitudes that are one-half or more of the voltage flicker due to reactive power fluctuation.

2.4 PERFORMANCE REQUIRED OF SUPPRESSION DEVICES

Fig.3 shows to what extent voltage flicker is reduced by the removal of various components in a simulation based on reproducing arc furnace currents recorded on a data recorder and removing the ΔQ, iN, and iH components with an electronic circuit. From this graph it can be seen that voltage flicker cannot be reduced significantly by only eliminating the reactive power components, and that it can be reduced significantly only if the components due unbalanced current and harmonic distortion current are also removed.
So, a suppression device capable of compensating for factors up to and including unbalanced current(iN) and distorted wave current(iH) is required. Due to their operating principles, the TCR-type suppression devices used up to now does not have rapid response enough to suppress distorted wave current. Thus it should be effective to apply active filter type flicker suppressin devices, since these devices provide rapid response to load variations and can suppress distorted wave current.

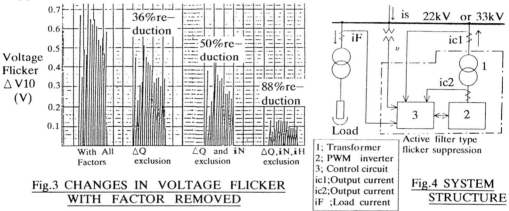

Fig.3 CHANGES IN VOLTAGE FLICKER WITH FACTOR REMOVED

Fig.4 SYSTEM STRUCTURE

3. ACTIVE FILTER TYPE FLICKER SUPPRESSION DEVICES

3.1 SYSTEM STRUCTURE

Fig.4 shows the structure of a system consisting of a fluctuation load (a steelmaking arc furnace) that is a voltage flicker source and a active filter type voltage flicker suppression is connected in parallel with the fluctuating load and compensates for voltage flicker factors.

3.2 MAIN CIRCUIT

The main circuit in the active filter type flicker suppression unit consists of a voltage-type PWM inverter and a transformer that is used to connect PWM inverter to power system. The PWM inverter is formed by a three-phase connection of single-phase bridge circuits that use GTO thyristor. Fig.5 shows the main circuit structure for a unit with a rated capacity of 18000kVA, and Table 1 lists the main ratings. The remainder of this section describes the features of this circuit.

(1) The voltage-type PWM inverter consist of two or three units each with a unit capacity of 9000kVA. Six single-phase bridge circuits are installed in a unit and these units are connected by series multiplexing on the transformer secondary windings. Futhermore, a three phase circuit is formed by delta-delta connection of transformer.

(2) Large-capacity GTO thyristors rated 6000V-3000A are used in main circuits. The overall size of the unit is minimized by increasing the capacities of the individual devices. The DC voltage of PWM inverter is 3000V.

(3) Each single-phase bridge circuit is controlled by PWM using a 390Hz carrier. The carrier frequency is effective increased to improve the response characteristics and reduce ripple current by phase sift of carriers between the series multiplexed single-phase bridge circuits.

(4) Since there is no transfer of real power between this unit and the power system, no power source is required in the DC side. This circuit includes DC capacitor that required to suppress the DC voltage fluctuations caused by suppressing unbalanced current and distorted wave current.

(5) A purified water recirculatory cooling system with a high cooling efficiency is used for the main components such as the GTO thyristor, and this also contributes to unit miniaturization.

(6) Concerning transformer, We succeeded in miniaturization the unit by using a three-phase four-legs iron core.

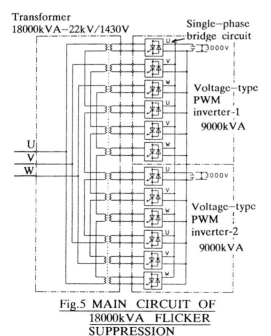

Fig.5 MAIN CIRCUIT OF 18000kVA FLICKER SUPPRESSION

Table.1 RATINGS OF ACTIVE FILTER TYPE FLICKER SUPPRESSION

System voltage	Three-phase 22kV or 33kV	
Rated capacity	18,000kVA	27,000kVA
Type	Series multiplexed voltage-type PWM inverter	
Series multiplexing	4	6
GTO switching frequency	390 Hz	
Inverter rateings (Ratings of unit)	Capacity 9,000kVA Structure Six single-phase bridge circuits Output voltage 1,430V Output current 1,050A DC voltage 3,000V GTO ratings 6,000V-3,000A Device configuration 1S-1P-4A Demention 3400W×2200D×3000H (mm)	
Inverter cooling	Purified water recirculatory cooling	

3.3 CONTROL CIRCUIT

This circuit compensates not only reactive power fluctuation, but also unbalanced current iN, and distorted wave current iH. The control circuit consists of an output current referenced calculation circuit, a current control circuit, DC voltage control circuit. Fig.6 shows the control block. This block has the following features:

(1) The current reference calculation circuit adopts a digital technique using a DSP (digital signal processor). It applies volatage flicker factor analysis technique and separates out each of the factors and detects them indevidually for optimum compensation.

(2) This circuit automatically controls the compeansation gain so that the maximum output within the rated range of the flicker suppression unit can be acquired in response to the size of the voltage flicker factor caused by the load. This means that the capacity of the unit can be used effectively and a high suppression performance can be achived.

(3) The current control circuit used feedback control to match the output of unit to a current reference. An analog circuit is used here, since it is also necessary to compensate for harmonic distortion current, which means that rapid response characteristics are required.

(4) Although the DC voltage is reduced by the loss in the unit, a voltage control circuit is used and the DC voltage is held at a fixed value. Thus the loss is supplier from the power system.

(5) The transformer core sometimes suffer from DC flux due to DC voltage in time of arc furnace shorting or rush current flowing. To prevent this, the output current of each single phase bridge circuit is controlled individually and the excitation current DC components flowing in the transformer windings is suppressed.

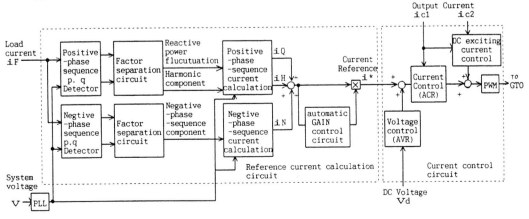

Fig.6 CONTROL SYSTEM BLOCK DIAGRAM

4. APPLICATION EXAMPLE

This suppression device has already been installed and put into service at plants belonging to several steelmaking companies. The following are representative sample applications.

4.1 VOLTAGE FLICKER SUPPRESSION

Fig.7 shows the volatge flicker when an 18000kVA suppresion device is applied to AC arc furnace(a 50MVA furnace transformer with a maximum reactive power of about 92MVA). A voltage flicker improved ratio of about 50% was achieved.

4.2 VOLTAGE FLUCTUATION SUPPRESION

This device is also effective at suppressing stedy-state volatge fluctuation. We applied an 27000kVA suppression device to a steelmaking plant that has several large-capacity arc furnaces. The 22kV voltage drop of 16% prior to improvement was suppressed to 10% by this suppression device. This allowed stable operation of equipments installed on same system.

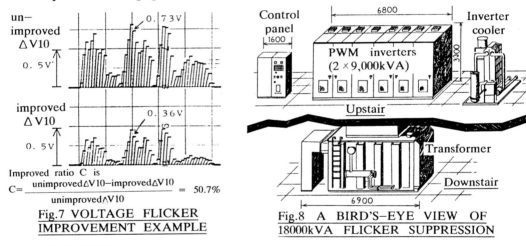

Improved ratio C is
$$C = \frac{\text{unimproved}\Delta V10 - \text{improved}\Delta V10}{\text{unimproved}\Delta V10} = 50.7\%$$

Fig.7 VOLTAGE FLICKER IMPROVEMENT EXAMPLE

Fig.8 A BIRD'S-EYE VIEW OF 18000kVA FLICKER SUPPRESSION

5. CONCLUSION

We have established a technique for voltage flicker factor analysis and demonstrated that in addition to reactive power fluctuations, which were previously considered to be a factor, unbalanced current and distorted wave current are also factors that cause voltage flicker.
We developed an active filter type vol-tage flicker suppression device that adopts a large-capacity PWM inverter in the main circuit and uses a control scheme that provides good suppression of voltage flicker.

REFERENCE:
[1] H.Inokuchi and F.Aoyama et al:'An application of the New Static Var Compensator Utilizing the PWM Converter,' UIE 11th Congress Paper,1988.

Activities of UIE-WG Power Quality

GUTIÉRREZ IGLESIAS, J. L.
ASINEL, Spain
Chairman of UIE Working Group: "Power Quality"

RÉSUMÉ

Le présent rapport est une synthèse des activités dans le sein du Groupe de Travail GT2 de l'UIE "Qualité de l'Alimentation".

ABSTRACT

This present report is a summary on activities of UIE Working Group 2 "Power Quality"

ACTIVITIES IN UIE-WG2 "POWER QUALITY"

UIE Working Group on Power Quality is specialized in research on Electromagnetic Compatibility (EMC).

A relevant group of 24 experts from 15 countries belongs to the Working Group.

Their task consists in covering the field of activities dealing with the assessment of voltage quality with relation to conducted disturbances, i.e., harmonics, flicker, unbalance and voltage dips. The WG produces a set of guides, based on research, which is largely used by standardization bodies.

At present it has been published relevant documents dealing with the problematic due to disturbances existing in industrial environment. These documents are technical reports that outline principles that may be used as the basis for determining the requirements for connecting distorting loads to public power systems, as well as to provide guidance for engineering practice to ensure adequate service quality for all connected consumers.

By other hand, activities of UIE WG Power Quality deal with basic research and relevant work on objective aspects of conducted disturbances. New measurement methods for evaluation and assessment of disturbances have been or are being developed, as well as adopted as possible worldwide standard.

PRESENT WORK AND RESEARCH

A short description is presented herein, explaining updated information resulting from recent research, as well as, relevant work on conducted disturbances.

At present, several Task Force Groups have been formed in this Working Group. They are specialized in different aspects of the problematic in power quality co-ordination taking into account the different types of disturbances, as mentioned above.

Task Force Group on Flicker. With respect to the well known UIE-IEC Flickermeter, the standard in Europe is being presented to North America where no standard measurement method exists. The UIE Flickermeter recently has been adapted for use in N-America and Japan where 120 V and 100 V lamps are used.

It is being tested in the USA and is being reorganized as a possible worldwide standard.

Data on flicker levels vs. complaints is shown for the UIE Flickermeter and U.S. Flickermeters.

Task Force Group on Voltage Dips. Voltage dips are among the most important disturbances, affecting the voltage quality of public networks and industrial installations. They are capable of annoying consumers, as well as causing substantial economic losses.

UIE has studied voltage dips and short interruptions for a number of years, and has elaborated a Guide on Dips. The paper gives a general description of this phenomena, and presents statistical data from co-ordinated measurements performed in different countries.

In addition, the application of mitigation measures such as SMES, are discussed.

Task Force Group on Harmonics. As the actual background levels of harmonic voltage distortion on supply networks due to residential equipment emissions increase, the connection of industrial consumer's electrical plant will become increasingly difficult and emission restrictions will become more severe.

This paper explains how a knowledge of the use of industrial equipment can facilitate its connection and how the emissions of some particular equipment can be reduced.

Information is based on the draft UIE Guide on Harmonics.

Task Force Group on Voltage Unbalance. Voltage unbalance is a well known disturbance existing in three-phase system. Nevertheless, at present, there has not been established evaluating and assessment methods in standards.

In UIE, and according to recent experience, it is being edited a Guide on Voltage Unbalance. The aim is to give guidelines for managing this type of disturbance.

In particular, a methodology is being proposed for measuring and evaluating voltage unbalance.

UIE-WG2 PUBLICATIONS ON EMC

Since the past decade, it has been published the following relevant documents:

- Flickermeter. Functional and design specifications. 1982.
- Flicker measurement and evaluation. 1st edition 1986. 2nd edition 1992. Addendum 1996.
- Connection of fluctuating loads. 1988.
- Measurement of Voltage Dips and Short Interruptions occurring in Heavy Industries. 1992.
- Guide to Quality of Electrical Supply for Industrial Installations. Part 1: General Introduction to Electromagnetic Compatibility (EMC), Types of Disturbances and Relevant Standards. 1994. Addendum 1996.
- Guide to Quality of Electrical Supply for Industrial Installations. Part 2: Voltage Dips and Short Interruptions. 1996.

Other new publications to be edited by end 1996 are the following:

- Guide to Quality of Electrical Supply for Industrial Installations. Part 3: Harmonics.
- Guide to Quality of Electrical Supply for Industrial Installations. Part 4: Voltage Unbalance.

A new an updated Guide to Quality of Electrical Supply for Industrial Installations. Part 5: Flicker, is to be edited by 1997.

The framework of all those publications basically consists on the description of disturbances, their origins and effects, measurement, evaluation and assessment methods, mitigation techniques and prediction studies for installations requirements.

Metal Processing and Heat Treatment

Contents

Tetronics tundish plasma heating 1
Iddles, D.M., Page, D.G., Liddiard, L.J. (Tetronics Reasearch & Development Co. Ltd), Omori, M. (Tetro International) UK

Development of large size cold crucible levitation melting 9
Kainuma, K., Fujita, M., Take, T., Ashida, Y., Tadano, H. (Fuji Electric Co R & D Ltd), Yamazaki, M. (Chubu Electric Power Co.) JAPAN

Controlled heating for the thixo-processing of aluminium 17
Jürgens, R. (AEG Elotherm GmbH) GERMANY

Induction heating for the wire industry 27
Kirkwood, R.J. (Radyne Ltd) UK

Installations for the continuous induction heat treatment of wires 35
Artuso, I. (ATE), Dughiero, F., Lupi, S., Partisani, S. (University of Padua), Facchinelli, P. (GCR Engineering S.p.A.) ITALY

Réalisation de moules de rotomoulage par électroformage sous courants pulsés 45
Aouaissia, C. (CTAA), Bercot, P., Pagetti, J. (Laboratoire de Corrosion et Traitement de Surface) FRANCE

Foundary automation 53
Booth, M., Wright, A.E. (Midlands Electricity plc), Steer, G. (Aston University) UK

Use of high frequency induction melting in cold crucible to produce crystalline materials without oxygen 61
Bezmenov, F.V., Kanaev, I.A., Ivanov, V.N., Raduchaev, V.A. (VNIITVCh) RUSSIA

Electric drying of monolithic steel ladles 73
Solmar, A., Danelius, J., Alavyoon, F. (Vattenfall) SWEDEN

Molten metal processing with an advanced electromagnetic pump 81
Katyal, A. (EMP Technologies Ltd), Bullard, H.W., Smith L., Hayes, P.J. (EA Technology Ltd) UK

ns 1996 MII 1

Tetronics Tundish Plasma Heating

D M IDDLES, D G PAGE, L J LIDDIARD, M OMORI*

Tetronics R & D Co Ltd, Lechlade Rd, Faringdon, UK
* Tetro International, Kawasaki, Japan

ABSTRACT

An improvement in steel quality can be realised by the introduction of plasma torch heating into a tundish. Case histories of several industrial units are discussed, together with a review of the respective merits in operation of single and twin DC plasma torch systems installed.

On peut améliorer la qualité de l'acier en introduisant le chauffage à la torche à plasma dans un avant-creuset. L'histoire de cas de plusieurs sites industriels est présentée, accompagnée d'un examen des mérites respectifs de l'exploitation des systèmes de torche à plasma DC simples et doubles installés.

INTRODUCTION

Currently more than 70 % of the worlds steel production is continuously cast[1], this figure is forecast to rise to 80-90 % by the year 2000. The emphasis within the steelworks community is on improving product quality, whilst maintaining cost competitiveness. The improvements in steel quality are being achieved throughout the process, but tundish heating is used prior to casting. A controlled increase in steel temperature prior to the steel entering the moulds allows the floatation of inclusions which gives greater uniformity in microstructure and sufficient superheat in casting to avoid premature freezing and/or void formation. However, too great a superheat to the steel allows the formation of centre segregated defects in the as cast structure.

The generation of superheat within the tundish is generally achieved by the use of either an induction coil or plasma process. These processes have their respective advantages and disadvantages:

Induction heating involves bulky equipment which is not easily retro-fitted to an existing tundish and causes large magnetic fields to be set-up surrounding the operators and control instrumentation. Plasma systems may be either AC or DC Transferred arcs and are easily retro-fitted to the tundish with minimal extra work. The use of either system will depend upon the customer and operational environment. In general, AC plasma torches tend to be noisier than their DC equivalents and can be prone to reduced service life.

Conventional DC plasma torch systems use a single torch with a return electrode placed in contact with the steel being heated, as can be seen in Figure 1. This single torch can be either cathodic or anodic in polarity.

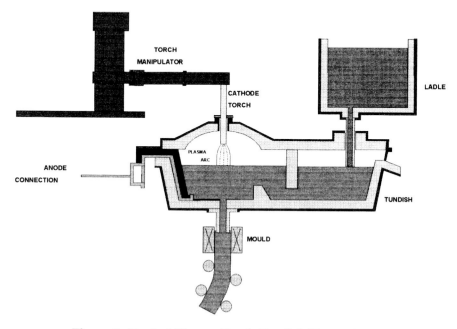

Figure 1: Typical Plasma Torch Tundish Heater Layout.

A new DC plasma torch system has been developed which does not require a return electrode to be placed in contact with the tundish. This system thereby obviates the requirement for the engineering associated with return electrodes and simplifies plant design and layout. The new system utilises twin plasma torches, of opposite electrical polarity, placed above the steel in the tundish, usually 250 mm to 300 mm from the melt surface. The current path is then considered to be from the cathode torch to the steel and back to the anode torch. Providing that there is a continuous electrical path within the tundish, the torches can be separated as far apart as the customer requires, thereby allowing heat transfer over a greater area.

Plants have been installed by Tetronics which are rated at 7000 A, 200 V for single plasma torch and 5000 A, 500 V for twin plasma torch. Some case histories are presented later.

TUNDISH METALLURGY

The temperature variation on first tapping the blast furnace, to continuously casting the steel is very significant, (tapping temperature of $\approx$ 1750°C at the blast furnace, mould entry temperature $\approx$ 1500°C - 1550°C, dependent upon the grade). This variation in mould entry temperature is sufficient to alter the casting operation and yield an inhomogeneous product for sale. Typical variation in the steel temperature within the tundish for a single ladle cast can be seen in Figure 2. The variation in mould entry temperature of the steel from the beginning to the end of the cast can be overcome by using a plasma torch, which results in the revised temperature profile during casting, also seen in Figure 2.

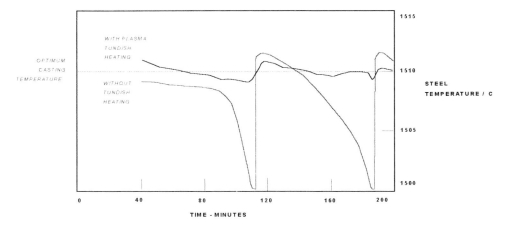

Figure 2: Steel Temperature Variation in the Tundish During Casting.

The influence of steel temperature superheat on the incorporation of inclusions and centre segregation within the cast structure can be gauged from Figure 3.

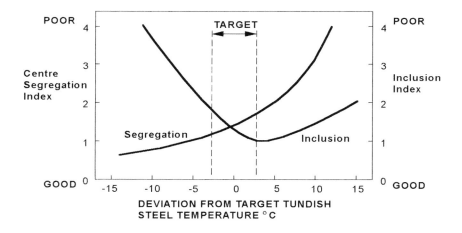

Figure 3: Influence of Tundish Superheat on Steel Quality.

TETRONICS PLASMA SYSTEMS

The Tetronics plasma torches are all mounted on to manipulators which enable the torches to be moved completely out of the tundish and casting operation when necessary. Safety for the operators is inherent in the manipulator design with an emergency air motor for driving the torch out of the tundish, in the case of loss of electricity supply and a water leak detection system on critical water coolant lines. All Tetronics tundish systems are PLC controlled to prevent arc initiation if for example, the water flows are too low, the water temperature is too high or there is no gas flow etc.

The plasma arc is initiated by a controlled high frequency discharge between the nozzle and the electrode. Once a transferred arc is detected, a contactor opens within the start supply and the non-transferred arc first formed, is stopped. The power transferred to the process can be varied by altering either the current, arc length or atmosphere within the tundish. The variation of arc voltage with arc length and atmosphere can be seen in Figures 4 and 5 respectively. By varying the power of the plasma arc within the tundish, the temperature of the steel entering the moulds can be controlled to ± 1°C by the use of a computer/algorithm control.

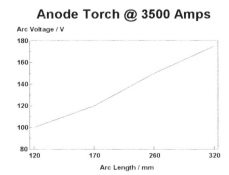

Figure 4: Arc Voltage Variation.

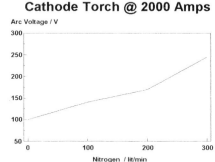

Figure 5: Arc Voltage Variation with Gas.

CASE STUDIES:

Pohang Iron and Steel Company.

A single plasma torch has been successfully installed at No 1 Continuous Caster (4 strand billet), Pohang Works, Korea. The plasma system is a 1.4 MW unit (7000 Amps, 200 Volts) water cooled cathode torch and ancillary power supply, water and gas services. The plasma system can either be operated manually or by means of remote computer control. The temperature variation of the steel during casting (100 tonnes per ladle, 3-4 ladles per cast) can be controlled to better than ± 5 °C and ± 1 °C, by manual and automatic control, respectively. The heating efficiency of the plasma torch in the tundish was > 70% and rises in the steel temperature of 20.7 °C min^{-1} were recorded. The nitrogen pick-up between ladle and billet was < 3 ppm, which was attributed to oxidation and air ingress into the tundish. For currents of < 5000 Amps, the electrode life is > 100 hours.

The tundish profile for the steel temperature within the ladle and tundish, as a function of natural convective/radiative conditions and the response when the plasma torch is applied, can be seen in Figure 6.

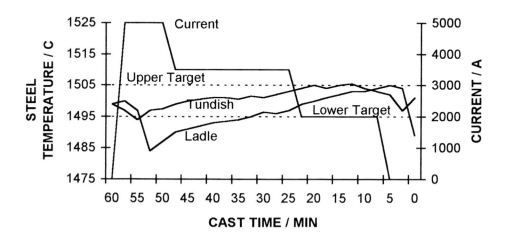

Figure 6: Tundish Characteristics for No 1 CC, Pohang Works.

Kawasaki Steel Corporation.

A twin DC plasma torch system, the worlds first, has been successfully installed at No 4 Continuous Caster (slab), Chiba Works, Japan. This new facility at Chiba Works has been built to be state of the art. The selection of the twin DC plasma system resulted in a easier engineering design for the plant and no maintenance of the return electrode in the tundish. The plasma system is a 1.2 MW (3500 Amps, 350 Volts) unit. The plasma torches for this application are based upon an adaptation of the conventional 70 mm diameter torch used in all other steelworks applications which employ water cooling and low gas (Ar) flow rates. The cathode electrode uses a thoriated tungsten tip whilst the anode electrode employs a copper button.

Through the use of the twin DC plasma torches at a power of 0.9 MW, the temperature drop in the liquid steel of 15°C during the last 25 minutes of casting can be compensated for. A typical plasma torch power profile during casting, is illustrated in Figure 7. The electrode lifes on both torches are in excess of 100 hours.

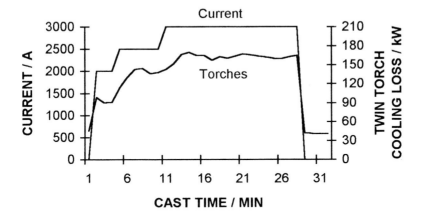

Figure 7: Tundish Characteristics for No 4 CC, Chiba Works.

CONCLUSIONS

Plasma heating has been shown to improve the quality of continuously cast steel by homogenising the casting temperature throughout the whole heating period which reduces grain size variation and void formation in the product. The additional heat from the plasma source in the tundish also enables the tap temperature at the furnace to be reduced by 10°C to 15°C, which improves the refractory life at the furnace tap-hole and hence reduced maintenance costs and lost operational time. An additional advantage of plasma heating is the ability to continue casting "cold heats", thus avoiding lost time at the rolling shop and the extra expense of re-melting the steel. A twin DC torch plasma system has been employed in a Japanese steelworks and shown to operate reliably at 1.2 MW, with torch lifes in excess of 100 hours. The twin DC plasma system has the advantage of not requiring a return electrode in contact with the liquid steel, which considerably simplifies the tundish and plant layout design.

For successful plasma torch operation in a tundish environment, the following criteria are required:
i) Low oxygen concentration, < 0.1 %.
ii) Clean steel, i.e no slag carry over into heating zone.
iii) Correct dam and weir design to ensure efficient stirring of the steel in the heating zone.

In this manner, heat transfer efficiencies of > 70 % can be achieved with noise levels of < 80 dB. The increase in the temperature of the steel from plasma heating is sufficient to compensate for the natural cooling losses.

Based upon the success of the twin torch system at Kawasaki Steel Corporation and the simplified plant design, further sales of the twin torch have been achieved, most notably a 2.5 MW (5000 A, 500 V) unit.

REFERENCES

1. Iron & Steelmaker, 1995, **22** (13), p 13.

Development of Large-Size Cold Crucible Levitation Melting

*K.KAINUMA M.FUJITA T.TAKE Y.ASHIDA H.TADANO
**M.YAMAZAKI

*Fuji Electric Corporate Research and Development Ltd., 5520 Minamitamagaki-cho Suzuka Mie JAPAN
**Chubu Electric Power Company Inc., 20-1 Kitasekiyama Ohdaka-cho Midori-ku Nagoya JAPAN

ABSTRACT

Cold crucible levitation melting (CCLM) furnaces can melt metals without contaminating them. Molten metals are levitated in a crucible by strongly induced electromagnetic forces. We have designed a duplex inverter configuration to obtain stable melting conditions. We have been able to melt 50kg of cast iron in a large CCLM in ten minutes. Molten metal can be tapped through a hole in the bottom of the crucible. During the development of the apparatus, we investigated eddy currents to evaluate the effects of electromagnetic force on molten metal and heat loss in the crucible. We investigated the free boundary of a mass of molten metal to estimate its shape. And we investigated heat transfer and fluid flow to evaluate the cooling ability of a crucible. In this paper, the design of a large CCLM and the results of experiments obtained with it are presented.

RÉSUMÉ

Les fours à fusion fluidifiée en creusets froids (CCLM) sont capables de fondre les métaux sans les contaminer. Les métaux en fusion sont fluidifiés dans un creuset sous l'effet de forces électromagnétiques de forte induction. Nous avons conçu une configuration d'inverseur duplex, afin d'obtenir des conditions de fusion stable. Nous avons réussi à faire fondre 50kg de fonte dans un gros appareil CCLM en dix minutes. Le métal en fusion peut être soutiré par un orifice pratiqué sur le fond du creuset. Dans cette expérience, nous avons étudié les courants tourbillonnaires afin d'évaluer les effects de la force électromagnétique sur le métal en fusion et la perte de chaleur dans le creuset. Nous avons étudié la périphérie libre d'une masse de métal en fusion afin d'en évaluer la forme. Et nous avons étudié le transfert de chaleur et l'écoulement de fluid afin d'évaluer la capscité de refroidissement d'un creuset. Dans cet articre, nous présentons la conception d'un CCLM de grosse capacité ainsi que les résultats des expériences.

I INTRODUCTION

Recently many products using electromagnetic technology have come into use. New methods of melting metal have been studied energetically.[1-3] Cold crucible levitation melting (CCLM) has lately attracted especial attention, because ① it allows metals to be melted without contamination, ② metals with high melting points can be melted with it, ③ a crucible can be used repeatedly, ④ metals can be melted rapidly and superheated, ⑤ molten metals become uniform in composition because they are strongly stirred in the crucible.

Levitation melting using only coils has been studied for 30 years. However, the results obtained in these studies have been unsatisfactory, because levitating molten metal is unstable when contained only by coils. Therefore, a new method of levitation melting using a cold crucible was recently developed. But, since it has been possible to levitate only a few kilograms of metal with this method, it has been used only in research and development. We have developed a CCLM which allows large amount of metals to be levitated and melted to apply to the industrial field. We have developed an apparatus which can levitate and melt 50kg of cast iron in ten minutes, and which has a hole in its bottom through which molten metal can be tapped. Therefore, this apparatus can be applied in actual casting processes.

II PRINCIPLE OF CCLM

Figure 1 shows the principle of the CCLM. A water-cooled crucible is made from copper segments. Coils are wound around the crucible and connected to high frequency power inverters. A high frequency current flows through the coils. An eddy current is induced in the crucible and metal to be melted. The metal in the crucible is levitated with electromagnetic repulsion forces. The eddy current generates Joule heat in the crucible and metal, and the metal melts. To realize stable melting conditions, we have designed a duplex inverter configuration, in which a lower frequency current (several kHz) is supplied to a lower coil and a higher frequency current (several tens kHz) to an upper coil. The lower coil is used mainly to levitate molten metals. The repulsion forces generated by coils is inversely proportional to the square root of frequency at a fixed power. The upper coil is used mainly to heat metals. With increasing frequency, the eddy current in a metal to be heated concentrates near the surface of the metal, and heating efficiency is improves.

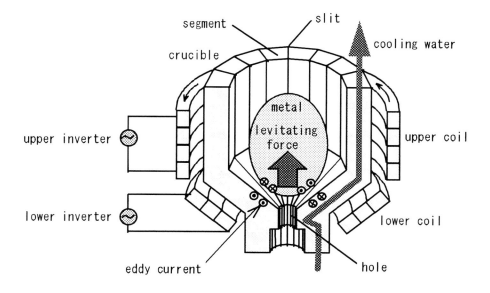

Fig.1 Principle of CCLM

III ESTABLISHMENT OF TECHNOLOGY FOR LARGE CCLM

The following were required to develop a large CCLM;

- methods of estimating electromagnetic force and heat loss.
- a method of optimizing a crucible and coil structures.
- technologies for high level cooling design.

The development of highly accurate such methods and technologies is described in the following sections.

3.1 THREE-DIMENSIONAL ANALYSIS OF EDDY CURRENT

In order to discover the amount and the distribution of electromagnetic force and heat loss, and in order to optimize a crucible and coil structure, we have applied three dimensional eddy current analysis.

With a CCLM, ① magnetic flux leaks over a wide space, and ② eddy current depth is shallow compared with conductor thickness. We used an integro-differential method of analysis with a thin plate element with current vector potential.[4] The unknown factor can be restricted to the surface of the conductor. In this analysis, we devised an original method of calculating eddy currents in coils.[5] We calculated the eddy currents induced by the upper and lower coils separately because the frequencies of the currents flowing through the coils were different, and estimated their characteristics by overlapping them.

Figure 2 shows the equi-current potential lines. Eddy currents flow along the lines. And the shorter the distance between the lines becomes, the greater eddy current density becomes. The distribution of eddy currents is horizontally uniform around the inside of a crucible. Eddy current loss concentrates around the hole in the bottom of the crucible. So the area around the hole must

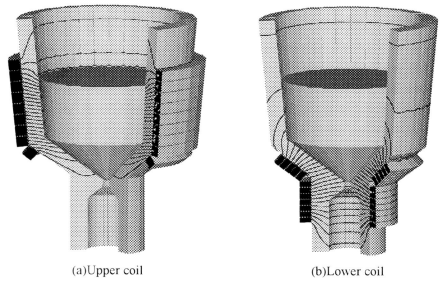

(a) Upper coil   (b) Lower coil

Fig.2 Equi-current potential line on the crucible

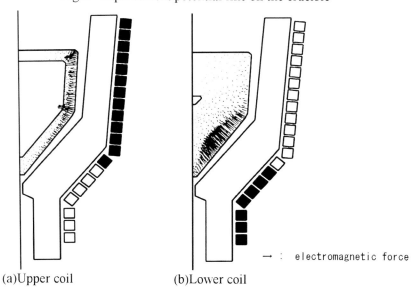

→ : electromagnetic force

(a) Upper coil   (b) Lower coil

Fig.3 Electromagnetic force distribution on molten metal

be cooled, to prevent temperatures from getting too high. Fig.3 shows the distribution of electromagnetic force exerted on the metal. The electromagnetic force generated by the lower coil is larger on the bottom of the molten mass than it is on the top. This makes levitating molten metal easier.

3.2 THREE-DIMENSIONAL ANALYSIS OF HEAT AND COOLING

Our research indicates that a large CCLM induces the great heat loss in coils and crucible. Thus cooling will be a very important item. So we analyzed heat transfer in the crucible to allow us to optimize the crucible cooling structure. The temperatures around the hole in the bottom of a crucible are higher than those at any other parts of the crucible, because eddy current density is higher there than elsewhere, and because the molten metal heats the crucible. To increase the cooling effect, we have developed a V-shaped cooling structure around the bottom of the crucible. With this V-shaped structure, the cooling water can flow near the hole and the cooling effect is increased. We then calculated heat transfer and fluid flow at the same time. In our analysis, the distribution of the heat source is calculated in a three-dimensional eddy current analysis.

Figure 4 shows the flow of cooling water and equi-thermal lines around the bottom of the crucible. The arrows plotted in the figure indicate the direction and velocity of cooling water. These results show that cooling water flow without stagnation around the bottom of the crucible.

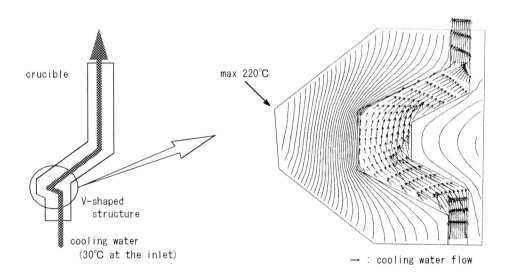

Fig.4 Distribution of water flow and temperature

The temperature of the crucible was calculated with a heat transfer coefficient obtained in a fluid flow analysis. Our results demonstrate that the temperature of the copper crucible is within a maximum temperature of 220°C.

IV ESTABLISHMENT OF THE TAPPING SYSTEM

Before using CCLM in industry, a way of taking out molten metal from a cold crucible must be developed. The results of our investigations have led us to adopt a method in which molten metal is tapped through a hole in the bottom of a crucible while controlling the balance of narrowing force $F1$, levitation force $F2$ and gravity force $F3$ (Fig.5). This method is superior to others, such as vacuum sucking and pouring by tilting a crucible, because of its simplicity and easy to maintain.

To verify the appropriateness of the method we chose, we analyzed the tapping of molten metal in a computer simulation. The electromagnetic force exterted on the molten metal was calculated by analyzing a magnetic field with the finite element method (FEM). The free boundary of the molten metal was calculated with the volume fraction method (VOF) considering gravity. In this method,

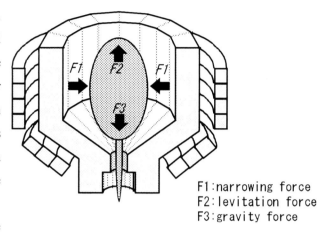

F1: narrowing force
F2: levitation force
F3: gravity force

Fig.5 Principle of Tapping molten metal

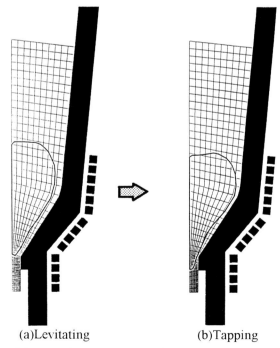

(a) Levitating   (b) Tapping

Fig.6 The results of free-boundary simulation

we calculated the fluid volume percentage of every element. The equi-ratio line (the ratio is 50%) represents the shape of the molten metal. Fig.6 shows the results obtained in the simulation. The results of our calculations suggested the correctness of the above principles.

V MELTING AND TAPPING EXAMINATION

Fig.7   Molten metal in CCLM (50kg of cast iron)

We made an experimental apparatus based on the analysis described above and conducted melting and tapping experiments with it. During the experiments, the power supply of the upper inverter was 450kW (30kHz), and that of the lower one was 400kW (3kHz). Fifty kilograms of cast iron were melted completely in about 10 minutes (Fig.7). And after that, the power supplies of the upper and lower inverters were changed to 400kW and 150kW, respectively. Then the molten

Fig.8   Tapping molten metal (cast iron)

metal was tapped through the bottom of the crucible (Fig.8), which took 10-30 seconds.

VI CONCLUSION

We have developed a large CCLM of 50kg-class. With it, we have been able to quickly melt fifty kilograms of cast iron, while keeping it almost completely without contact to the crucible both during melting and tapping. So this apparatus has reached the level where it is applied to industrial fields.

ACKNOWLEDGMENT

This research was conducted in a joint investigation with the National Research Institute for Metals and Chubu Electric Power Company Inc. The authors are grateful for the advice provided by Dr. Tsuboi (Fukuyama University) on the three-dimensional analysis of eddy current.

REFERENCE

1. K.Sakuraya, et al.: `Levitation Melting by Duplex Configuration of Inverters with Different frequency`, Simulation and Design of Applied Electromagnetic Systems, 1994, 483-486

2. A.Fukuzawa, et al.: `Cold Crucible Type Levitation Melting by Supplying Two Frequencies`, ITEM Nagoya ISIJ, 1994, 172-177

3. H.Tadano, et al.: `Levitation Melting of Several Kilograms of Metal with a Cold Crucible`, IEEE Trans., vol.**30**(6),

4. H.Tuboi, et al.: `Eddy Current Analysis of Thin Conducting Plates Taking Account of Source Currents`, T.IEE Japan, vol.**110**A(9), 1990

5. Y.Ashida, et al.: `Eddy Current Analysis of Cold Culpable Levitation Melting Model`, IEE Japan, No.**1010**(1994)

CONTROLLED HEATING FOR THE THIXO-PROCESSING OF ALUMINIUM

by ROBERT JUERGENS, AEG ELOTHERM GmbH, Remscheid, Germany

ABSTRACT

The last ten years have witnessed the introduction of a new manufacturing process for aluminium alloy formings, the thixocasting process. More recently, the car industry has implemented this process to produce car parts submitted to high quality and safety standards. The material to be formed by this process is induction heated beyond solidus. The desired liquid part for the subsequent forming operation generally amounts to between 30 and 40 %. Due to the physical properties of aluminium alloys, heating the thixo bars according to a precisely pre-defined energy curve is required. Incrementing the energy per period of time with great accuracy is decisive in the success of making high quality formings. Such accuracy is achieved by the implementation of a patented process, enabling precise measurement of the power induced into the material to be heated. Monitoring electrical parameters such as current, voltage and frequency enables on-line process control in the course of the heating operation. Thus, the heating operation becomes a controlled process. Changes in the quality of the material may be detected and compensated within the heating plant. Thanks to this breakthrough, one is now much nearer the objective of using 100% of the material to be formed within a thixocasting or thixoforming plant.

RÉSUMÉ

Il y a une dizaine d'années que s'est développé un nouveau procédé de fabrication de pièces en alliage d'aluminium, le procédé de "thixo-moulage" (thixocasting). Plus récemment, l'industrie automobile a commencé à utiliser ce procédé pour fabriquer des pièces soumises à des critères de qualité et de sécurité particulièrement exigeants. Ce procédé implique le chauffage par induction des pièces brutes jusqu'à un niveau de température supérieur au liquidus. La part liquifiée que l'on souhaite obtenir en vue de l'opération de formage à effectuer en aval est généralement comprise entre 30 et 40%. Compte tenu des propriétés physiques du type d'alliage d'aluminium utilisé, la barre cylindrique en thixo-aluminium doit être chauffée selon un profil d'énergie précis, préalablement défini. L'obtention d'un niveau de qualité élevé passe par une incrémentation très précise de l'apport d'énergie par unité de temps. Cette précision est atteinte par la mise en oeuvre d'un procédé breveté qui permet de mesurer la puissance induite dans la pièce elle-même. La surveillance des paramètres électriques, intensité, tension et fréquence, permet le contrôle direct du processus de chauffage en cours. Les éventuelles variations de la qualité du matériel brut peuvent être décelées et prises en compte pour corriger de façon appropriée le processus de chauffage. Grâce à cette percée technologique, nous nous sommes grandement rapprochés de l'objectif consistant à utiliser dans les installations de thixo-moulage (thixocasting) et thixo-forgeage (thixoforging) 100% du matériel à transformer.

MII 18

THE MANUFACTURING PROCESS OF THIXO FORMINGS

The car industry is gradually designing and manufacturing improved and lighter car parts. One must say that aluminium alloy components have been constantly used for car building for quite a long time. The thixocasting process which has now been under development for approximately ten years, is used, for example, to manufacture large series of pressuretight injection distributors. Components made of hypereutectic aluminium-silicon alloys, manufactured by means of a thixoforming process (thixocasting or thixoforging), show excellent materiel properties. But what is the reason that such components considered as vehicle safety parts are not yet manufactured in series by the thixoforming process to be massively integrated in cars? At the time being it is the insufficient reliability of the process when manufacturing large series. For the demonstration of this fact, let us break down the entire manufacturing process of the thixo part into its different stages and examine each of these stages individually:

Process stages	Machine type	Remarks
Manufacturing of the raw material	Continuous casting with intense inductive stirring during solidification	Presently variations in quality, charge (batch) dependent, problems when several casting lines are operated in parallel
Cutting to the appropriate length	Sawing of precisely weighted sections	No problems, reliable
Heating into a partially liquid state	Induction heating plants	Currently in turntable plants, more or less effectively dependent on supplier
Loading the part into the forming chamber	Handling devices, robots or transfer loader	No problems
Forming in partially liquid state	Modified pressure-casting machines	Operating in a narrow process window
Finishing	Milling and drilling machines	No problems
Heat treatment	Ovens	No problems

Table 1 Manufacturing process of thixo formings

In the following, we proceed to an in-depth analysis of the heating process suitable for thixo alloys and the lay out of a machine design likely to ensure a reliable and reproduceable heating process for thixo bars.

IT IS THE MATERIAL TO BE HEATED WHICH DEFINES THE HEATING PROCESS PARAMETERS

To better understand the difficulties encountered when heating aluminium alloys up to the temperature limit area between the solid and liquid phase, we first consider the initial condition and the physical properties of the material when in the temperature range situated between solidus and liquidus.

Suitable alloys for thixo forming processes have a fine nodular structure. Grain size distribution differs along the radius of the cross section of the cylindrical thixo bar. In most cases, the closer the grains are to the core region, the more the grain size increases.

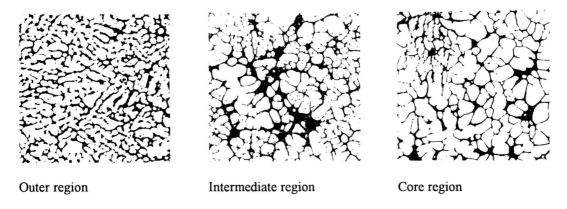

Outer region   Intermediate region   Core region

Figure 1.: Structure of the alloy AlSi7Mg0,3 suitable for the thixoforming process

The grain size directly influences the physical properties of the alloy, since among others such relevant material-specific parameters as electric and thermal conductivity depend from it. Heating tests have proved that the well known relationship between electric conductivity and thermal conductivity also applies to thixo alloys. Based on this important observation, one may use the known electric conductivity to control the heating process. A detailed description of the method used for determining the electric conductivity of the round thixo bar will be given further below in this paper. The constitution diagram, enthalpy curve, thermal conductivity diagram and curve of the effective specific thermal capacity give us a hint to the difficulties we will encounter when trying to heat a thixo alloy up to a liquid part of 40%.

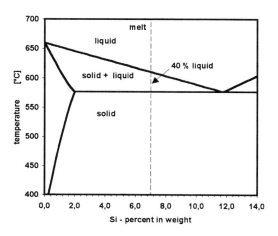

Fig. 2 Constitution diagram of an AlSi alloy.

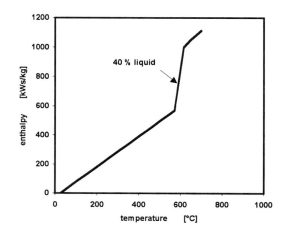

Fig. 3 Enthalpy of an AlSi alloy.

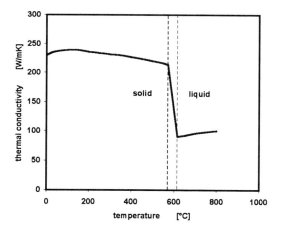

Fig. 4 Thermal conductivity of an AlSi alloy.

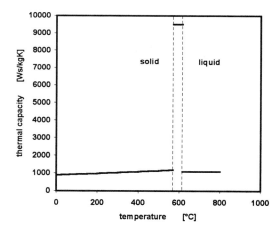

Fig. 5 Thermal capacity of an AlSi alloy.

We do not exactly know the properties and the behaviour of thixo alloys in the intermediate state between solidus and liquidus. Therefore, exact calculation of the temperature distribution or, even better, of the energy distribution within the thixo bar, is not possible. Nevertheless, a rough estimate performed by means of a FEM calculation gives us a valuable approximation.

For cylindrical thixo bars, the following differential equation is to be solved:

$$\frac{\partial \vartheta}{\partial t} = \frac{\lambda(r)}{\gamma \cdot c_P(r)} \cdot \left(\frac{\partial^2 \vartheta(r)}{\partial r^2} + \frac{1}{r} \cdot \frac{\partial \vartheta(r)}{\partial r} \right) + \frac{w_r(r, \rho, f)}{\gamma \cdot c_P(r)} \qquad 1$$

Definitions:
- λ thermal conductivity
- c_P effective specific thermal capacity
- γ specific weight
- ρ electric resistivity
- f frequency of the induced current
- w_r thermal source density

If we consider the existence of a straightforward skin effect and an energy introduction penetrating through the cylindrical surface being equivalent to the induction heating power generated within the material, the thermal source density contained in 1 may be described by formula 2:

$$w_r = \frac{2 \cdot q_H}{\delta} \cdot \frac{r_o}{r} \cdot \frac{e^{-\frac{2}{\delta}(r_o - r)}}{1 - e^{-\frac{2}{\delta}(r_o)}} \qquad 2$$

Definitions:

$\delta = \sqrt{\dfrac{\rho}{\mu \cdot \pi \cdot f}}$ Penetration depth of the current

q_H Density of the heating power introduced through the surface

A FEM programme has been used to calculate the non stationary temperature distribution according to the thermal source density distribution as defined in Fig. 6. For the purpose of this calculation, the introduced power has been decreased according to several pre-defined steps. The thermal source density distribution has been adjusted to each step. Variations in material characteristics have been replaced within the calculation by constant transitions which take place within a very small temperature interval. The surface and core temperature curve shown by Fig. 7 may be reached only by substantially reducing the input power in the course of the heating process. By the end of the heating, in our calculated example, the input power only amounts to 4% of the initial power introduced on start-up of the heating operation.

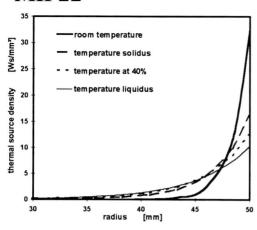

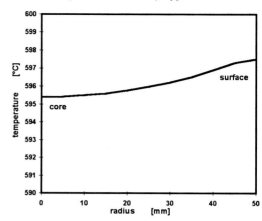

Fig. 6 Thermal source density distribution

Fig. 7 Temperature curve in time with stepped power input

Fig. 8 shows the calculated temperature curve along the radius. Assuming that the temperature range comprised between the solidus and the liquidus lines amounts to 45 K, a difference in temperature of 2 K would consequently lead the effective liquid part to differ by 4.5% from the expected liquid part. In our example, this would result in a liquid part of 40% in the peripheral area and a liquid part of 35% in the core region. The calculated temperature curve leads to the following conclusion: to ensure that the temperature does not exceed the upper limit at any moment, the power input has to be regulated with great precision. In the course of the heating operation, there should not be the slightest deviation from the set values.

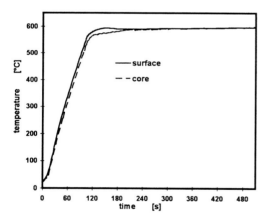

Fig. 8 Temperature distribution within the thixo bar after induction heating

The result is that time-dependent control of the power input is the sine qua non condition for the controlled heating process of thixo bars.

PROCESS CONTROLLED HEATING BY IMPLEMENTATION OF A PATENTED TECHNIQUE ENSURING MEASUREMENT OF THE EFFECTIVE POWER INDUCED INTO THE MATERIAL

The usual method of measuring the generator output power and time integration of this power does not make it possible to reach the required precision of dosing the energy to be induced into the thixo bar. This becomes obvious when taking a quick look at the efficiency and the possible metering errors within the partial load range. The electric efficiency of induction heating of individual thixo bars is lower than 35%. In other words, only 35% of the generator output power effectively heat the thixo bar. If we now consider that the power rating used during the major part of the total process duration amounts only to between 5 and 10%, application of the conventional method would lead to an inadmissibly high error rate. When using measurement appliances of precision class 0,2, according to VDE 0410, we obtain the following result:

$$\frac{0.002}{0.05 \cdot 0.35} \geq \text{Error} \geq \frac{0.002}{0.1 \cdot 0.35}; \text{ i.e. approx. } 11\% \geq \text{Error} \geq 5\%$$

This error rate may be considerably reduced by metering the power effectively induced into the material. Fig. 9 shows the basic circuit enabling such measurement.

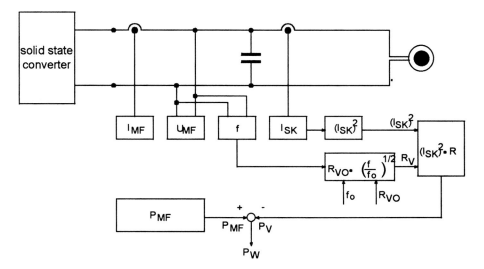

Fig. 9 Basic circuit enabling the measurement of the effective power induced into the material

Unlike the usual effective power measurement, this device additionally measures the current of the resonant circuit of the heating plant. The following steps may now be taken on this basis:

- Determination of the parameter representing all the losses depending on frequency and current. This is achieved by a single calibration test without material.

- Subsequent storage of the parameter within the measurement appliance.

- Calculation of the effective losses of the resonant circuit during the heating operation.

- On-line computing of the effective power induced into the material by subtraction of the generator power and the resonant circuit losses.

The above-described method enables us to evaluate precisely and at any moment the energy contained in the material by integration of the effective power induced into it. Therefore, the method also makes us aware of the energy-to-time historical curve as it takes place within the thixo bar. Measuring directly the energy condition of the material thus allows process controlled heating up to the solid-liquid state. Consequently, the indirect energy measurement through reading the temperature of the body is no longer required.

Continuous acquisition of the generator current and voltage, the resonant circuit current and frequency, all used for calculating the parameters generator power, effective power and energy induced into the material, provides additional means for process control, of which we finally would like to enumerate only the following:

- Permanent comparison of the above-mentioned data with reference data stored in a master file provides improved quality control throughout the heating process.

- Variations in the raw material are detected in the course of the heating operation. This feature allows one to compensate for such variations by suitable regulation of the heating process.

- In case of subsequent faults in the forming operation, suitable fault processing routines could be performed, taking into account the actual energy condition of the material.

All these measures have only one objective: the achievement of a reliable heating process of thixo bars with regard to the subsequent forming process. One should always bear in mind the necessity of properly taking into account all possible variations in the raw material so as to allow the subsequent process to produce 100% quality parts.

References

1. B. Wendinger and G. Schindelbacher: 'Verarbeitung und Eigenschaften von thixotropen Aluminiumlegierungen', *Gießerei-Praxis,* **1994**, 11/12, 317-326.

2. J.-P. Gabathuler and J. Erling: 'Thixocasting: ein modernes Verfahren zur Herstellung von Formbauteilen', *conference at ETH Zürich,* **1994**, ISBN 3-907967-05-4.

3. K. P. Young and R. Fitze: 'Semi Solid Metal Cast Aluminium Automotive Components', *3rd Intl. Conference on Processing of Semi Solid Alloys and Composites,* **1994**, Tokyo

4. H. Geisel: 'Wärmequellenverteilung und Temperaturverlauf beim induktiven Erwärmen', *VDI-Zeitschrift,* **1963**, 105 Nr.31 November

5. H.-G. Matthes: 'Qualitätssicherung beim Induktionshärten durch Erfassung der Wirkleistung im Werkstück', *Härtereitechnische Mitteilungen,* **1993**, Nr.49 Januar

Induction Heating For The Wire Industry

ROBERT J KIRKWOOD

Product Group Manager - Heat Treatment
Radyne Ltd Wokingham Berkshire England

Following a short introduction into the basic principles of induction heating applied to wire heating, several applications are reviewed.

Examples of single wire, multi wire and rope strand processes are discussed including in line hardening, tempering and coating operations. Specific equipments are identified for given wire sizes, speeds and temperatures.

Une brève introduction, sur les principes de base du chauffage par induction utilisé pour le chauffage des fils, est suivie de l'étude de plusieurs applications.

Des exemples de procédés monofilaires, multifilaires et à toronnage multiple sont analysés, y compris les opérations de durcissement, trempe et revêtement en ligne. Des équipements spécifiques sont indiqués pour des grosseurs de fils, des vitesses et des températures données.

Basic Electrical Principles of Induction Heating as Applied to the Wire Industry

An electrically conducting wire within an induction heating coil is heated due to the resistance of the wire to an alternating current induced in that wire by the coil (at the same current) in accordance with Ohm's Law. In magnetic materials the heating effect is amplified by hysteresis loss, the effect of which disappears at the Curie temperature (approximately 760°C). The current flows in the surface of the conductor to a depth dependent on the frequency of the alternating current, the resistivity and the effective permeability of the conductor. Heating deeper than the current flow is by conduction. As a general rule the depth of inductive heating should be somewhat less than half the conductor radius although with some compromises the depth is exceeded in practice. Where the inductive heating depth approaches or exceeds the conductor radius the induced currents cancel each other out nullifying the heating effect. Raising the frequency of the alternating current (typically 3 - 200 kHz) decreases the inductive depth of heating for a given material. Materials of higher resistivity and/or lower effective permeability raise the inductive depth. The effective depth of current penetration (p in mm.s) is given by

$$p = \frac{1}{20\pi} \sqrt{\frac{\sigma \times 10^{-2}}{\mu F}}$$

σ = resistivity in microhm centimetres
F = frequency in cycles per second
μ = effective permeability(=1 for non magnetic materials)

With wire of a known diameter and material properties and current penetration at the optimum one quarter diameter, frequency can be determined. Frequencies in excess of 200 kHz reduce the viability of the application with current technology.

These principles in conjunction with some limitations of available equipment tend to prescribe the range of wire heating applications. Carbon steel wires above 0.8 mm diameter heated to c750°C maximum generally present no problems. Above 750°C the carbon steel minimum wire diameter increases to c4.0 mm. Minimum diameters increase and maximum temperatures decrease for most other materials to be viably induction heated. Advances in induction heating giving increased power at higher frequencies for reducing capital costs and increasing operating efficiencies are bringing more applications within the scope of induction heating.

Wire Heating Applications

Induction heating is applied to a wide variety of wire processes treating either single wires, multi wires running in parallel, or wires stranded into rope.

Applications have included hardening, tempering, annealing, low relaxation processes, coating processes, and preheating, among others.

Compared to alternative heating methods lower maintenance costs, lower scrap rates, and improved quality have been cited as benefits of induction heating by existing users. All applications are instant switch-on with no equipment warm up times, no heat transfer to handling structures and are suitable for gas atmospheres. Being in line reel to reel, processes batching can be eliminated reducing work in progress. Heating lengths can be shorter than many alternative methods, varying generally between 1.0 and 5.0 metres with applied power controlled from line speed; adjusted for wire diameter.

Hardening and Tempering

In line hardening and tempering of carbon steel wire using induction heating is widely carried out with a variety of end users for the wire. Typically the wire is heated to, say, 950°C and quenched to harden; reheated to, say, 350°-650°C and re-cooled to temper. Heating rates, temperatures and cooling methods are all variable to suit required properties of the end product.

The heating equipment for a typical line is shown in fig.1. In this instance the equipment was for the treatment of deformed bar for reinforcing concrete structures to give a low relaxation high yield strength wire.

Figure 1
3 Stage Heating
For In-Line
Harden&Temper

Wire diameters range from 5.0 to 13.0 mm diameter with speeds up to 70 metres-minute and from 0.6 to 2.0 tonne-hour subject to diameter. For this output heating for hardening is carried out in two stages - stage 1 up to approximately 750°C determined by the Curie temperature for the material, and stage 2 above Curie to the required maximum temperature. Stage 1 is a 250kW power unit at 10 kHz, stage 2 a 180kW power unit at 50 kHz or 200 kHz. Heating for tempering is carried out with a 150kW unit at 10 kHz. Immediately after the second and third heating stages the wire is sprayed by high pressure water jets which are an integral part of an in line recirculating water system reducing the wire temperature to c40°C. Subsequently the wire is air dried with in line air wipes. Total line length from entrance to first stage heating coil to exit of second air wipe is c20 metres.

Typical heating coil lengths for each stage are 1.8 metres. At the 10 kHz heating stages a single size coil bore gives good heating efficiency over the full wire size range. For the post Curie heating stages coils are interchangeable with bores matched to certain wire diameters. Figure 2 shows a typical coil assembly arranged on its mounting frame to suit a 1000 mm wire pass height. With quick release power and cooling water connections coil changes can be effected within 2 minutes. Each coil assembly includes wire guides and an internal ceramic liner with approximately 7.0 mm minimal radial clearance to the wire.

Figure 2 - Typical Coil Assembly

The whole installation is P.L.C. controlled, with applied powers and wire tension for a given wire size, material specification and required end product computed over a wide speed range during commissioning. All parameters are adjusted automatically in response to line speed changes during running to ensure a constant end product. Running data including wire temperatures is constantly monitored and compared with known requirements.

This specific application is representative of the features which may be incorporated in single wire lines for a variety of processes. Wire sizes, materials, temperatures and line speeds are not limited to those indicated. Heating coils may be varied in respect of length, number of turns, and diameters to cater for a wide range of requirements including maintaining temperatures for limited times subsequent to raising to a preset temperature. Control flexibility is such that, with appropriate mechanical handling, processes can be reel to reel without line stoppages. Parameters applicable to batch heating processes are not necessarily applicable with the faster heating rates obtainable via induction heating.

Multi Wire Applications Including Tyrecord Diffusion

Induction heating is applied to a number of wire processes where typically 10-24 wires run in parallel and are simultaneously heated within a single flattened oval heating coil. Applications include wire coating, heating prior to galvanizing and, with some provisos, annealing.

In general wire diameters are small, typically 0.8 - 2.0 mm but ranging up to 6.0 mm. Since most applications of this type involve below Curie temperatures current cancellation is not generally a problem with operating frequencies of around 20 kHz. With this frequency current cancellation is used to advantage with wires up to 2.0 mm diameter preventing temperatures in excess of Curie however long a (broken) wire might be within the heating coil.

The most widespread multi wire application is tyre cord diffusion. Wires pre-coated with copper and zinc are heated to c550°C melting the coatings to form brass which diffuses into the base wire. Wires are run with a typical inter axis dimension of 15.5 at 25.0 mm. Best results are obtained with a profiled heat input with a high initial rate followed by a slower final rate. Typically for 18 wires at 15.5 mm inter axis with throughput based on DV=70 (D = diameter) in millimetres and V = speed-metres per second) 80kW and 40kW output power units have been used for the respective stages.

Figure 3 shows a typical installation. 2 coil assemblies each 2.5 metres long heat the wire. Each coil assembly is connected via busbars to power unit spositioned below the coils in a basement. Downstream insulated muffle zones slow the rate of heat loss. Between the coils and muffles gas tight boxes permit easy wire threading and are entre points for gas atmosphers if required. For inter axis dimensions in excess of 15.5 mm each wire passes through an individual ceramic liner. Figure 4 shows a coil cross section. Inter axis dimensions below 15.5 mm are possible and in fact result in higher heating efficiencies but the individual liners are removed. Threading up consequently is more difficult but this does not necessarily preclude inter axis dimensions down to 6.0 mm.

Figure 3 - Typical Diffusion Installation

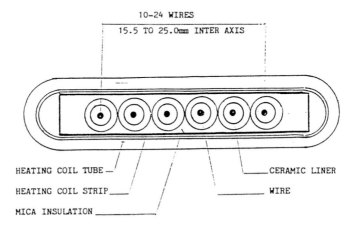

Figure 4 - Typical Multi Wire Coil Cross Section

Low Relaxation

This process requires the heating of single wire or rope strand to approximately 450°C while under tension and is essentially similar to the temper stage described in the first application. Variations on equipment are subject to the nature of the wire or strand handling equipment. If the direction of travel of the product is reversible the heating coil is static, as previously described. With non reversible travel of the product the coil assembly is mounted on a motorized carriage.

Figure 5 illustrates a moving coil assembly coupled to a static power unit viewed from the strand input end. Not shown downstream of the coil at the end of the track is a static quench trough in which the strand is rapid cooled after a short air cooled zone. In the event of a line stoppage a product in this air cooled zone is likely to be inadequately processed. Prior to restart up this product is cooled to ambient by quench troughs mounted at each end of, and moving with, the coil assembly. On restart up the coil assembly is moved progressively further from the static quench at a rate corelated to strand speed and stopping at a distance related to normal strand running speed.

Typical equipments consist of either a 500kW 3 kHz power unit coupled to a 2.0 metre coil or a 500kW 10 kHz power unit coupled to a 4.0 metre coil dependent on product size and ultimate mechanical properties required. With the former rates of 150, 115, 100 and 75 metres-minute for strand sizes 9.5 mm (0.375 inch) 12.7 mm (0.5 inch) and 15.2 mm (0.6 inch) and 17.8 mm (0.7 inch) respectively are attainable with a 50 mm bore diameter to the coil liner.

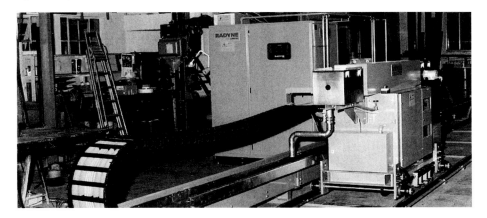

Figure 5 - Typical Low Relaxation Installation

Other Applications

The previous examples are all applications involving carbon steel wire. For the same material similar equipments are available for pre-heating prior to galvanizing, drying, curing of coatings and annealing (either of the final product or interstage for successive reductions). In some instances to overcome limitations of small diameters heated above 750°C, induction heating is used up to that temperature followed by heating in conventional furnaces above that temperature with consequent shorter line lengths.

Wires of other materials can be heated but generally at lower efficiencies. Figure 6 shows a limited comparison of inductive efficiencies for some materials heated to different temperatures with specific coil lengths, diameters, and operating frequencies. This table illustrates the significance of wire diameter to coil diameter above the Curie temperature. This is further magnified when power unit costs for given frequencies are superimposed on the table. As a guide for basic power units at 100kW costs increase by factors of approximately 1.2, 1.5 and 2.3 for 10 kHz, 25 kHz and 50 kHz respectively relative to 1 kHz.

MATERIAL	TEMP	DIA	SPEED	COIL LENGTH	COIL DIA.	FREQUENCY	OVERALL EFFICIENCY
	°C	MM		MM	MM	kHz	
STEEL	20-600	0.8	DV=50	2000	20	25	.47
		0.8	DV=100	2000			.38
		0.8	DV=100	5000			.55
		2.0	DV=50	2000			.73
		2.0	DV=100	2000			.68
		2.0	DV=100	5000			.74
	20-750	5	1T-HR	2000	20	10	.72
		9	"	"	20	10	.84
		5	"	"	25	10	.68
		9	"	"	25	10	.81
	750-900	5	"	"	20	50	.24
		9	"	"	20	50	.57
		5	"	"	25	50	.20
		9	"	"	25	50	.52
NON MAGNETIC STAINLESS	20-600	5	"	"	20	50	.25
BRASS	20-600	5	"	"	20	50	.23

Figure 6 - Material Efficiency Comparisons

Installations for the Continuous Induction Heat Treatment of Wires

ARTUSO I.* - DUGHIERO F.** - LUPI S.** - PARTISANI S.** - FACCHINELLI P.***

* - ATE - Applicazioni Termo-Elettroniche - Vicenza (Italy)
** - Dept. of Electrical Engineering - University of Padua (Italy)
*** - GCR Engineering S.p.A. - Milano (Italy)

Abstract

In the production of high-quality steel wires an increasing role is played today by continuous induction heat treatments, which allow to achieve very high production rates with reduced installation space requirements, minimum energy consumption and better working environment.

Among various heat treatments, the stress relieving process for stabilisation of wires and strands is considered in the paper. This process requires through heating of steel wires up to 400-600 °C with a final uniform temperature distribution in the wire cross-section with a tolerance of about 3 %.

The attention is focused on the choice of the main installation parameters which assure the temperature uniformity within the required tolerance.

The analysis has been performed by the use of modern mathematical modelling programs, giving as an example a set of diagrams with typical results.

An important conclusion of the calculations made on existing lines, which have proven to assure optimum metallurgical results, is that a considerable reduction of the length of the heating line can be obtained making reference to the temperature uniformity obtainable by the soaking process in air occuring between the exit of the inductor and entrance of the water quenching unit.

Finally a modern industrial installation of an italian manufacturer, designed on the basis of the above criteria and characterized by very high conveying speeds (up to 420 m/min) is shortly described.

Résumé

Dans la production de fils d'acier de haute qualité, les traitements thermiques par induction en ligne ont de nos jours un rôle grandissant, puisque ils permettent d'obtenir des productions très élevées tout en occupant des espaces très réduits, avec une consommation d'énergie minimum et des conditions de travail améliorées.

Dans ce travail en particulier le processus de stabilisation des fils d'acier est pris en consideration.

Ce processus exige le chauffage continu des fils jusqu'à 400-600 °C avec une distribution finale uniforme de la température et une tolérance d'environ 3%.

On a considéré en particulier le choix des paramètres principaux du projet qui permettent de garantir la distribution finale uniforme de la température.

L'analyse, développée en utilisant des programmes numériques de calcul, a donné un set de diagrammes utiles pour le projet.

Une importante conclusion des calculs effectués sur des lignes en service, qui donnent des résultats métallurgiques optimaux, est que l'on peut obtenir une considérable réduction de la longueur de la ligne, en se basant sur le processus d'homogénéisation de la température qui a lieu entre la sortie de l'inducteur et l'entrée de l'unité de refroidissement.

Enfin on décrit une moderne installation construite per un constructeur italien, conçue sur la base des critères sus-mentionnés et caractérisé par des vitesses d'avancement très elevées (jusqu'à 420 m/min).

1. Introduction

The main technological problems arising in the heat treatment of steel wires up to temperatures of 400-600 °C have been already dealt with in some papers. [1-4]

Although the results given in the above papers constitue a good general basis for the design of this type of installations, they have been obtained in years in which the development of HF solid state converters was at a very beginning and numerical packages for the mathematical modelling of thermal transients in the induction heating of ferromagnetic workpieces were not sufficiently tested and currently available.

In this paper the choice of some important design parameters is reconsidered in the light of the present technological acheevements, giving - as an example of typical results available today - a series of data related to the induction stabilization process of steel wires (for diameters ranging from 3 to 9 mm) performed with heating up to 400 °C.

The data have been used for designing an installation for the continuous treatment of steel wires UNI-C70 with production speed up to 420 m/min. A short description of the installation is also included.

2. Stabilization process of wires and strands

To improve the mechanical properties of wires and strands a special treatment has been developed that is called stabilization. Essentially, this treatment consists of applying "plastic tensile strain" during the stress-relieving heating operation. Variations in the tensile strength are considerably influenced by the "straining temperature". The combination of strain and heat causes significant changes in the mechanical properties when compared with those of wire that has simply been drawn and stress-relieved. As known, stress-relieving is a heating operation, followed by cooling, which serves to remove internal stresses. Residual internal stresses are normal during wire drawing because the deformation is essentially transformed into heat. The wire is cooled on the capstans by transfer of the heat from the wire to the capstan and/or to the fan-cooled air. Consequently, because the phenomenon occurs in very brief time spans, there is a temperature gradient between the core and the surface of the wire. This results in residual tensions. Deformation during drawing causes orientation of the structural grain, with particular reference to the crystals. This contributes to hardening, which translates into an increase in the tensile strength. The relation between deformation and increased tensile strength has been and is the subject of different theories that we will not go into here. Sufficient is to say that the relation between deformation (reduction of the area) and tensile strength is influenced by many factors, such as carbon content, cross-section reduction per draft, overall reduction and, to a lesser degree, drawing speed, cooling capacity, lubrication, etc.

During heating of a drawn wire, variations in the mechanical characteristics depend in large part on the carbon content and preceding degree of work hardening (percentage of area reduction). In the case of wire and strand for prestressed concrete, we are dealing with steel wires or strands having a carbon content of 0.65-0.80% and overall reduction of 80-85%. Through heating to about 400°C for a time that varies from 0.5 to 2.0 seconds, increased tensile strength is obtained accompanied by considerable increase in the elongation at break. Other mechanical properties are also influenced by stress-relieving at this temperature, such as torsions, bends, etc. For the specific application of wire or strand for prestressed concrete, the wire must be able to resist high stresses for long periods of time without increasing solicitations. This means higher resistance to relaxation. In substance, the effect of stress-relieving is to modify the elastic properties of the wire. If, during heating for stress-relieving, the wire or strand is subjected to pulling strain near the elastic limit of the element at that temperature, as stated above, "stabilization" is achieved. The applied strain, together with the heating, also makes it possible to obtain perfectly straight wire and achieve tensile strengths that are similar or superior to those of the initial product, depending on the process temperature. No theory has yet been developed that explains with absolute certainty the mechanism that regulates the phenomenon. Metallographic research has shown variations in the crystalline interface, with complex dislocations which, however, cannot by themselves justify the phenomenon. In any case, the practical results of the treatment are evident and can be measured, and they are extremely interesting for the specific application of the product in question. The combination of temperature and strain causes a profound modification of the proportional limit, which increases by up to 80-85% of the ultimate tensile strength, and elongation can reach 6-7%. Even more fundamental in this specific case is the enormous increase in resistance to relaxation which, for stabilized material, is 4 to 5 times superior in respect to wire that has been drawn or simply stress-relieved. On modern production lines, "stabilization" is done as part of a continuous, straight through process that combines stretching of the wire or strand with heating that is controlled in induction heaters.

3. Main parameters of the induction heating process
- *Frequency choice*

It's well known from the induction heating theory that the choice of the optimum frequency must be done taking into account the conflicting requirements of achieving maximum efficiency in energy transfer and limited thermal gradients between the surface and the axis of the wire.
Such requirements are met with frequency values satisfying the following conditions:

(1) $$m = \frac{d}{\sqrt{2}\,\delta} = 2.5 - 4.5$$

with: d - wire diameter; $\delta = \sqrt{\rho/\pi\mu f}$ - induced currents penetration depth; ρ, μ - electrical resistivity and magnetic permeability; f - frequency.

In the stabilization process of steel wires, an accurate prediction of thermal gradients during heating can be done only taking into account the variations of electrical resistivity and permeability with temperature and - as regards permeability - with the local magnetic field intensity.
As a consequence the final temperature differential ΔT between surface and axis is a function - for the different values of m - also of the magnetic field intensity H_0 produced by the exciting coil at the surface of the wire.

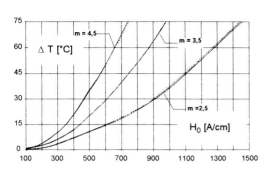
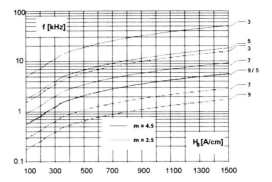

Fig.1 - *UNI-C70 wire: temperature differential as a function of H_0 for various values of m and diameter d*

Fig.2 - *UNI-C70 wire: frequency values as a function of H_0 for various values of m and diameter d*

In figure 1 such dependence is given as an example as a function of H_0 for the stabilization process at 400 °C of UNI-C70 steel wire.

The diagram allows the designer to specify the range of values of H_0 satisfying conditions (1), for a prefixed maximum admissible value ΔT. Starting from the same range of values of H_0, the corresponding range of frequencies can be specified for the different diameters d from diagrams similar to figure 2.

- **Inductor efficiency**

A parameter of fundamental importance in the installation design is also the efficiency η of the energy transfer from the inductor coil to the wire.

The well known relationship for η :

(2) $$\eta = \frac{1}{1 + \alpha \sqrt{\frac{\rho_i}{\rho \mu}} \frac{A_i}{\sqrt{2} P}} = f(m, H_0, \alpha)$$

with: ρ_i, μ_r - inductor resistivity and wire relative magnetic permeability; A_i, P - active power coefficients of inductor and load; α - coupling coefficient between inductor and wire, shows that for well designed inductors, when the type of steel constituting the wire is specified, the efficiency depends not only on m and H_0 but also - to a great extent - on the coupling coefficient α.

The diagrams of figure 3 give, for the same examples considered in figures 1 and 2, the values of η as a function of the above parameters. They show the following:

- the heat treatment of ferromagnetic steel wires below Curie point are characterized always by relatively high efficiency values

- in the range of H_0 previously defined, the highest efficiency can be obtained by the lowest values of H_0 and α and the highest value of m.

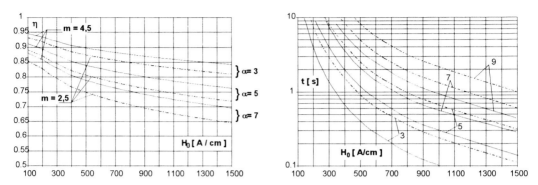

Fig. 3 - *UNI-C70 wire: efficiency values as a function of H_0 for various values of m and coupling coefficient α*

Fig. 4 - *UNI-C70 wire: heating time as a function of H_0 for various values of m and diameter d*

- **Heating time**

The analysis of the heating transients gives also, for the different wire diameters, the time values necessary to reach the prefixed final temperature with specified values of H_0 and m.

For the case considered previously the heating times are obtainable from the diagrams in figure 4.

- **Inductor length**

All other parameters being specified, the inductor length ℓ is strictly dependent on the production rate M. Figure 5, which gives the production rate per unit inductor length as a function of H_0 in the optimum range of m, allows the designer to determine the inductor length according to the required value of M.

- ***Example*** - Stabilization of UNI C70 wire: d=5 mm; coupling coefficient: α=5; production rate: M=2000 kg/h; (see table I).

The example clearly shows, on one hand, that a great limitation of the line length ℓ can be obtained allowing relatively high temperature differentials at the exit of the inductor.

ΔT [°C]	20		40	
m	2.5	4.5	2.5	4.5
H_0 [A/cm]	700	400	1.100	540
f [kHz]	3.0	5.5	4.5	7.5
η	0.78	0.85	0.74	0.83
t [s]	1.3	2.0	0.6	1.0
M/ℓ [kg/h m]	400	250	950	500
ℓ [m]	5	8	2.1	4

Table I

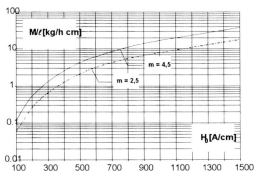

Fig. 5 - *Production rates per unit inductor's length as a function of H_0 for various m*

On the other hand, the experience done on existing lines which have proven to assure optimum metallurgical results allows to state that the temperature uniformity required at the entrance of the quenching unit can be easily obtained relying on the transient soaking process in air occuring between the exit of the inductor and the entrance of the quenching unit itself. This means that is possible to reduce considerably the length of the heating line allowing a relatively greater temperature differential at the inductor outlet and positioning conveniently the quenching unit in order to assure the desired temperature uniformity.

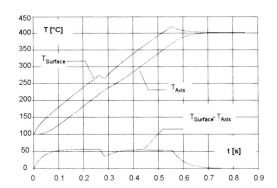

Fig.6 - *Transient temperature distribution in UNI-C75 wire*

An example of this criterion is shown in figure 6, which refers to the transient temperature distribution during the stabilization heat treatment from 100 to 420 °C of a 5 mm UNI-C75 steel wire, obtained in a line comprising an induction heater constituted by two inductors (1300 mm length; 150 mm spacing). The test done on the wire after stabilization have confirmed the optimum metallurgical results.

4. Stabilizing line for production of prestressed concrete (p.c.) steel wires

The line described here is designed for the production of low relaxation prestressed concrete steel wires with diameters ranging from 3 to 9 mm and max. conveying speed 420 m/min.

As shown in figure 7, the line includes the following equipment: 1-motorized pay-off stand; 2-wire straightening group; 3-induction furnace for p.c. wire; 4-water cooling and air drying unit; 5-haul-off capstan; 6-pinch wheel pulling unit; 7-guide channel, flying rotating shear, guide tubes to coilers; 8-basket coilers.

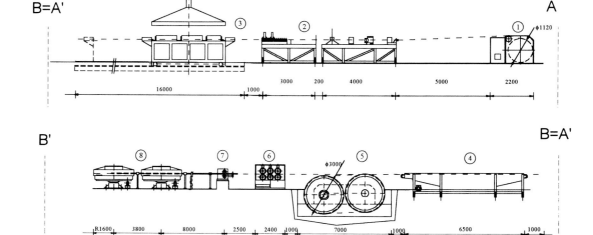

Fig.7 - *Layout of the line*

Technical data of the line: wire diameters: 3 to 9 mm; tensile strength: 160-240 kg/mm²; working speed: 200-420 m/min; max. straining force: 5000 kg; capstan motor power: 200 kW; max. coils weigth: 3000 kg; stress relaxation(1000 hrs): max. load loss 2.5% at 70% of actual breaking load.

Induction heating equipment

The purpose of this continuous operation induction heater is to stabilize, or "stress-relieve" prestressed concrete wires of UNI-C70 steel by heating them at a maximum temperature of 420 °C. The induction heating equipment is constituted by:

- one heating unit, composed by a set of 3 equally spaced inductors (coil length: 3x1100 mm) as shown in figure 8. The unit is mounted on a trolly which allows a longitudinal displacement of 10-12 m to recover the part of wire not treated during stopping and starting of the line. The unit contains: -a medium frequency (M.F.) inverter group with modular water-cooled SCR assemblies; -a firing and regulating unit with PLL frequency control; -a low loss, water-cooled M.F. capacitor bank; -connecting busbars; -starter unit; -measuring and protection devices; -inlet/outlet water cooling thermostats and flow-meters.
- a solid state frequency converter rated 600 kW/8 kHz
- a heat-exchanger water cooling assembly
- an operator control desk with instrumentation board and PC. A regulation algorithm is implemented which changes the power reference level according to the wire conveying speed and the material temperature measured by optical pyrometers near the heater outlet. For each

product to be processed a working schedule is provided which includs the relevant operating parameters from which the control system elaborates the power reference level of the frequency converter. The measured temperature value is used as a further control to obtain fine setting. The result is a highly precise control of the final temperature of the process, both in normal and transient (stopping and starting) working conditions. The operator control desk with functions controlled by microprocessor allows to set and monitor the line parameters (diameter of wire or strand, temperature, efficiency, etc.) and record the production data (e.g. line speed, temperature, pull/elongation) for subsequent statistics and quality assurance.

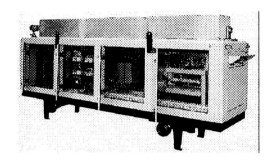

Fig. 8 - *Heating unit*

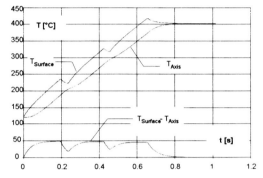

Fig.9 - *Transient temperature distribution in UNI-C70 wire*

Technical data of the induction heater

Wire diameters: from 3 to 10 mm; Max. conveying speed: 420 m/min (for 4 and 5 mm wire); Max. throughput: 6500 kg/h (9.4 mm); Total installed power: 800 kVA; Medium frequency output power: 600 kW; Frequency: 6-8 kHz; Coil length: 3x1100 mm; Max. treatment temperature: 420 °C; Temperature tolerance (in steady state conditions): 2%.

The design transient temperature distribution for the wire with diameter d=7 mm corresponds to the diagrams of figure 9.

References

1. **M. NEIRYNK, N.V. BEKAERT**: "L'Electrothermie en tréfilerie d'acier", *VIII Congrès Int. d'Electrothermie*, Liège, 1976, sect. II-c, n.4
2. **J. REBOUX, B. LAPOSTOLLE, J.C. BRUNE**: "Contribution du chauffage par induction à la modernisation des lignes de traitement thermique dans les tréfileries de fil d'acier", *Journées d'Etude*, Versailles, 5-6 Avril 1978
3. **J.P. METAIL**: Induction moyenne fréquence appliquée au chauffage de fil - Etude des paramètres de rendement énergétique", *Journées d'Etude*, Versailles, 5-6 Avril 1978
4. **L. MAIFREDY, F. FOUGERES, M.THEOLIER**: "Influence d'un chauffage très bref à mi-chaud par induction moyenne fréquence sur la structure et sur l' aptitude au tréfilage des fils d'acier eutoctoide déjà fortement écrouis", *Int. Conference on Induction Heating and Melting*, Liège, 2-6 October 1978, n.31

RÉALISATION DE MOULES DE ROTOMOULAGE PAR ÉLECTROFORMAGE SOUS COURANTS PULSÉS

PRODUCTION OF ROTOMOULDED MOULDS BY ELECTROFORMING WITH PULSED CURRENTS

Camel AOUAISSIA
Responsable R & D
CTAA FRANCE
17 rue des Prés Heyd
90200 GIROMAGNY
FRANCE

Patrice BERCOT, Jacques PAGETTI
LABORATOIRE DE CORROSION ET TRAITEMENT DE SURFACE
32 rue Mégevand
25030 BESANCON
FRANCE

Le rotomoulage ou moulage par rotation est une technique permettant l'obtention de corps creux de toute dimension à partir de matières thermoplastiques ou thermodurcissables. La mise en forme de ces dernières dépend étroitement de la qualité du moule, aussi, le choix du mode de fabrication de cet outillage, de même que la nature du métal ou alliage pour le réaliser, sont fondamentaux si l'on veut obtenir des pièces rotomoulées de qualité.

Dans notre cas, les moules sont réalisés par la technique de l'électroformage qui est un procédé permettant de produire ou de reproduire un objet métallique ou plastique par voie électrolytique à partir d'un modèle que l'on élimine en fin d'opération.

Cette technique, très ancienne (première référence signalée en 1838), conduit en utilisant un courant continu, à des pièces électroformées présentant deux défauts caractéristiques qui sont les "effets de pointe" et les "effets de creux".

Pour intervenir sur ces deux phénomènes, nous avons retenu la technique des courants pulsés avec inversions anodiques qui permet de réduire, de façon spectaculaire, les "effets de pointe" tout en contribuant à réguler les lignes de courant de l'installation électrolytique, minimisant ainsi les "effets de creux".

Rotomoulding or rotational moulding is a technique which allows to obtain hollow part of every kind of dimensions from thermoplastic or duroplastic material. The shaping of the latter depends on the quality of the mould. The choice of the technique of manufacturing, the nature of the metal or alloy are very important if you want to obtain good quality rotomoulded parts.

In our case, the moulds are manufactured by the electroforming technique which is a process which allows the production or the reproduction of a metallic or plastic object by electrolytic way, from a model which is eliminated at the end of the process.

This very old technique (first reference known in 1838) leads, by using a direct current, to electroformed parts presenting two caracterictical disadvantages : the "dent effects" and the "hollow effects".

To operate on these two phenomenons, we chose the technique of pulsed currents with anodic inversions which allows to reduce spectacularly, the "dent effects" conducing to regulate the current lines of the electrolytical installation, minimizing the "hollow effects".

Le rotomoulage ou moulage par rotation est une technique mal connue concurrençant l'injection pour la production de petites et moyennes séries. Ce procédé permet d'obtenir des corps creux dont le volume peut varier d'un à plusieurs milliers de litres dans certains cas.

Décrivons succinctement le cycle de production. La matière à transformer est introduit en quantité déterminée dans un moule qui animé de mouvements de rotation selon deux axes perpendiculaires sous faibles vitesses (10 à 20 tours par minute) va donner naissance après des temps de chauffe et de refroidissement appropriés, à une pièce prenant la forme de l'aspect de l'intérieur du moule. Ces différents cycles de production sont explicités sur la figure 1.

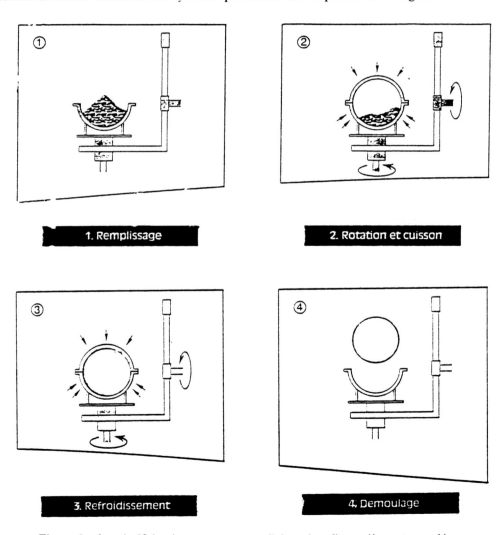

Figure 1 : descriptif des étapes permettant l'obtention d'une pièce rotomoulée.

Ce mode de transformation des matières plastiques dépend étroitement de la qualité du moule qui est à la base de la transformation du matériau. Le choix du mode de fabrication de ce moule ainsi que la nature du métal ou alliage le constituant, sont fondamentaux si l'on veut obtenir des pièces rotomoulées de qualité.

Dans notre cas, nous avons à produire des pièces en PVC plastifié dont l'aspect de surface doit copier fidèlement le cuir, c'est pour cette raison que nous avons choisi de produire nos moules ou outillages par électroformage de nickel. Le choix de ce matériau s'explique par ses propriétés mécaniques, thermiques et sa capacité de reproduction d'un motif avec une extrême fidélité.

Intéressons-nous à présent aux mécanismes de l'électroformage et de son application aux courants pulsés.

L'électroformage est un procédé permettant de produire ou reproduire un objet métallique ou plastique par voie électrolytique à partir d'un modèle (mandrin) que l'on élimine en fin d'opération.

L'un des grands avantages de cette technique est la reproduction extrêmement fidèle de l'état de surface du modèle initial.

Le fait d'utiliser un mandrin fusible ou non permet d'obtenir des objets aux formes complexes qu'il serait très difficile d'obtenir avec un autre procédé.

Les principaux métaux électroformés sont le cuivre et le nickel. Il est également possible d'utiliser les métaux précieux tel que l'or ou l'argent. Les applications sont extrêmement nombreuses et nous ne citerons ici que des secteurs d'utilisation comme l'aérospatiale, l'imprimerie, l'industrie du disque (vinyle et CD), la bijouterie, . . . et l'automobile.

Parmi les exemples pouvant exister dans ce dernier secteur, nous mentionnerons la réalisation d'électrodes destinées à la fabrication des moules d'injection par électroérosion et la fabrication des moules de slush moulding et de rotomoulage.

Dans le cadre de cette étude, nous nous sommes plus particulièrement intéressés à ce dernier point. En effet, réalisant en interne nos outillages de rotomoulage, nous sommes confrontés régulièrement aux problèmes liés à la technique de l'électroformage à savoir les manques d'homogénéité des épaisseurs des parois des moules électroformés.

Par conséquent, il y a là naissance à d'importants coûts dûs aux poids de matière à transformer ne peuvent pas être revus à la baisse de même que les temps de finition particulièrement importants (usinage, renfort) si l'on veut mécaniquement optimiser ces outillages.

Afin de réduire tous ces travaux, en aval de la fabrication des moules, nous avons décidé d'examiner, via une collaboration Entreprise-Université, les solutions pouvant exister en amont et durant la fabrication de ces outillages. Après des études préalables, nous nous sommes aperçus que si nous agissions directement sur le moule pendant le temps d'électrolyse, nous pourrions résoudre les problèmes d'épaisseurs de parois. Une des pistes prometteuses fut de travailler sur la nature même du courant électrique indispensable pour réaliser une électrolyse.

Les premières références de pièces électroformées ont été signalées en 1838 et depuis cette date, les utilisateurs de cette technique ont toujours employé le courant continu. Ce dernier possède un certain nombre de désavantages qui se traduisent par deux défauts caractéristiques antagonistes intitulés "effets de pointe" et "effets de creux".

Examinons, de façon plus détaillée, ce que l'on entend par "effets de pointe" et "effets de creux".

Les "effets de pointe", comme l'indique la figure 2, se manifestent dans la cellule électrolytique dès lors qu'il existe sur la pièce placée en cathode, des angles saillants. Les lignes de champ vont se canaliser sur ces pointes et l'épaisseur en métal déposé va augmenter extrêmement rapidement au détriment des autres zones de la pièce. Il n'est pas rare d'obtenir des excroissances de plusieurs dizaines de centimètres pour une épaisseur moyenne de 2 à 3 millimètres. Notre but est d'obtenir une pièce électroformée régulière en épaisseur, aussi, est-il nécessaire, une fois cette dernière sortie du bain, d'effectuer un traitement mécanique (usinage) pour éliminer ces épaisseurs localisées. Le coût de ce traitement vient s'ajouter à celui du métal perdu par cette opération.

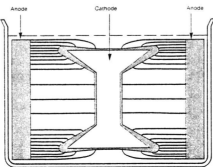

Figure 2 : distribution des lignes de courant entre l'anode et la cathode : "effets de pointe"

Les "effets de creux", comme l'indique la figure 3, se manifestent dans les zones où les lignes de champ rencontrent des angles rentrants. La conséquence de ce phénomène est que localement, l'épaisseur du dépôt électrolytique sera très faible (quelques dixièmes de millimètres au lieu des 2 à 3 millimètres moyens). La pièce électroformée va donc présenter des zones de fragilité et il sera nécessaire de la renforcer localement (brasage). Cette opération, d'un coût non négligeable, aura des répercussions sur les transferts thermiques du moule à la matière lors de l'opération de rotomoulage.

Les "effets de creux" seront d'autant plus importants que les "effets de pointe" se manifesteront.

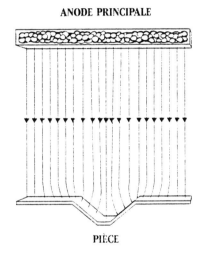

Figure 3 : distribution des lignes de courant entre l'anode et la cathode : "effet de creux"

Comme nous l'avons mentionné plus haut, il existe des artifices pour remédier à ce type de problèmes. Nous les citerons ici pour mémoire, cette liste n'étant nullement exhaustive.

Utilisation d'écrans conducteurs ou non (voleurs de courant) pour atténuer les "effets de pointe" et utilisation d'anodes auxiliaires ou d'électrodes bipolaires pour réduire les "effets de creux".

Après examen et étude de ces diverses techniques, nous nous sommes aperçus qu'en agissant sur les "effets de pointe", les lignes de champ à l'intérieur de la cuve électrolytique étaient mieux régulées et, par voie de conséquence, autoriseraient une meilleure contribution à la résolution des "effets de creux".

Pour intervenir sur ce phénomène, nous avons retenu la technique des courants pulsés dont nous allons expliciter la nature.

Si l'on examine la figure 4, nous nous apercevons de l'existence de deux zones. L'une caractérise le dépôt (phase cathodique) et l'autre le polissage électrolytique (phase anodique). La surface délimitée par l'aire anodique est inférieure à celle de l'aire cathodique, par conséquent, nous aurons bien un dépôt électrolytique au cours du temps, mais ce dernier verra ses "effets de pointe" très atténués.

En effet, les courants pulsés avec inversions anodiques permettent d'obtenir un polissage électrolytique sur les excroissances caractéristiques provoquées par les "effets de pointe" et ainsi les lignes de champ, mieux régulées, permettent d'obtenir un dépôt beaucoup plus régulier en épaisseur.

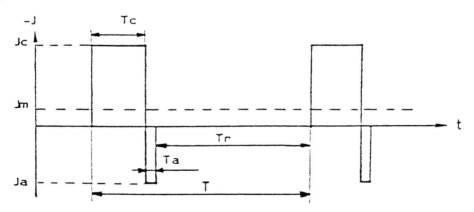

T : période
T_c : Temps d'imposition du courant cathodique
J_c : Densité de courant cathodique
J_m : Densité de courant moyenne
T_r : Temps de repos
T_a : Temps d'imposition du courant anodique
J_a : Densité de courant anodique

Figure 4 : courants pulsés avec inversions anodiques

Les résultats obtenus dans une cuve expérimentale de nickelage de 100 litres, puis dans un bain industriel de 5000 litres ont été les suivants :

- diminution des épaisseurs de l'ordre de 70 à 80 % dans les zones présentant des "effets de pointe"
- augmentation des épaisseurs de l'ordre de 25 % dans les zones présentant des "effets de creux"

Ces valeurs découlent d'un comparatif établi entre une pièce réalisée en courants pulsées avec inversions anodiques et une pièce témoin réalisée en courant continu tel qu'on peut le voir sur la photographie ci-après.

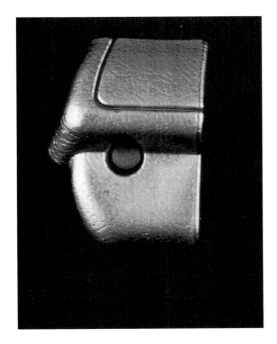
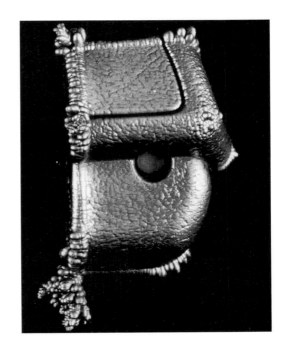

Courants pulsés avec inversions anodiques　　　　　　　Courant continu

Les moules ainsi produits sont plus réguliers en épaisseur. Cette homogénéité se traduit par la production de pièces rotomoulées avec un poids abaissé. Les essais menés sur le site industriel ont permis de gagner jusqu'à 30 % de la matière transformée.

Les coûts de revient de la fabrication des pièces rotomoulées peuvent donc être abaissés tant sur les durées de finition des moules que sur la quantité de matière transformée.

Bibliographie :

- C. AOUAISSIA, Thèse, Besançon.
- R. J. CRAWFORD - Rotational Mouding of plastics - RSP - 1992
- INCO - Nickel Electroforming - 1991
- P. J. NUGENT, Ph-D, Belfast - 1990
- P. BERCOT, Thèse, Besançon - 1988
- INCO - Guide to Nickel Plating - 1988
- INCO - Guide du nickelage - 1973
- P. SPIRO - Electroforming - Ed. R. Draper - 1971
- R. ROUSSELOT - Répartition du potentiel et du courant dans les électrolytes - Ed. Dunod - 1959

Foundry Automation

MALCOLM BOOTH*, TONY WRIGHT*, GLEN STEER**

*Midlands Electricity plc., Halesowen, UK.

**Aston University, Birmingham, UK.

ABSTRACT

This paper reviews recent electrical process optimisation and control system developments within the UK foundry industry, which have led to wider opportunities for process automation. The benefits resulting from these developments include improved energy efficiency, production flexibility, improved quality control, and load optimisation with associated opportunities for energy cost, and energy demand, reduction.

Case study results are currently being analysed and demonstrate the generic improvements referred to in this paper.

ABSTRACT

Cet article résume les récents développements des systèmes de commande et l'optimisation des procédés électriques de l'industrie métallurgique britannique qui ont ouvert de plus riches débouchés à l'automation des processus. Parmi les advantages de ces développements, un rendement énergétique optimisé, la souplesse de la production, un contrôle qualité plus efficace et l'optimisation des charges qui, ensemble, permettent de réduire le coût comme la demande d'énergie.

Les résultats d'études de cas sont actuellement sous analyse et démontrent les améliorations génériques énumérées dans le présent article.

INTRODUCTION

To date the level of automation within the foundry industry has been lower than that of many other industrial sectors. The dominant reasons are the hostile environment within which foundries operate, with extremes of dirt, temperature, and magnetic fields, and also the deeply ingrained traditional working practices.

The emergence of the UK foundry industry from recession has brought with it the realisation that, in order to maintain present high order levels, the issue of increased automation must be addressed. In addition to this external factors in the shape of Government legislation, concerning health and safety and working practices within foundries, have added weight to the arguments for increased automation.

During the last 25 years Midland Electricity plc (MEB) has been assisting foundries to establish more efficient processes. This paper reviews recent applications of automation within a foundry environment in the form of :-

- Autopour systems
- Production scheduling
- Energy management and control

AUTOPOUR SYSTEMS

Mould production rates of up to 300 per hour are forcing foundries to assess the potential for automating the pouring process. This process requires :-

- Precise location of the dispensing nozzle
- Accurate shot weight control
- Precise temperature control
- Small footprint to facilitate retrofit to existing equipment

Most existing automatic metal dispensing systems are based on pressurised induction heated refractory bodies *(fig.1)*, to which is attached filling and pouring spouts. A stopper and nozzle arrangement is fitted to the pouring spout to dispense metal from a constant metal head. This type of pouring equipment has no technical size limitation but, however, is difficult to install in retrofit situations.

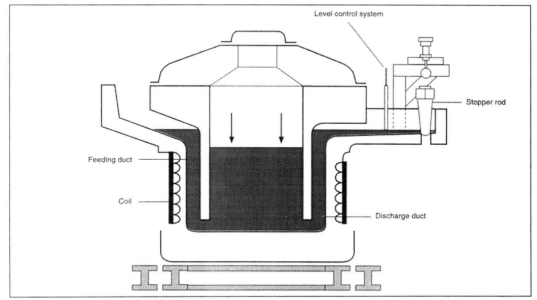

Figure 1 - Existing automatic metal dispensing system

A more compact alternative system, developed by EA Technology in collaboration with a West Midlands foundry, is the vacuum lift heated metal dispensing process[1]. In this process (*fig.2*) the volume of metal added from the transfer ladle is raised into the refractory tube which is heated by an induction coil. Power is applied at medium frequency (approx. 1kHz) and the heated metal is then returned to the refractory bath as metal dispensing occurs.

The vacuum autopour system has been successfully installed to existing automatic moulding lines[2]. Ongoing development and installation of automated casting lines is presently being undertaken at a large foundry in conjunction with MEB.

PRODUCTION SCHEDULING

Production scheduling aims to maximise throughputs at a minimum cost. Within the foundry environment daily and weekly patterns of work are governed by the need to meet delivery dates and, as a result, the processes of melting, holding and casting are often considered in isolation. Where the work in the foundry is varied, and production runs are relatively small, there is considerable scope to relate manufacturing costs to specific castings and therefore optimise daily/weekly production cycles.

However, the success of production scheduling requires the collection and analysis of historical data which needs to be linked into the entire process. For example, the casting process has to be linked to the availability of cores, moulds, hot metal, and energy costs. Only by linking such data is it possible to optimise daily/weekly work patterns and achieve maximum throughput.

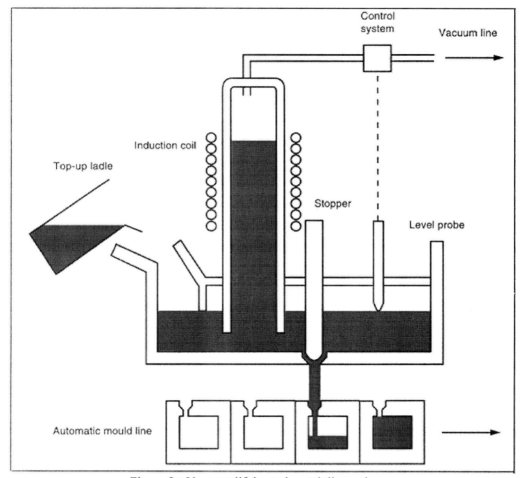

Figure 2 - Vacuum lift heated metal dispensing process

Foundry production scheduling software has been developed on behalf of MEB by EA Technology Ltd. Initial algorithms compared historical energy and production information to predicted half hourly energy costs. Trial schemes based upon this information, using pool price electricity costs, have proven that significant direct energy savings can be made when the working pattern provides opportunities for change. For 'non-pool price' customers the benefits of production scheduling are also available through obtaining a higher load factor, leading to reduced maximum demand charges. Also, with a flatter load profile, the customer may be able to negotiate more favourable supply contract terms with its electricity supplier.

ENERGY MANAGEMENT AND CONTROL

Detailed analysis of metal throughputs and associated energy costs are required to undertake any form of production scheduling or furnace efficiency trials. In many instances customers are unable to provide information related to the specific running costs/efficiency of individual pieces of plant and the only energy costs recorded are those of the main billing meters. Acquiring the relevant data for an energy audit requires an understanding of the associated industrial process and full commitment from the participating company. Unfortunately, in many instances when energy audits are conducted neither of these parameters are met!

During the past 18 months MEB, in association with local industries, has developed techniques for monitoring energy use within iron foundries. These techniques involve recording detailed half-hourly energy usage, including the identification of power factor, of individual furnaces with energy monitoring equipment. Recorded data is then analysed and presented to the customer. Such was the value of information, in both reducing energy consumption and improving productivity, that the customers expressed their desire for an equally accurate on-site energy management system. This system has enabled the customer to adopt a furnace sequencing procedure in order to achieve the optimum scheduling of the individual furnaces and therefore maximise efficiency. This procedure has led to a 20% increase in furnace outputs.

In order to provide this service a series of tests were performed on various commercially available solid state energy monitors. The majority of monitors were rejected after trials highlighted problems of poor accuracy and/or poor communication reliability in the demanding foundry environment. Hardware and software was subsequently developed by MEB to communicate with the selected energy monitors and the existing billing meters *(fig.3)*. Algorithms ran concurrently within the same software in order to process the real time data and provide the information that the foundry workforce required. Selective data was also input by furnace operators to assist them in controlling the furnaces. A large overhead display then provided real time data for furnace operators to work to. The multi-drop monitoring and control system also provides the production controller and energy manager with real time information of furnace operations and daily/weekly energy and efficiency reports.

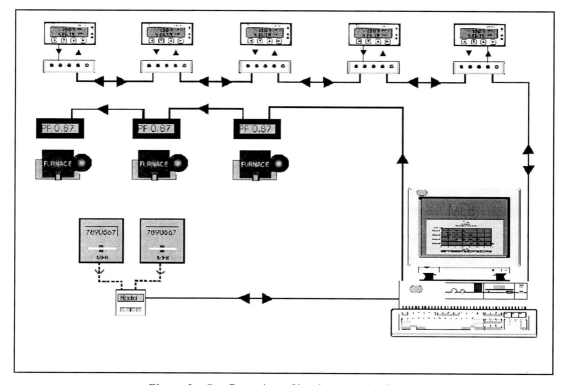

Figure 3 - Configuration of hardware and software

Practical difficulties were experienced with the installation of the monitoring system, mainly attributable to the hostile environment. These difficulties included magnetic distortions from furnace operations, dirt, and voltage spikes, with associated harmonics, from inverters. In order to overcome these problems high levels of filtering were implemented as well as other modifications.

Initial results indicate that the pay back period for such equipment based on energy costs alone will be substantially less than 6 months for a typical installation. Feedback from companies who have installed such equipment indicates that accurate real time data from the furnaces has enabled them to manage maintenance regimes more efficiently, reduce downtimes and optimise furnace lining lifetime, leading to improved furnace efficiency at all times.

CONCLUSIONS

The project has enabled MEB to identify several processes within foundry operations which offer scope for automation. Some of these areas, such as moulding lines which have adopted autopour technology, have already benefitted from such automation. MEB is now in a position to exploit the potential of automation in other associated processes such as casting and will continue to do so through a close working relationship with its foundry customers.

ACKNOWLEDGEMENTS

The authors wish to thank the many companies associated with the foundry industry, especially G.Clanceys Ltd, and Glynwed Foundries, Sinclair Works, who have contributed in much of the work referred to in this paper. The special contribution of EA Technology in support of process and product developments is gratefully acknowledged.

REFERENCES

1. C.F. WILFORD and N.B. WILLIAMS, "A new development for the automatic dispensing of molten metal", paper 12, Proceeds of the BCIRA Int. Conf., Warwick, March, 1990.

2. T.J. SMITH, "Experiences with vacuum lift heated autopouring of cast iron", Proceeds of the IBF Conf., Harrogate, June, 1991.

Use of High-Frequency Induction Melting in Cold Crucible to Produce Crystalline Non-Oxygenous Materials

F.V.BEZMENOV, I.A.KANAEV, V.A.RADUCHAEV, V.N.IVANOV

VNIITVCH, St. L.Tolstoy 7, St.Petersburg, 197376, Russia

ABSTRACT

The work was dedicated to the development of technologies and equipment to produce high-melting non-oxygenous compounds and, first of all, boron carbide ($B_{12}C_3$). Boron carbide crystals thanks to their high temperature of melting, hardness, chemical resistance are perspective materials for manufacturing chemical utensils, wear resistant elements of different devices, high temperature thermocouples, indicators for measuring hardness at high temperature, thermoelectric temperature transducers.

A method of induction melting and crystallization of high-melting metals in "cold crucible" has been suggested. The paper presents theoretical reasoning and the design of the installation for melting boron carbide in cold crucible at frequencies 3,5 and 5,28 MHz.

Technology of growing semi-crystalline ingots and also technology of growing boron carbide monocrystals in induction melting in cold crucible from the gaseous phase have been developed.

RÉSUMÉ

L'ouvrage est consacré à la creation des technologies et de l'équipement pour l'obtention des composés de haute fusion sans oxygène, premièrement du carbure de bore ($B_{12}C_3$). Grâce à leur température élevée de fusion, à leur dureté et à leur stabilité chimique les cristaux du carbide de bore se présentent come particulièrement prometteurs pour la fabrication de la verreri de laboratoire, des éléments résistants à l'usure pour toutes sortes d'appareils, des thermocouples de haute température, des indicateurs de durete sous température élevée, des convertisseurs thermoélectriques.

On propose une méthode de fusion par induction et de cristallisation des métaux de haute fusion au "creuset froid". L'article renferme la justification théorique et la structure de l'installation de fusion de carbide de bore au creuset froid sous des fréquences de 3,5 et de 5,28 Mhz. On a réussi à créer une technologie permettant d'obtenir des lingots polycristallines, ainsi que des monocristaux du carbide de bore lors de la fusion par induction au creuset froid à partir de la phase gazeuse.

In the recent years high melting compounds have acquired more and more importance in science and industry thanks to unique combination of thermo-electrophysical, mechanical and chemical properties. High melting compounds have not only high temperature of melting (1400-4000 C) but high hardness, chemical resistance, high-temperature strength, low rate of evaporation and vapour pressure, small coefficient of thermal expansion etc. Among high melting materials a special place is occupied by non-oxygenous materials including carbides, borides, nitrides of transition rare-earth metals (Ti, Zz, Hf, V, To, W, Mo) and non-metals (Si, B).

The present paper is dedicated to the development of technology and equipment to produce high melting non-oxygenous compounds, first of all, boron carbide ($B_{12}C_3$ or sometimes designated as B_4C). Boron carbide crystals thanks to high temperature of melting, hardness, chemical resistance are perspective materials for manufacturing chemical utensils, wear resistant elements of different devices , high temperature thermocouples, indenters for measuring hardness at high temperature, thermoelectric temperature transducers, crucibles for melting metals and alloys with high temperature of melting. High abrasive property of boron carbide crystals is used in treatment of manufactured corundum, glass, precious stones, in finishing cutting tools out of hard alloys, in dressing of grinding wheel. Being a good insulator in its pure form, boron carbide acquires the properties of semiconductor due to doping and can be used in manufacturing non-linear high-ohmic resistances.

When selecting a method of growing boron carbide the following well-known methods were analyzed: solid phase (recrystallization), gaseous phase, liquid phase (non-crucible zone melting, Verneuil's method, modified Czochralski's method) and solution method. Liquid phase methods were recognized as the most perspective ones, and the method of induction melting and crystallization of high melting materials in "cold crucible" has been suggested, its diagram is given in Fig. 1.

Cold crucible - 1 is made of water cooled and isolated from each other metal sections and coupled together in such a way that crucible can keep the charge of high melting material in cold and melted states and, on the other hand, this crucible does not prevent from the propagation of electromagnetic field of the inductor - 2 inside the crucible.

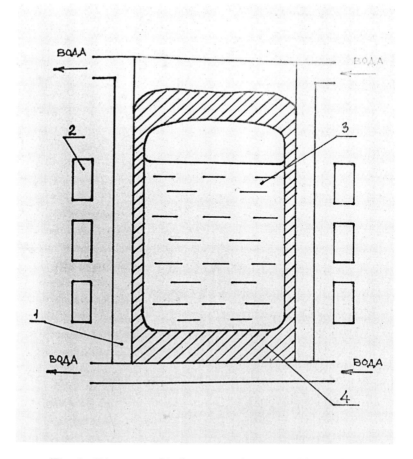

Fig. 1. Diagram of induction melting in cold crucible.
1 - "cold crucible" made of copper pipes which form a rigid reservoir for powder but in the induction zone they do not form a closed circuit and permit the field of an inductor to penetrate into the crucible
2 - an inductor
3 - melt
4 - powder

The charge of high melting material is placed into the cold crucible. A washer with the dia 7-10 mm made out of melted or baked compact boron carbide is placed in the center of the crucible at the level of the upper induction turn.

Boron carbide powder has a low electric conductivity, therefore the charge is transparent for high-frequency field. Compact boron carbide in the center of the charge having sufficient electric conductivity begins to get heated and melted. Heat from the zone of started heat expands to the neighbouring charge layers which after melting begin themselves to absorb the energy of electromagnetic field and heat the next layers. The melting wave moves towards the water cooled walls of the cold crucible. The charge layers which are in close vicinity of the walls get intensively cooled and heat energy coming from the melted charge is captured by the water cooled sections of the crucible. The heat equilibrium takes place and slag is formed in which melt is preserved - 3.

When developing the experimental installation for melting boron carbide the commercial installation for melting oxide materials of type "Crystal 401 - Phyanite" (power 60 kW, frequency- 5,28 MHz, voltage at the inductor - 6 kV, melting - in the air) was used.

However, thermal, electric and chemical properties of boron carbide considerably duffer from oxides properties. Therefore, the installation was redeveloped.

Melting of boron carbide should be conducted in inert gas (argon or helium), therefore, the water cooled crucible was placed in a sealed working chamber which was equipped with systems of air evacuation, inert gas supply, pressure control in the chamber and pressure release in an emergency. Inert gases have low electric strength, especially at high temperatures when their thermal ionization is possible. Therefore, the heating unit for melting boron carbide was redeveloped, the gaps between the leads were increased, their sharp angles were smoothed, the inductor turns were placed in parallel. To eliminate the break-down, the surface of fluoroplastic flange fixing the leads was made corrugated. For convenient extraction of melted boron carbide having low density the knock-down construction of cold crucible was designed.

The least reliable elements of a heating unit of high-frequency installation are water cooled sections of the crucible which are burn-out at break-down. This can result in water getting into the hot zone, its evaporation and sharp increase of pressure in the chamber. Therefore, the chamber is provided with pressure release device.

To select the most reliable heating unit we made and tested copper and aluminium sections and aluminium

sections with strong dielectric coating out of aluminium oxide having the thickness of 0,03-0,05 mm (Fig. 2). Increase of electric resistance between the sections is rather important since boron carbide slag has considerably higher electric conductivity at working temperatures than the oxide slag. Therefore, the eddy currents induced in the metal sections could get closed via the slag. This could result in reducing the power released in the melt and spontaneous crystallization.

Experiments showed that sections made of aluminium with insulation coating have significant advantages.

Boron carbide has high heat conductivity (13 W/m.grad at melting temperature); keeping it in a melted state requires high heat release. By this reason, autotransformer connection of the inductor was used which permitted to increase the voltage at the inductor from 6,0 to 8,5 kV.

Experimental melts at frequency 5,28 MHz showed the difficulty to keep the melt in cold crucible. At high electric conductivity of boron carbide the depth of current penetration is small. Power is released in a thin layer on the melt surface and thanks to heat conductivity spreads both in the melt and through the slag into the cold crucible. Therefore, on the experimental installation the oscillator and the heating unit were re-adjusted for frequency 3,5 MHz.

A series of melts was conducted on the experimental installation. After loading 500-800 g boron carbide charge into the cold crucible made of aluminium sections coated with aluminium oxide and placing compact boron carbide in the center of the crucible, the working chamber was sealed and air was pumped out of it. Then, the chamber was filled with argon up to the excess pressure of 5-10 mm of Mercury column; in the process of melting the pressure was maintained constant in the chamber. The excess gas supply was controlled by hydraulic back-pressure valve.

After melting the boron carbide its specific resistance reduced and the working regime of valve oscillator changed (Table I). Speed of melting wave moving towards the walls of the cold crucible was 0,1 mm/s.

TABLE 1
PARAMETERS OF COLD CRUCIBLE AND REGIME OF VALVE OSCILLATOR IN THE BEGINNING OF MELTING AND AFTER CHARGE MELTING AND SLAG FORMATION

Material of sections	Coating material	Inner dia	Length	Beginning of melt			End of melt		
		mm	mm	kV	A	A	kV	A	A
Aluminium	Aluminium oxide	70	250	7-8	2,5-3,0	1,8-2,0	5-6	5-6	1,2-1,4

After keeping the melt in a stationary state (homogenization state) its crystallization was performed by taking the cold crucible out of the inductor (when growing monocrystals) or reducing the voltage at the inductor (when growing semi-crystalline ingots).

Structure and properties of semi-crystalline samples of boron carbide depend on the speed of cooling (Table 2).

Samples grown at slow reduction of voltage (0,3 kV/min) have large grain structure and high microhardness in combination with increased brittleness (pressure giving imprint without forming cracks is minimal).

TABLE 2
STRUCTURE AND PROPERTIES OF SEMI-CRYSTALLINE SAMPLES OF BORON CARBIDE GROWN AT DIFFERENT SPEEDS OF COOLING DUE TO REDUCTION OF VOLTAGE AT THE INDUCTOR

No.	Voltage reduction rate at the inductor	Size of grains	Microhardness	Maximal pressure on indenter when measuring microhardness
	kV/min	mcm	kg/mm^2	kg/mm^2
1	0,3	1300-1600	4400-4500	100
2	8,0	200-400	4000-4300	150
3	500 (hardening)	20-30	3300-3500	200

Fig. 2. Heating unit of modified experimental installation for melting boron carbide.

For the structure (Fig. 3) are typical the release of eutectic formed by two phases - boron carbide (light grains) and graphite in the form of dark plates.

At quick reduction of voltage (8 kV/min) the structure with smaller grains is formed (Fig. 4); eutectic is practically absent. Only separate inclusions of graphite are noted. Microhardness and brittleness of samples is slightly reduced. (Table 2).

In the regime of hardening at instantaneous switching off the voltage at the inductor and cooling the boron carbide with water the single-phase product is grown practically. Sizes of grains are reduced to dozens of microns.

Brittleness and microhardness of the material go down. Thus, the regularity in decreasing the grain sizes and reduction of brittleness and microhardness with the increase of cooling speed are observed.

In the process of induction melting and crystallization of boron carbide its purification from impurities takes place (Table 3).

TABLE 3
CONTENT OF IMPURITIES IN THE ORIGINAL CHARGE
AND IN THE MELTED PRODUCT

Impurity	Content of impurities	
	in original charge	in melted product
C free	0,10	0,10
Mg	0,88	0,02
O_2	1,80	0,70
N	0,19	0,10
Fb	0,33	0,19

Boron carbide is most intensively purified from magnesium - the main technological impurity.

At slow cooling of the melt due to taking the crucible out of the inductor at the speed of 10 mm/min, isometric crystals of boron carbide up to 3 mm were grown. The absence of clear outer cut is typical for them. Parallelism of opposite faces is observed.

Along with isometric crystals, boron carbide crystals of other morphological types - wafer, prismatic and columnar were grown by the method of recrystallization through a gaseous phase in cold crucible. In gaseous phase growing of crystals by using induction melting in cold crucible the main task is the formation above the melt of so-called adiabatic roof out of solid boron carbide having multiple cavities in which crystals grow on spontaneously appeared nuclei by recrystallization through gaseous phase. Sizes of grown crystals are:

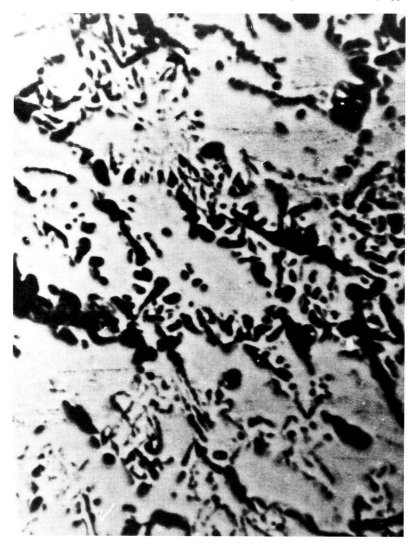

Fig. 3. Eutectic of B12 C3 (B4 C) and C (magnification x 635; photographing in the reflected light). Samples are grown at 0,3 kV/min

Fig. 4. Oscillating-columnar crystal X-ray photograph of boron carbide

- wafer: 3x3x0,1 mm;
- columnar:: 0,5x1,5x5 mm;
- prismatic: 3x3x3 mm.

Growing time is 15-20 min.

Belonging of prismatic and columnar crystals to trigonal syngamy was established and parameters of their lattices were defined:
- prismatic - $a = 5,41 A$ and $= 60$
- columnar - $a = 5,6 A$ and $= 65,4$

Clear outer cut of crystals is the evidence of perfection of their inner structure. High quality of grown crystals is confirmed by the results of X-ray structural analysis too (Fig. 4). Good division of doublet on the oscillating-columnar crystal X-ray photograph shows a low level of internal stresses in a crystal and uniformity of its melt.

CONCLUSION

An experimental installation was developed and boron carbide melting was conducted in cold crucible at frequencies 3.5 and 5,28 MHz.

Technology was developed for growing semi-crystalline ingots by reducing the voltage at the inductor after melting the charge, and monocrystals by taking the cold crucible out of the inductor. It was shown that after melting and crystallization, boron carbide is purified from impurities.

Also, the technology was developed for growing boron carbide monocrystals in induction melting in cold crucible from the gaseous phase. Crystals grown from the gaseous phase have the structures of high uniformity and perfection as per the results of investigations.

Electric Drying of Monolithic Steel Ladles

ARNOLD SOLMAR, JOHN DANELIUS, FARID ALAVYOON
Vattenfall, Sweden

ABSTRACT

Liquid raw iron and steel are transported by means of a ladle inside the steelworks. The ladle wall is traditionally made from layers of bricks and an outer shell of steel. Drying the brick layers is usually done with poorly controlled burners without any exhaust stack. Steel engineers have identified a possibility for improvement if bricks could be replaced by a monolithic material. The layer could be made by a moulding technique instead of manual, time-consuming bricklaying. The aim of producing a clean steel without impurities also indicates a demand for a more resistant contact surface with steel. Environmental reasons are forcing the steelworks to reduced emissions from badly controlled fossil fuel burners and also to decrease the amount of worn-out brick lining that is disposed in land-fill sites. Electric heating is proposed as the best method of drying.

RESUME

Le fer et l'acier brut á l'état liquide sont transportés à l'intérieur des aciéries dans des poches de coulée dont la paroi est habituellement constituée de couches de briques réfractaires et d'une enveloppe externe en acier. Le séchage des assemblages de briques est communément effectué à l'aide des brûleurs mal contrôles et sans récupération des gaz. Les ingegénieurs de secteur sidérurgique ont mis en évidence une possibilité d'amélioration qui consisterait à remplacer l'ensemble des briques par un matériau d'une seule pièce. Les enveloppes réfractaires pourraient alors être fabriquées par moulage, procédé plus rapide que l'empilage manuel des briques. L'utilisation de matériaux ne réagissant pas avec l'acier est rendu nécessaire par la volonté de produire un acier propre, sans impuretés. D'autre part les aciéries doivent réduire à la fois les émissions provoquées par les brûleurs mal contrôlés utilisant des énergies fossiles et la quantité de briques usées mises en décharge. Le chauffage électrique est proposé en tant que meilleure méthode de séchage.

BACKGROUND

The electricity generating companies are interested in identifying new sales opportunities for their electricity. The steel industry uses fossil fuel-based ladle drying and heating plants, which are suitable for conversion to electrical operation. The introduction of new heating technology has now become topical at a time when new lining methods are beginning to be introduced. Rationalisation of maintenance and a simultaneous increase in the service life of ladles are achieved by replacing bricks with moulded linings. Heating must be accurately controlled and uniform throughout the ladle in order to ensure proper drying, when approximately 5% of water must be removed from the moulded material. Electrical heating is expected to meet the technical requirements for the desired heating principle.

COMPUTATIONAL MODEL

In order to be able to assess the dimensioning of an item of electrical drying equipment, calculations of the power requirement and the temperature distribution in the outer wall of the ladle during drying have been made. The modelling method and results are described below.

GENERAL FORMULATION

Much interest has been shown in developing general-purpose codes for simulating fluid flow and heat/mass transfer problems in the last couple of decades, thanks to the increasing availability of powerful computers. Such codes are based on the use of computational fluid dynamics (CFD). The aim of these general-purpose CFD packages is to obtain reliable solutions that satisfy engineering demands for a wide range of applications, and at the same time have an acceptable degree of accuracy. The reliability and accuracy of CFD codes are usually evaluated by applying them to well-defined problems where experimental data is available.

The commercial CFD code PHOENICS version 2.1[1], has been used for the calculations in this paper. PHOENICS solves the equations for the time-dependent, or steady-state conservation of momentum, mass and heat, etc., in one, two or three spatial dimensions. The code is based on the finite volume approach[2], in which the geometrical domain of the problem is divided into a sufficiently large number of small control volumes. The partial differential equations to be solved are then integrated for each control volume. One thus obtains a system of algebraic equations, which is then solved using an appropriate numerical algorithm. In PHOENICS, problem geometry can be created in Cartesian, polar or body-fitted co-ordinates (BFC). A computational grid is generated in the geometrical domain by dividing each co-ordinate direction into a number of intervals. Each control volume is a volume element defined by adjacent grid lines. In the recent versions of PHOENICS, it is possible to use the so-called multiblock technique. In this approach, a complex geometry is first divided into a number of simple parts. Each part is then modelled separately. The equation solver in PHOENICS can then connect these parts to produce the complex geometry in which the equations are to be solved. This method has been used in the present work. The geometry is divided into about 7000 control volumes.

SIMPLIFYING ASSUMPTIONS

The mathematical model which is used for the calculations in this study is based on the following assumptions:

a) The problem is two-dimensional. Only a vertical cross-section is modelled.
b) Heat-up and evaporation of the water in the pores of the porous parts of the container are disregarded. This is a restrictive assumption which is made for simplicity. In order to compensate for this simplification, the power needed to heat up and evaporate the water is calculated afterwards.

c) The container is made of different materials, with different material properties. Heat transfer through the materials is assumed to be only through conduction. The temperature-dependence of the material properties is accounted for in the model. The equation that is solved to determine the evolution of the temperature field as a function of time reads:

$$\frac{\partial(\rho c_p T)}{\partial t} = \nabla \cdot (k \nabla T)$$

where ρ, c_p, k and T respectively denote density [$\frac{kg}{m^3}$], heat capacity [$\frac{Ws}{kgK}$], conductivity [$\frac{W}{mK}$] and temperature [C]. Temperature T is a function of time and spatial co-ordinates, while heat capacity and conductivity are, for some of the materials, functions of temperature.

d) Heat exchange between the container and the surrounding air is modelled as:

$$q = \alpha(T_{out} - T)$$

where q [$\frac{W}{m^2}$], α [$\frac{W}{m^2 K}$] and T_{out} [C] denote heat transferred to (from) the container from (to) the surrounding air, heat transfer coefficient and temperature of air in the immediate vicinity of the container walls. In the case studied here, it is assumed that $\alpha = 100$ in the container and $\alpha = 8$ outside the container. The temperature inside the container is prescribed as a function of time according to Fig. 1. The outer temperature is kept constant at $T_{out} = 20 C$.

RESULTS

An electric heating device has been designed to meet the technical specifications of a Swedish steel producer. The size of the ladle is about 100 tons. The drying procedure involves a temperature rise from 20 °C to approximately 400 °C over about 35 hours. Estimates of the temperature fields show that the inclusion of a layer of insulating fibres drastically reduces the external temperature of the ladle. In theory, not more than 200 kW are required to cover the transmission and drying heat requirement. Ventilation losses depend on the specification from the respective lining supplier. Only minimal ventilation of the ladle is required since the quantity of water given off is very small, and at the same time the drying time is long.

ELECTRICAL HEATER DESIGN

There are two types of electrical ladle drier currently in use in Scandinavia. Both types consist in principle of electrical resistance elements suspended from a lid which fits the ladle opening. Experience has been gained from the use of both suspended SiC elements for higher temperatures and radiation tubes for lower temperatures. There is also experience with the use of belt elements mounted on the external surface that is lowered into the ladle. The two main principles are illustrated in Figure 4. Temperature control is performed with the use of thyristor technology. Temperature sensors can be positioned both in the electrical heater package and in the lining. Archiving of the recorded temperature in the lining during drying is recommended. This will provide a factual basis for any subsequent discussions concerning the drying sequence.

EXPERIENCE

The intention in providing this contribution to the conference was to present the operating experiences gained from the drying of the ladles included in the numerical calculation of the temperature field. Thanks to an extremely favourable production situation, the steelworks in question has not yet had the opportunity to install an electrical heater, although this will take place in 1996. However, experiences with the continuous use of electrical heating since 1986 can be reported from Scandinavia. The ladle size in this case is approximately 50 tonnes, and the ladle drier has an output of 300 kW. The average power consumption is 3 000 kWh for a 48-hour cycle with a maximum temperature of 650 °C.

REFERENCES

1. B. SPALDING: 'The PHOENICS-2.1 Companion', CHAM TR313.
2. S.V. PATANKAR: 'Numerical Heat Transfer and Fluid Flow'. Hemisphere, Washington D.C., 1980.

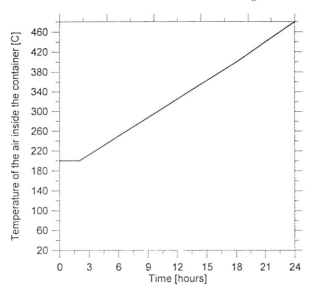

Fig. 1 Drying specification, i.e., the temperature of the air adjacent to the inner walls of the container.

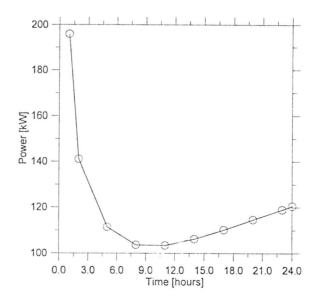

Fig. 2 Heat flux to the ladle during the first 24 hours of drying.

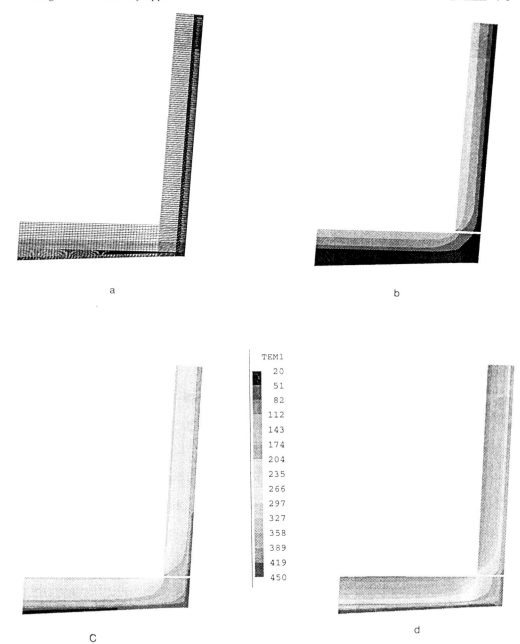

Fig. 3 a) The computational grid, and the calculated temperature field for b) time=2 hours, c) time=14 hours, and d) time=24 hours.

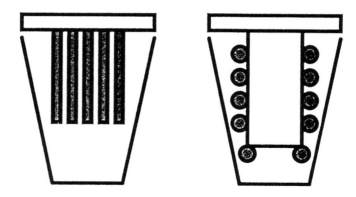

Fig. 4 Principle designs for electrical ladle heaters.

Molten Metal Processing Using an Advanced Electromagnetic Pump

ANIL KATYAL*, HARRY BULLARD**, LES SMITH** AND PAT HAYES**

*EMP Technologies Ltd, Derbys, UK
** EA Technologies Ltd, Capenhurst, UK

The surge for optimising aluminium melting operations has led to the innovative development of a simple and cost effective method of controlled subsurface stirring of liquid metal using non-submersible electromagnetic pump. The pump with no moving parts can be retro-fitted to both small or large, tilting or stationary, open well or close type reverberatory furnaces.

The paper describes the electromagnetic pump hardware, retro-fitting of pump to furnaces, applications and experiences of aluminium melting operations with the electromagnetic pump. Specific examples of metal processing with the electromagnetic pump to improve metallurgical quality in aluminium casthouses and a novel method of making aluminium metal matrix composites are discussed in the paper.

Optimiser la fusion d'aluminium est une des priorités actuelles. Ainsi, une méthode simple et bon marché visant à contrôler le mouvement du métal en fusion se trouvant sous la surface a été mise au point grâce à une pompe électromagnétique non-submersible. La pompe, sans aucune partie en mouvement, peut être adaptée à n'importe quel type de fours (petits ou grands, basculants ou fixes, fours de fusion maintien avec ou sans aujet).

L'exposé décrit le matériel, son installation, son fonctionnement et les différentes expériences menées avec la pompe en matière de fusion. L'utilisation de la pompe électromagnétique pour améliorer la qualité métallurgique dans les fonderies et une nouvelle méthode pour la fabrication des alliages d'aluminium (MMC) sont aussi abordées dans cet exposé.

INTRODUCTION

The benefits of liquid metal stirring in metallurgical processes is well known however efficient means of stirring are limited. Recently an electromagnetic pump was developed jointly by EA Technology and Calder Aluminium Ltd for aluminium scrap melting which is now an established tool for metal circulation in reverberatory type of furnaces.

ELECTROMAGNETIC PUMP

The pump coil assembly shown in Figure 1 is constructed of a multi-layered induction coil supported between two stainless steel flanges and is held together by stainless steel tie rods. . The coil is water cooled and has a sophisticated earth leakage monitor. The pump operates by induced eddy currents in molten metal by the magnetic field generated by the pump coil. The interaction between the magnetic field and eddy current generates an electromagnetic force propelling the molten metal.The pump performance curves are shown in Figure 2. The pumping velocity of the metal can be infinitely varied and the direction of flow reversed by the touch of a button. The pump system comprises of four components:

1. a circular induction pump coil assembly
2. a three phase 380/415V, 50/60Hz remote power supply
3. a closed circuit water cooling system
4. a system control panel and a workstation for operator

Electrical schematics of the electromagnetic pumping system are shown in Figure 3. The pumping and stirring is controlled through the use of a PLC. The alarm system and fault indicators are independently powered to maintain the integrity of system.

PUMP SPECIFICATIONS

Length (nominal)	:	750 mm
Flange Dimensions	:	500X500 mm
Refractory Pipe	:	Non-wetting silicon carbide
Nominal Bore	:	110 mm
Rating	:	50 kW (nominal)
Flow rate	:	upto 8 tonnes/min Aluminum at rated power
Connected Power	:	Mains frequency, 380/415V, 70KVA.
Control	:	Reversible flow, infinitely variable
Pump Priming	:	50 mm nominal at reduced flow
Maximum Operating Temp.	:	1000^0 C

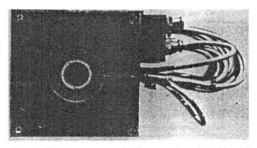

Fig.1: Electromagnetic Pump showing Liquid metal path and cooling water pipes

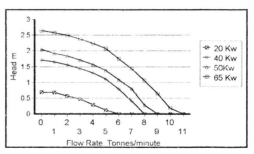

Fig. 2 : Pump Characteristics Aluminium Head Vs Flow Rate

DESIGN OF MELTING APPARATUS

The melting apparatus[1] is a 900mm internal diameter refractory lined vessel, with tangential inlet and outlet ports connected to the pump and furnace are shown in Figure 4. The pump inlet and exit ports are installed at the hearth level allowing the pump to function with minimum metal heel of 50-60mm. The connection between the furnace, pump and the charge well is specifically constructed in non-wetting refractory to avoid any alumina/oxide build-up. A slight slope is constructed towards the furnace in the pump refractory connections, this enables the drainage of the pump tube required during alloy changeover or the works shutdown period. A vertical component to the rotating fluid flow in the well is formed which results in rapid immersion of light gauge scrap in to the molten stream with little opportunity for oxidation. Furthermore the velocity in the skim well (not shown in Figure 4) is allowed to reduce, enabling oxides to float to the surface for skimming, thereby minimising oxide entry to the furnace. Thus the system offers the following advantages to metal melters:

- thin gauge materials can be fed into the molten stream without being exposed to direct flame.
- the subsurface flow ensures minimal disturbance of metal surface thereby no additional source of oxidation being introduced to the system.
- baled, as received, infeed can be melted.
- the heat transfer to the molten bath is homogeneous and there is virtually no corrundum build-up inside the furnace.
- the melting system allows charging of both lightly and densely baled materials without debaling, although there are some exceptions to this rule.
- the furnace doors remain shut, minimising heat loss, and increasing refractory life.

Fig. 4: Refactory Lined Vessel with tangential ports connected to the pump and furnace

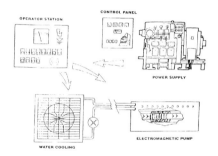

Fig. 3: Electrical Schematics of the Electromagnetic Pump

- increased production as no additional time is required for stirring which is part of the continuous process.
- ingots and alloying additions may be charged through the well.

SYSTEM OPERATING PERFORMANCE

The melting rate is influenced by the type of scrap, its bulk density and size of charging well; other things being constant. For 900mm diameter charging well typical melting rates for loose swarf are shown in Table 1. Computer flow modelling[2] of the system however suggests that a larger size charge well will provide adequate vertical and horizontal components to the flowing fluid, necessary for the rapid immersion of light gauge scrap. This may be of interest to foundries melting low density chips, swarf and UBC's at 4-5 tonnes an hour.

Bulk Density $Kg\ m^{-3}$	Melt Rate $T - h^{-1}$
200-400	1.5-2.0
500-700	2.2-2.7
700-1000	2.5-3.0

Table 1: Melt Rate of Loose Swarf from a Standard Size Well

Energy Consumption $KW\ t^{-1}$		Energy Cost £t^{-1} ingot	Furnace Utilization %
Pump	Gas		
21.9	1038	9.94	92.1

Table 3: Average Energy Consumption over nineteen weeks

Typical metal recoveries from the system obtained over long periods at Calder Aluminium are shown in Table 2 and these figures indicate that metal recoveries are equivalent to that obtained from induction furnaces from identical materials. Metal recoveries are on ingot out basis, excluding recoverable metal from the dross. Energy consumption data to melt and hold metal in a furnace equipped with cold air, natural gas burners without any heat recovery is given in Table 3.

Table 2 : Typical Melting and Recovery Rates with Electromagnetic Pump

Material	Melting Rates t h^{-1}	% Recovery
Swarf loose	1.5	87-88
Swarf baled	2	92-93
Lithographic plate	2.0-2.5	91-92
Clean foil	2.5-3.0	98
Shredded scrap	2.5-3.0	85-87
Profile section cuts	1.5-2.5	88-92
Graded metallics	1.5	72-75
Alloying Elements, Si	2	-

OPERATING REGIME

The furnace when started from "absolute cold" conditions follows a typical preheating cycle to soak the refractories with heat. The pump, the connecting pipes and the charge well are also heated with a top hat burner placed over the charge well. The pump is primed by melting solid scrap within the reverberatory furnace. As soon as 50mm of metal has reached in the pump the top hat burner over the charge well is removed. From now on the charge well seeks heat from the reverberatory section of the furnace, as metal is in continuous circulation in the system and the flow can be controlled as desired. Alloy changeover require emptying the furnace however as it is simultaneously filled with further metal no preheating of the charge well is required.

Calder Aluminium operates five days a week around the clock however the furnaces remain hot during the weekend, with a residual heel and the pumps are set at a low power level 8-10 kW to maintain circulation. It has been found the cost of preheating from a cold start is more than leaving the furnace hot. To better appreciate the pump capability, one pump was subjected to periodic heating and cooling. The pump and the furnace were on for three days and off four days. This heating and cooling cycle was observed for approximately twelve months without adverse affect on the pump or pump refractory. During preheating care is to be observed not to overheat the pump tube as overheating can cause oxidation of thin metal film adhering to the wall of the tube. The blistered oxide can then restrict the flow of metal from the furnace. This build-up can be eroded by purging an inert gas for 10-15 minutes in the direction of the restriction, however the best practice is to control the preheating temperature . To avoid overflowing the charge well in the event of a blockage in the exit port, a level sensor automatically trips the power supply off.

PUMP MAINTENANCE

The pump requires preventative maintenance. The quality of water is very critical to obtain a good life of the pump. For efficient operation the water conductivity should be checked on weekly basis despite a close circuit water system. Other electrical parts should be air cleaned at regular intervals as accumulation of dust can build stray currents and cause overheating of power supply. It is recommended to inspect the silicon carbide refractory pipe at works shut down periods or six month intervals. By experience one may find much longer service from the silicon carbide refractory pipe. In fact at Calder Aluminium plant one of the silicon carbide refractory pipe was in operation for twenty months. Replacement of pump refractory pipe may be effected with a hot furnace. After draining the furnace, a stand by pump may be installed and operation resumed within eight hours. If the same pump is used then resumption will take about 10 hours. Among other features of preventative maintenance are the temperature of the pump cooling water. This temperature should not be cooled to dew point as the coil acts as a heat exchanger and can cause condensation when the furnace is off or during restarts after a shut down. This can cause power tripping off due to earth leakage and its consequence should be first verified by first lowering the power supply to minimum and then gradually raising it as the condensed vapour on the refractory dries off after a while with the heat.

REMOVAL OF ALKALI ELEMENTS

Aluminum recycling operations have a constant need for removing magnesium from the melt. Other contaminants in the alloy can be sodium, calcium and lithium. All these elements can be removed by purging chlorine and the electromagnetic pump assists in the gas dispersion into the melt resulting in high chlorine usage. Magnesium removal rate for a nominal 14t aluminum melt, approx. 350 mm deep bath, purged with chlorine at a rate of 1kg per minute are shown in figures 5 & 6. During the test no slippage of free chlorine from the system was detectable by Draeger tubes. However, the preferential reaction of chlorine with magnesium declined in favour of aluminum as the magnesium content fell to 0.05%.

Tests to remove magnesium with fluxes have also been conducted. A few proprietary magnesium removal fluxes have been successful, without any operational problem and removing magnesium at a rate of 0.2% per hour.

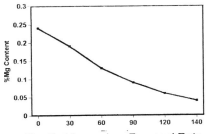

Fig. 5 : Magnesium Removal Rate with Chlorine from Aluminium Alloy

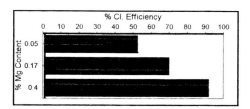

Fig. 6 : Chlorine Usage to remove Magnesium from Aluminium Melt

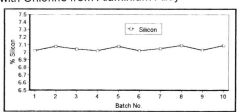

Fig. 7 : Batch Composition Variation

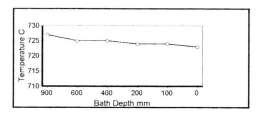

Fig. 8 : Thermal Gradient Across Bath Depth

FURTHER APPLICATIONS

The practicality of applying the electromagnetic pump to the recovery of light gauge scrap is now established. Table 4 shows a list of EMP installations. Productivity increases between 12-30% have been reported by the users of the electromagnetic pumping and stirring system. Stirring with the EMP system results in uniform temperature and chemical homogenity as Figures 7 & 8 show the variations of a 33 t capacity furnace. The system is also suitable for furnaces for heavy infeed materials, for example sows. A schematic arrangement of a close reverberatory furnace fitted with the electromagnetic pump is shown in Figure 9. The rigid construction of the electromagnetic pump coupled with the

Table 4. Existing EMP Installations

Company	Date	Furnace Size
Calder Aluminium, GB	8-92	+ 30 tonnes
Aldevienne, FR	9-94	30
FE Mottram, GB	8-95	20
Nottingham Metal Recyclers, GB	9-95	50
H Landseer-Bailey Ltd, GB	2-95	7
Asian Secondary Smelter	3-96	20
UK Primary Aluminium User	10-95	25
UK Secondary Smelter	12-95	20
Ingot/Billet Producer, EEC	12-95	25
Aluminium Extrusion Co, EEC	12-95	50
Aluminium Foundry, GB	1-96	6

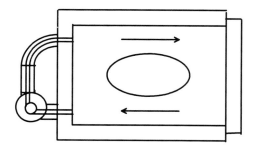

Figure 9: Typical installation of electromagnetic pump for stirring in a closed reverberatory furnace

use of flexible connections the pump is equally suited for installation on tilting type furnaces. The pumping and stirring system has also been applied to produce Al-Cu and Al-Si master alloys containing 35% copper and 27% silicon respectively at 980-1000°C.

METAL MATRIX COMPOSITES

Metal matrix composites (mmc's) are alloys often reinforced with a ceramic phase. One of the problems in producing MMC'c is to keep the heavier ceramic composite in suspension without segregation. The methods of mechanical stirring invariably cause quality problems due to gas bubble or oxide films. Coreless induction furnaces have an inherent stirring action depending on the applied frequency, coil dimensions and power density applied. However this method suffers from uncontrolled temperature rise during holding periods. Ideally, independent Control of Heating And Stirring (CHAS) is required. A new technique[3] using electromagnetic pumping and stirring these composites has overcome the drawbacks.

CONCLUSIONS

With several pumps now under continuous operation in Europe, the EMP Technologies pumping and stirring system is now an established and reliable technology for aluminum melting and refining applications. Subsurface stirring gives an effective means of melting thin gauge materials at recoveries equivalent to that obtained from an induction furnace.

REFERENCES

1. US Patent 5,350,440
2. Winwood, R., Computational Fluid Dynamics Model of the EMP Technologies Furnace Systems; EATL report no. V1635, Jan.1995.
3. P J Hayes, "Minimising Segretation in Metal Matrix composites" BCIRA International Confernece, Warwick University, 19-21 April 1994.

ACKNOWLEDGEMENTS

The authors acknowledge the commitment and dedication of their colleagues in developing this technology.

Strip Heating

Contents

Recent applications of high power transverse flux induction heating in the metal industries 1
Walker, R., Hayward, A. (Davy Corporation) UK

Towards intelligent use of induction heating in steel strip processing lines 9
Delaunay, D. (Stein Heurtey) FRANCE

Induction heating of flat metal bodies 17
Bukanin, V., Nemkov, V. (St Petersburg Electrotechnical University) RUSSIA, Dughiero, F., Forzan, M., Lupi, S. (University of Pauda) ITALY

Complementary profile heating 27
Jackson, W.B. (EA Technology) UK

Development of a direct resistance heating with transformer coupling 35
Yao, Y., Fukuyama, M., Ikuta, F., Tanino, M. (Neutren Co Ltd), Onada, M., Hirota, Y., Saito, Y., Nagase, T. (Nippon Steel Corporation) JAPAN

Non-deforming induction heating of metal discs 43
Zgraja, J. (Technical University of Lodz) POLAND

High flux induction for the fast heating of steel semi-products in line with rolling 51
Prost, G., Hellegouarc'h (CELES), Bourhis, J-C. (EDF-DER), Giffray, G. (IRSID) FRANCE

The simulation non-linear electromagnetic heating and deformation fields in the induction heating system 59
Vasiliev, A., Zarevsky, V., Lablonskaj, O. (St Petersburg Electrotechnical University) RUSSIA

The design and performance of multi-megawatt variable width TFX induction heaters for the ultra rapid heat treatment of metal strip 69
Gibson, R.C. (EA Technology) UK

The Application of High Power Transverse Flux Induction Heating in the Metal Strip Industries

RONALD A. WALKER
Davy International, Prince of Wales Road, Sheffield, U.K.
ADRIAN HAYWARD
Davy International, Prince of Wales Road, Sheffield, U.K.

Davy International have developed a Variable Width Transverse Flux Induction Heating system for strip metals, registered trade name TFX, based on licensed technology from EA Technology - Capenhurst. Evolution of the current design has spanned two 1700 mm wide multi-megawatt Production installations on aluminium, a 350 mm wide Pilot Line for copper and brass and a 500 mm Pilot Line for a range of carbon steels. The Variable Width Inductor design uses a rigid, solid coil/core assembly which has proven in service to be necessary for long term reliability and adjustable duple modifier packs on each pole to control the distribution of magnetic flux at the strip edges giving a uniform cross strip temperature profile.

A modern, high technology, environmentally friendly, energy efficient furnace option for small to medium capacity continuous annealing lines for the metals industry.

L'équipe Davy International a mis au point un système de Chauffage par Induction à Flux Transversal à Largeur Variable, pour le traitement des métaux en bande, dont le nom déposé commercial sera TFX, fondé sous licence technique de EA Technolgy - Capenhurst. L'évolution de l'étude de conception actuelle a embrassé deux installations en Production à mégawatts multiples à une largeur de 1700 mm, une Ligne Pilote d'une largeur de 350 mm pour le cuivre et le laiton, ainsi qu'une Ligne Pilote de 500 mm pour une gamme d'aciers au carbone. Le modèle d'Induction à Largeur Variable utilise un ensemble bobine/noyau rigide d'état plein, dont la fiabilité en cours de production l'a rendu indispensable. Il détient également des dispositifs modificateurs en double réglables sur chaque moitié de chaque pôle pour commander la distribution du flux magnétique aux rives de bande, dans l'objectif de fournir un profil thermique constant à traverse la bande.

Une option moderne pour l'industrie métallurgique d'un forneau de pointe à consommation efficace qui ne nuit pas à l'environment.

In the early 1970's continuous annealing of strip gauge steels was introduced by the Japanese, gaining acceptance and reaching Europe by the late 1970's. It was recognised that continuous strip annealing (CSA) gave greater consistency of mechanical properties, better shape and improved surface cleanliness over traditional batch oven methods. The Japanese designed CSA Lines gave very good results but were costly and inflexible and as such installations were, and still are, justified on very high capacity, some exceeding one million tonnes per annum, with few product changes.

Recognising the potential for CSA lines from the early Japanese work, the then Electricity Council Research Centre at Capenhurst, U.K. (now EA Technology), began development of an electrically powered alternative with a view to a more compact flexible technology which could be justified on much smaller annual tonnages. They pioneered CSA research and development with Transverse Flux Induction Heating taking advantage of the advent of the thyristor which had made medium frequency power supplies available in the megawatt ratings required to meet viable production capacities; had thyristors been available in the nineteenth century no doubt Faraday would have beaten them to it!

In 1982 Davy acquired rights to this pioneering technology and currently hold a sole worldwide licence for the Product which they market under their registered trade name TFX. The early design of Inductor had a very limited capability in coping with variations in strip width and as such installations required a magazine containing several inductors to cater for the Product width ranges required. Two multi-megawatt systems of this type were supplied to the Aluminium Industry between 1983 and 1986, a 1.8MW to Japan and a 2.8MW to Belgium. The larger of these systems had a suite of five inductors covering product strip widths between 900 and 1600 mm in gauges from 0.3 to 2.5 mm. It covered heat treatable and non-heat treatable alloys from 1000 to 6000 series, temperatures up to 550°C and speeds up to 200 m.min^{-1}. These installations provided invaluable knowledge of product behaviour under the rapid heating conditions afforded by TFX and of the beneficial metallurgical properties to be obtained. They also provided much information on magnetic flux behaviour under varying conditions and the stresses imposed on Inductor components by the induced forces. All this data was used to amend, enhance and refine the already sophisticated mathematical models used in all aspects of TFX application design. Both systems are in full production today, between them, among other things, turning out auto body and lithographic material with improved tensile/ductility ratios and superior fine grain surface properties.

A three stage inductor system was supplied to a Customer in Sweden in December 1990 with a combined power rating of 250kW. This was installed in a Pilot Line to develop the process knowhow on copper and brass strip in a protective atmosphere at temperatures between 150 and 750°C. The strip width was fixed at 350 mm with gauges from 0.2 to 1.25 mm running at speeds up to 50 m.min^{-1}. The successful outcome of the Project has seen a very clean product with finer grain size than obtained in conventional furnaces and the decision to build the inductors into a narrow strip Production facility in Finland.

Following this in December 1992, a joint development effort with British Steel saw a two stage, 800kW inductor system installed in a Pilot Line in South Wales for processing a wide variety of low carbon steel strips. Additional funding for the work was provided by the European Coal and Steel Community and by EA Technology and National Power Plc. The line had a fixed 500 mm strip width and was capable of heating gauges between 0.4 mm and 1.2 mm up to 850°C at speeds up to 50 m.min^{-1} in a protective atmosphere. The project not only addressed the thermal cycle requirements of steels ranging from ultra low carbon mild to extra-high strength types but enabled the behaviour of magnetic strip below Curie temperature to be observed and influenced under effectively production conditions. The Project is nearing its conclusion having successfully demonstrated the potential of TFX in the annealing of steel strip products; a Production facility based on this work is now actively under discussion.

All basic transverse flux inductors with cross strip coil configurations produce a characteristic flux shape and hence temperature profile across the strip width. A key part of transverse flux inductor design technology is to produce as flat a thermal profile as possible with typically $\pm 2\%$ deviation from the mean as the target (see Fig. 1). TFX technology achieves this with remotely adjustable flux modification components located appropriately on both sides of the centre of each inductor coil. These devices as fitted to the early 'fixed width' inductors were relatively simple pieces of magnetically permeable material with limited lateral movement whose sole purpose was to influence magnetic flux distribution around the strip edges where the characteristic perturbations occur.

To meet two of the key objectives of TFX, i.e. a very fast heating rate and very compact inductor design, necessitates generation of power densities up to 1.5 MW.m^{-2}; this in turn produces considerable forces within an inductor which is oscillating at twice the supply frequency. It had been confirmed, from the aforementioned installations, that to achieve the mechanical reliability demanded of todays production machinery, the primary inductor coil/core assembly needed to have no moving parts, be exceptionally rigid and to employ specially developed coil forming and insulation techniques unlike those appropriate to conventional induction heating equipment.

The ten years to 1992 had produced a wealth of operational experience with transverse induction equipment in both production and pilot line environments with ferrous and non-ferrous products. The Davy confidence in the performance and mechanical integrity of the evolved TFX product was consolidated. There remained, however, one vital step to complete in the development of TFX, the design and proving of a single inductor adjustable to a wide range of strip widths - a Variable Width Inductor codenamed VWI.

Work on a common VWI concept had started back in 1991 on two fronts. The Davy Licensor, EA Technology, had begun a research project sponsored by East Midlands Electricity Plc, Midlands Electricity Plc, Eastern Electricity Plc, Yorkshire Electricity Group Plc and National Power Plc. Davy themselves had completed some computer performance simulations, mechanical feasibility studies and preliminary designs. In September 1993 the two projects were combined under a collaborated agreement which culminated in June 1995 with the successful demonstration of a VWI. This was achieved using a converted one megawatt inductor installed in a Pilot Line at the EA Technology laboratories and was conducted on both aluminium and steel strip products ranging in width from 600 mm to 1200 mm.

The Development teams were justifiably pleased with the result because not only had they met the Variable Width objective but they had accomplished it without compromise to the coil/core design fundamentals established from the earlier work. The principle revolved around the flux modification devices which had been enhanced to not only redirect flux from above mean to below mean segments of the thermal profile but to also cancel out surplus flux. This gave better control over the temperature profile and with careful design enabled a 2:1 turndown on the maximum strip width capability of any given inductor (see Fig. 2). There was of course a penalty to pay in reduced efficiency at the narrow strip width end of an inductor range but the very high efficiency of TFX transverse flux induction heating equipment meant that even at its lowest point the efficiency was considerably higher than that achieved in traditional furnaces; typically on steel applications one could expect an efficiency range of 70 to 90%. This approach was considered far better than any potential design alternative which may have offered higher overall efficiency at the expense of long term reliability and hence operating costs.

Figure 1

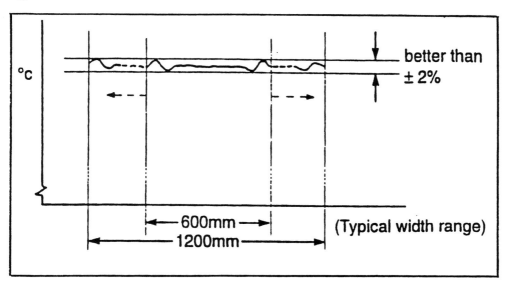

The temperature profile across the strip is uniform to within ±2% for any width within the VWI range.

Figure 2

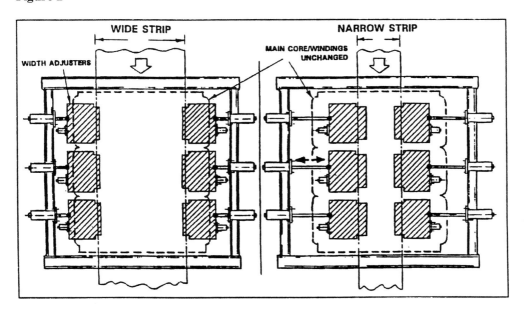

The VWI adapts to strip width by positioning of the flux modification devices without disturbance of the main core or windings.

Transverse Flux Induction Heating has now come of age and can take its place along side the well established Conventional (or Axial flux) induction heating technology to provide modern environmentally friendly, energy efficient alternatives for a great many of the Metals Industries heating requirements. The Axial Flux devices offer a lower cost solution for magnetic strip materials in the thicker gauge ranges. They are not suitable for non-magnetic materials including ferritic steels above their Curie temperatures and, within sensible limits of power supply frequency, soon become impractical in overall physical length as strip gauges get thinner and/or production capacities increase. Transverse Flux Induction heaters on the other hand are more costly to produce but operate equally successfully on magnetic and non-magnetic materials. They operate with much lower supply frequencies, usually well below the radio band, are more efficient to operate at temperatures up to 1250°C and can process relatively thin strip gauges. Because of their high power density capability they are also far more compact in achieving realistic production throughputs. There are some carbon steel applications which are arguably more cost effective using a combination of Axial and Transverse flux equipment for the below and above Curie parts of the cycle.

Davy have developed a PC based integrated control package for the TFX-VWI which makes the most of the flexibility and controllability of the Product. The fact that in a TFX furnace only the strip gets hot makes line speed and grade changes quick and easy under fully automatic control. This makes TFX of particular advantage for products which do not have tight time dependent elements in their process cycles, such as overageing. Even on processes which are restricted to fixed line speeds to afford strict thermal regimes, TFX is a more efficient and compact alternative to direct or indirect gas fired CSA furnaces on medium size lines where annual tonnages do not exceed say 250k tonnes of steel or 100k tonnes of aluminium. For standard products with higher annual capacity requirements the Japanese CAL and CAPL equipment comes into its own.

To complement the TFX furnace, Davy have evolved, through parallel development, an atomised water spray quench system which is capable of taking heat out of the strip very evenly across the width at rates up to $100°C.s^{-1}$ This is aimed at aluminium products and those steel products which require an essentially sawtooth cycle. Gas cooling options can be incorporated in TFX Line installations where required which would be sourced from proprietary suppliers of such equipment.

Typical applications that are foreseen for TFX are as follows:-
1)   Compact annealing lines for aluminium alloys capable of half hard, three quarter hard, fully soft

and solution treatments across the spectrum of alloys as appropriate. Because of the non-time dependent nature of these processes full advantage can be taken of TFX flexibility. Speeds can be reduced to around 10 m.min^{-1} for coil changes and ramped to full production speeds for processing which allows for very small capacity strip accumulators, single coil batches and minimal scrap losses (see Fig.3).

2) Compact annealing lines for IF steels, high strength steels and niche market steels which require simple heat treatment cycles. The process will be contained in a protective atmosphere which, coupled to the very short time at temperature afforded by rapid TFX heating, will result in very light oxidation even when combined with a rapid water mist quench; pickling will therefore be much simplified. Lines can incorporate temper rolling mill facilities if required.

3) Vertical Bright Annealing (VBA) lines for stainless steels. These are by nature extremely tall constructions conventionally made so by the long radiant tube heating stages employed. TFX can reduce the height by up to 40% because of its compactness giving the obvious savings on civil and building aspects as well as more efficient operation.

4) Compact annealing lines for copper and brass strip again in a protective atmosphere with rapid water quenching. There is also the potential to boost the throughput of existing conventional furnace lines by supplementing or replacing one or more sections with a TFX unit and to avoid the Beta phase issue by using the TFX to raise the temperature earlier in the cycle giving more time at higher heat levels.

5) Curing ovens for laminated or other polymer and paint coatings in high speed process lines where advantages of finish quality and adhesion integrity can benefit from heating the substrate to cure from below as opposed to direct radiant heaters from above.

These are just a few of the opportunities envisaged for the TFX Product which it is believed offers an environmentally acceptable, energy efficient solution for small to medium size strip metal continuous heat treatment processes in the future using electricity.

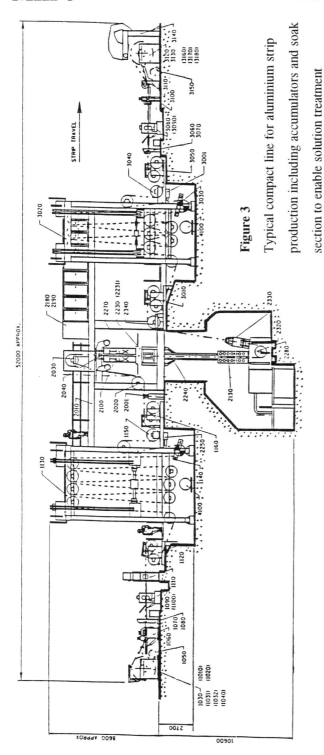

Figure 3

Typical compact line for aluminium strip production including accumulators and soak section to enable solution treatment

ITEM	DESCRIPTION
1000	ENTRY SECTION
1010 x	COIL STORAGE SADDLES
1020 x	ENTRY COIL CAR & RAILS
1030	UNCOILER WITH CENTRE GUIDE
1040	SPOOL UNLOADING
1050	PEELER & BREAKER ROLL
1060	PINCH ROLL & FLATTENER
1070	ENTRY SHEAR & PINCH ROLLS
1080	SHEAR SCRAP DISPOSAL
1090	EDGE TRIMMER (OPTION)
1100 x	SCRAP BALLER (OPTION)
1110	STRIP JOINER
1120	No.1 BRIDLE AND DRIVE
1130	ENTRY STORAGE LOOPER
1140	STEERING ROLL
1150	No.1 TURN ROLL
1160	No.2 BRIDLE & DRIVE (B)
2000	PROCESS SECTION
2001	STRIP THICKNESS GAUGE
2010	SUPPORT STRUCTURE STEELWORK
2020	No.1 DEFLECTOR ROLL
2030	No.2 DEFLR ROLL & TENSIOMETER
2040	ANTI-FLUTTER ROLLS (IND ENTRY)
2100	INDUCTOR
2110	INDUCTOR TRAVERSE
2120	OFF LINE STATION.
2130	STRIP QUENCH SYSTEM
2140 x	COOLING SYSTEM (ELECTRICS).
2150	TFX FUME EXTRACTION
2160	STEAM EXTRACTION SYSTEM
2170 x	HV/LV TRANSFORMER
2180	FREQUENCY CONVERTER
2190	O/P TRANSF, CAPACITOR CUBICLE BUS BARS/FLEXIBLE CABLE.
2200 x	LV/AC DISTRIBUTION BOARD.
2210	INDUCTOR GUIDANCE SYSTEM.
2220 x	CONTROL SYSTEM COMPUTER.
2230	PYROMETERS
2231 x	PAINTED STRIPE CALIBRATION
2240	THERMAL SOAK SYSTEM
2250	STITCH DETECTOR
2260	CONTROL BOX-INDUCTOR TRAVERSE
2270	ACOUSTIC ENCLOSURE
2280	QUENCH ROLL & DRIVES
2290	QUENCH ROLL REMOVAL EQUIPT.
2300 x	QUENCH CIRC INCLUDING SOFTENING
2310 x	COOLING TOWER
2320	QUENCH SQUEEGEE SYSTEM
2330	AIR KNIFE SYSTEM
2340	No.3 DEFLECTOR ROLL, WITH TENSION RELIEF.
3000	EXIT SECTION
3010	No.3 BRIDLE AND DRIVE (B) No.2 BRIDLE AND DRIVE (A)
3020	EXIT STORAGE LOOPER
3030	STEERING ROLL
3031	EXIT STITCH DETECTOR (B)
3040	No.2 TURN ROLL
3050	No.4 BRIDLE AND DRIVE
3060	EXIT SHEAR AND PINCH ROLLS
3070	SHEAR SCRAP DISPOSAL
3080	EDGE TRIMMER (OPTION ONLY)
3090 x	SCRAP BALLER (OPTION ONLY)
3100	TEST SAMPLE PRESS
3110	PINCH R/DEFL & THREAD-UP TABLE
3120	RECOILER WITH EDGE GUIDE
3130	BELT WRAPPER
3140	SPOOL LOADING & STORAGE
3150 x	EXIT COIL CAR & RAILS
3160 x	COIL CAR STORAGE SADDLES
4000	ANCILLARY EQUIPMENT
4060	LOOPER ROLL REMOVAL RIG

x ITEMS NOT SHOWN ON THIS DRG.

Towards intelligent use of induction heating in steel strip processing lines

D. DELAUNAY

STEIN HEURTEY - Z.A.I. du Bois de l'Epine - 91 130 RIS ORANGIS - FRANCE

ABSTRACT :

Induction heating is increasingly used in continuous processing lines on account of its advantages. Its flexibility in particular enables to minimise transient states between two successive coils. Through two concrete examples we show that this potentiality can be more advantageously utilised when induction is the only heating mode. However, the cost of electric energy generally dissuades from choosing this configuration, especially in high-productivity lines. In this case nevertheless, using a low-power booster makes it possible to significantly reduce the constraints of transient situations when the booster is controlled in connection with the main furnace.

RESUME :

Le chauffage par induction est de plus en plus utilisé dans les lignes continues, en raison de ses avantages. Parmi ceux-ci, la flexibilité permet de réduire les transitoires thermiques d'une bobine à l'autre. A travers deux exemples concrets de chauffage avant galvanisation, nous montrons que cette potentialité est plus facile à exploiter lorsque l'induction est l'unique mode de chauffage. Cependant le coût de l'énergie électrique dissuade généralement d'opter pour cette configuration, en particulier pour des lignes à grande productivité. Dans ce cas l'adoption d'un booster de puissance modeste permet néanmoins, s'il est piloté en couplage avec le four principal, de repousser les contraintes des transitoires de manière significative.

1. INTRODUCTION

For many years, induction heating has been used in the steel industry. As regards continuous heating of thin strips, the first industrial installations of really modern design were built in coating lines in Germany as from the early sixties [1]. These first applications of inductive heating were favoured by the low cost of electric energy. Afterwards, the development of industrial applications was still linked to progress made in induction power sources, solid state supplies replacing motor power generators and tube oscillation power sources. At the same time as the performance of semi-conductors improved, the number of steel strip processing and coating processes increased in order to meet metallurgical requirements or enhance the corrosion resistance of sheets.

Induction heating has naturally found a place in these processing lines since it fully satisfies the needs of such installations thanks to its well known advantages:

- Higher product quality due to greater heating accuracy and especially to reproducibility. The heating rate can be much more easily calculated and modelled as power is injected directly into the strip. The strip surface finish -particularly its emissivity- does not affect the heating efficiency.

- Enhanced productivity, since thermal transients are eliminated. There is indeed no thermal inertia.

- Greater flexibility in line operation, ON-OFF control, no strip overheating in furnace in case of shutdown of line speed.

- Compact design due to very high flux density so installation is easier in an existing line.

- Low maintenance costs.

- High thermal yield.

2. INDUCTION DEDICATED TO THE PROCESS

Taking advantage of induction characteristics, STEIN HEURTEY, jointly with CELES, has equipped several processing lines with induction heating[2]. These facilities cover most of strip annealing and coating processes; induction is used either as booster or as the main heating mode.

Ours applications of induction heating are illustrated by a first example. It concerns an all-electric annealing furnace, prior to galvanising, operated under nitrogen and hydrogen atmosphere. The yearly capacity of this line amounts to 90,000 metric tons.

The specific features of the line, commissioned in 1989[3], are summarised in table 1 (case 1). The heating chamber is composed of an electric resistor preheating section (1,660 kW), followed by the induction heating section (3,000 kW), and final heating and soaking with 840 kW electric resistors. The induction part comprises 5 in-line inductors and enables to raise the strip from 350 to 700° C. The inductors work in a tight enclosure, the heat and power losses are minimised thanks to the special design of the furnace-inductor assembly.

Power control of the inductors is performed by a computer which applies a set-point preset by the production parameters and partly corrected by a value measured by pyrometer.

With this control mode, inductors are process-bound, just like the sections heated by electric resistors.

3. INDUCTION OPTIMISES PRODUCTION

In the preceding example, induction, though predominant, is controlled as any heating device, i.e. the electric resistors. Consequently, it is not possible to fully utilise the potentialities of inductive heating, and especially the flexibility resulting from the absence of thermal inertia.

Table 1

Characteristics of induction heating section

	CASE 1	CASE 2
Maximum strip width	1,350 mm	750 mm
Strip thickness	0.5 to 1.5 mm	1.5 to 6 mm
Maximum line speed	90 m/min	90 m/min
Maximum output	28 t/hr	30 t/hr
Initial temperature	350° C	20° C
Final temperature	700° C	500° C
Quantity of inductors	5	2
Frequency	8 kHz	6 kHz
Total rated power	3,000 kW	3,000 kW
Total heating length	18.6 m	7.2 m

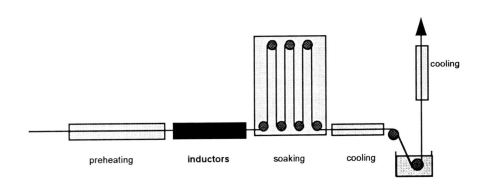

"Fig. 1"

Case 1 : Process section lay-out

To demonstrate this potential interest of flexibility, we will describe an induction unit recently started up by STEIN HEURTEY and CELES. As in the preceding case it concerns a heating installation before galvanising. However, this line features a new concept of hot rolled narrow strip galvanising[4], which replaces a conventional batch process.

In this line, the strip is currently heated from 20 to 500° C by induction only. The yearly production amounts to 150,000 tons. The maximum strip width is 750 mm, and the strip is heated by two identical inductors. The total connected power, the maximum strip travel speed and the hourly production rate can be compared to those of the preceding line on table 1 (case 2), with a more compact arrangement (figure 2).

In its design the induction furnace includes several innovative concepts :
- magnetic shields offering high electrical efficiency and use of conventional materials in the vicinity of inductors (rolls, casing)
- impedance matching for increasing power available for narrow strips
- complete tightness as inductors are capable of heating the strip to 500° C in an atmosphere which could contain up to 100 % hydrogen.

While the absence of thermal inertia is now utilised with its full potential, the objective here is not only actual ON-OFF operation of the line that can be operated without any problem in 2 shifts for instance, but also on-line management of transient states, thereby enabling the strip thickness or width to be doubled.

This management of transients is obtained by inductors control at supervisory level. It consists in adjusting the largest size to 100 % of the inductor power and to make the power decrease (or increase) speed-dependent so as to maintain a constant temperature. Figures 3 and 4 illustrate thickness decrease and increase respectively.

Moreover, this level 2 model also integrates constraints resulting from zinc wiping in order to control the coating thickness according to the strip travel speed for instance.

These two combined functions enable to achieve actual flexibility and to obtain noticeable production and quality gains during transients.

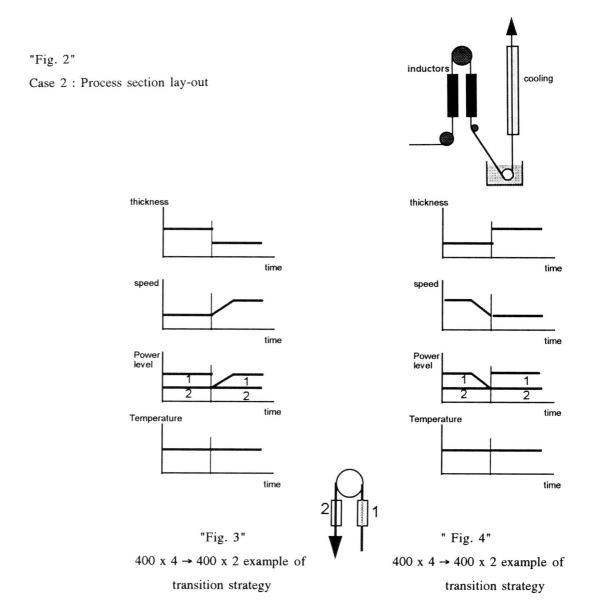

"Fig. 2"

Case 2 : Process section lay-out

"Fig. 3"

400 x 4 → 400 x 2 example of transition strategy

" Fig. 4"

400 x 4 → 400 x 2 example of transition strategy

4. TAKING ADVANTAGE OF INDUCTION AT LOWER COST

As we explained, a suitably controlled all-induction heating therefore makes it possible to get the most of this heating mode, if it needs to be proved. Of course, the economical aspects (investment and energy costs) restrict the induction use to small capacity plants, in spite of a better thermal efficiency. However, high returns can be obtained with a specific utilization of induction within the furnaces. It is the booster concept [5].

The booster is an inductor arranged in a conventional processing line, either a new or an existing one. It is used to make up for the furnace thermal inertia, the heat demand due to a change in operating conditions, such as thermal cycle or strip size, being instantaneously compensated.

In case of increased heat demand, the inductor supplies the additional energy required until the furnace has reached its new settings : it is the positive booster (figure 5). In case of decreased heat demand the booster maintains the initial power while the energy of the main furnace is lowered to its new value: it is the negative booster (figure 6).

With this concept, the inductor and the furnace must obviously be controlled in parallel. The line control strategies are then applied to all of the process actuators: inductor, zinc wiping if any, and so on [6]. Combined with this optimising system, the inductor provides the process with greater flexibility and lower connected power.

Typically, the booster represents 10 % of the connected power in the main furnace and approx. 2 % of the energy consumed. The production gain calculated for a concrete example represents two thirds of the production losses due to transient strips, when conventional strategies are applied such as dummy coil, constant over heating or "laissez faire" approach.

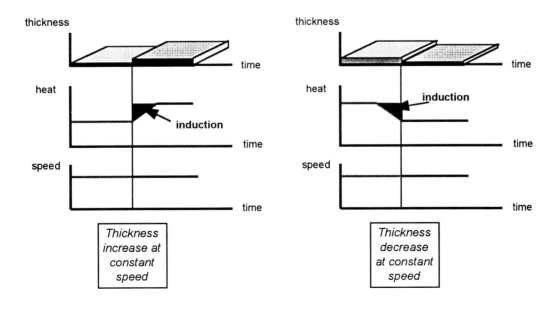

"Fig. 5" "Fig. 6"

Positive booster Negative booster

5. CONCLUSION

To get the most out of induction heating at its lowest cost is necessary to make induction popular. The intelligent control of the induction within the whole process actuaors can allows to get out more than 80 % of induction advantages with less than 20 % of total installed power.

REFERENCES

1. B. MEUTHEN: 'Coil Coating in Western Germany', *Proceedings of the NCCA conference,* Scottsdale, Arizona, May 14-17, 1978.

2. R. WANG, Th. DERIAUD: 'Induction Heating for Continuous Heat Treatment of Steel Strips', *Proceedings of the European Congress on Induction*, Strasbourg, France, March 20-22, 1991.

3. G. RAMBAUD, Th. DERIAUD, Y. RIOLLET: 'All Electric Equipment for the Process Section of a Continuous Galvanizing Line', *Proceedings of Galvatech'92 conference*, Amsterdam, September 8-10, 1992.

4. W. PUSCH: 'Down Stream Processing of Rolled Strip. Response to a Trend toward Value-Added Products'. *Proceedings of the International Symposium on Flat Steel Products Technology,* Cairo, Egypt, October 23-25, 1994.

5. M. BOYER: 'Procédé de Traitement Thermique de Bandes Métalliques', *French Patent 92.03311*, 1992.

6. P. DUBOIS, M. BOYER: 'Global Approach to Solve CGL Process Inertia Troubles and Application to Mathemtical Models' *Proceedings of the Galvatech' 95 Conference*, Chicago, September 18-20, 1995.

Induction Heating of Flat Metal Bodies

BUKANIN V.[*] - DUGHIERO F.[**] - FORZAN M.[**] - LUPI S.[**] - NEMKOV V.[*]

[*] - St. Petersburg Electrotechnical University (Russia)
[**] - Dept. of Electrical Engineering - University of Padua (Italy)

Abstract

The problem of the induction heating of flat metal bodies is still open in spite of a number of theoretical researches and experimental studies. In this paper a short analysis of the present situation in the induction heating of strips and slabs has been developed. The results obtained by the authors in the frame of an international Russian-Italian project are discussed. The main attention is paid to transverse flux systems with a short overview on longitudianal flux and multi-phase travelling wave induction heating of flat metal workpieces.

Résumé

Le problème du chauffage par induction des corps métalliques plats est toujours actuel malgré les nombreuses recherches théoriques et expérimentales. Dans ce travail deux aspects ont été développés. D'une part une brève analyse sur les connaissances actuelles du chauffage par induction des bandes et des produits plats et d'autre part les résultats obtenus par les auteurs dans le cadre d'un projet international Russe-Italien. On traite en détail le chauffage en flux transverse avec quelques aperçus sur le chauffage en champ progressif des produits métalliques minces.

1. Introduction

The induction heating of flat metal workpieces is applied in a variety of industrial applications and different types of inductors (e.g. longitudinal flux, transverse flux, travelling wave inductors) are used according to the process requirements.[1] In the present paper the analysis is focused on the transverse flux heating (TFH), with an overview to the other types of inductors.

The following significant achievements in induction heating technique took place in the last years:
- industrial application of high-efficiency transistor generators for frequencies up to 500 kHz
- development and practical use of 2D and 3D computer codes .

These achievements have influenced strongly the choice and design of inductors for heating of flat metal workpieces (e.g. slabs, sheets and strips).

2. Longitudinal Flux Heating (LFH)

The system geometry corresponds to the schematic of figure 1; its main inherent features are:
- simple and reliable coil design, based on well established technologies
- relatively simple and well predictable distributions of the eddy currents, the resulting power sources and temperature patterns
- non-critical dependence of efficiency and temperature distribution on limited variations of workpiece properties, size and positioning inside the coil

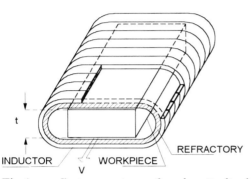

Fig.1 - *System set-up for longitudinal induction heating. [t-workpiece thickness; v-direction of movement]*

- good uniformity of power distribution in the flat workpiece width for proper frequency choice.

These features of LFH make it attractive and widely used for flat bodies heating.[1,2] Practical applications are in a wide range from low frequency high power heating of big slabs to local high intensive heating of blades and knife edges at very high frequencies up to 27 MHz.

- **Slab heating**

Correct frequency choice is of a great importance for slab heating. Theoretical studies show that the most uniform distribution of power sources in the width of the slab and hence the most rapid heating correspond to a ratio "slab thickness/penetration depth" $t/\delta \cong 3$.

For $t/\delta <3$, the efficiency falls down and underheating of the edge zones takes place. For $t/\delta >3$, the coil efficiency remains approximately constant while an excess power is observed in the edge zones due to the so-called edge-effect of "rectangular" cross-section workpieces.

Since a certain excess power can be useful for compensation of thermal losses from the edges, the most quick and effective heating of slabs occurs with $t/\delta = 3.5 - 4$.

Line frequency may be therefore effectively used for LFH of slabs with thickness greater than 250 mm for steel, and 50 mm for aluminium alloys.

The limiting factors of slab heating speed are:
- overheating of edge zones for "thick" slabs, when the frequency is higher than optimal
- high temperature gradients in steel slabs due to poor thermal material conductivity
- thermal stresses which can provoke internal cracks at the "cold" stage of the process

- big electrodynamic forces (up to 250 N/kW), especially for aluminium slabs heated at line frequency.

These forces affect the induction system components causing strong vibration, noise and possible faults of thermal and electrical insulation as well as mechanical damages.

The above factors must be taken into consideration along with economical analysis.

The most favourable application of the induction slab heating is the in-line temperature field correction in the continuous casting-hot rolling process. Both line and medium frequencies up to 1000 Hz may be used in this case. Medium frequency installations, even if more expensive, provide better power control, uniform three-phase loading of the supply line, more accurate temperature pattern correction and lower electrodynamic forces.

- **Strip heating**

The other threshold case of LFH is thin strips heating: radiofrequencies are required in this case in order to obtain a ratio $t/\delta \geq 3$ and good coil efficiency.

Modern transistor generators have not only high efficiency (η = 85-90%) compared with tube oscillators (η = 0,6), but much less weight and low floor space demand. A relatively low output voltage (400-1000 V) allows in many cases to avoid the use of matching transformers, giving additional gains in the installation efficiency.

These factors strengthen the position of LFH in its old competition with TFH. The overall electrical efficiency of the installation may be around 65 - 70% for strip thickness $t \geq 2$ mm in the case of stainless steel, titanium alloys or plain steel above Curie point, while these values are reduced to 35-40% and 0,4 mm respectively for aluminium alloys.

For more thin strips the use of tube generators is required with consequent considerable decrease of the efficiency. At the same time bare and metal-coated magnetic strips may be heated with high efficiency in longitudinal field in a wide range of thicknesses.[1,3]

3. Transverse Flux Heating (TFH)

TFH is an effective method for heating thin flat workpieces (e.g. sheets and strips).[1-4]

The main advantages of TFH are:
- high electrical efficiency, not achievable with LFH. It is of a special importance for materials with low resistivity such as aluminium or copper alloys
- possibility to use much lower frequencies (50 Hz - 10 kHz) compared to LFH.

This method, however, has also some disadvantages:
- non-uniformity of power and temperature distributions with overheating at the edges

- strong dependence of heating uniformity upon coil design and variations of strip material properties, size and position inside the coil
- big electromagnetic forces which may be critical for a successful application of the process
- complexity of coil design due to the multivariable nature of the problem.

The minimum number of variables for a classic TFH inductor is four if the strip material is non-magnetic. A much greater number of variables must be taken into account if field correction means are used.

A lot of field correction methods and means such as variation of coil winding form, gap value, magnetic circuit geometry or their combination have been proposed. They give good results for particular applications; some of them permit a coil adaptation to strip width variation[3], but there are no general guidelines for solution of the problem of uniform and efficient TFH till now.

A long-term theoretical research and experimental tests, carried out by the authors in the frame of a joint Russian-Italian project, can contribute to TF coil design for different applications.[5,6]

Different geometries of coil windings are used in TFH but all of them may be considered as a combination of two threshold cases.

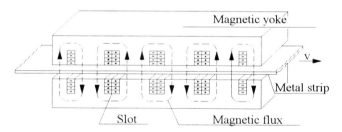

Fig. 2 - *TFHTC- Transverse Flux Heater with Transverse Conductors*

The classic TFH inductor (figure 2) has slots with current carrying conductors located perpendicular to the length of the strip and to the direction of its movement. It is a transverse flux heater with transverse conductors (TFHTC).

A similar inductor, rotated by 90°, constitutes a transverse flux heater with longitudinal conductors (TFHLC). In this system the magnetic field is mainly 2D and the eddy current pattern is more favorable for strip edges heating control.

In single phase coils, the TFH provides periodical waving distribution of power in the strip, with zero values under the poles axis. These oscillations do not influence strongly the temperature pattern in TFHTC since they are "equalized" by the strip movement, while the contrary happens in TFHLC and special means must be provided to reach temperature uniformity.

Multi-phase TFH inductors

Multi-phase inductors allow to obtain eddy current patterns not available for single-phase coils; as a result, they open new ways for field and temperature profiles control.[5]

The other significant advantage of multi-phase TFH consists in the reduction of the alternative component of the electromagnetic forces affecting the system elements. It decreases strongly vibration and noise of the coil structure. This feature of multi-phase coils is of special importance for heating large aluminium strips and slabs.

A threshold case of multi-phase heating, with the exciting currents distributed uniformly along the coil surface with phase changing linearly in distance, is the so-called travelling wave induction heating.[1,5] In this case the resulting normal component of the alternative force affecting the coil structure is equal to zero but a tangential force acting either on the strip or the coil, appears and must be taken into consideration in the inductor design.

4. Simulation methods

Modern sofware packages allow a 3D calculatation of electromagnetic fields and a prediction of temperature distributions.[4,6] However a big number of variables and high computation times make a direct 3D simulation time- and cost-consuming, especially for coupled electromagnetic and thermal problems. Good preliminary approaches are necessary for the effective use of 3D codes. A set of software codes for TFH simulation have been developed, adapted and tested by the authors for this purpose.

- **2D** *Analytical methods*

The codes based on these methods refer to a simplified representation of TFH inductors with arbitrary distributed current sheets of different width, as sketched in figure 3.[5,8] The currents may have different magnitudes and phase, with the only limit that the algebraic sum of all currents must equal zero. Fourier series tranformation has been used for field calculation.

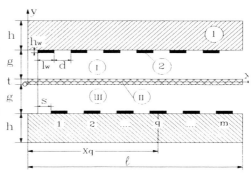

Fig. 3 - *Idealized 2D model of a double-side TWIH system [I,III and II-airgaps and load; 1,2-magnetic yoke and exciting current sheets; l, l_w-inductor and current sheets lengths; d-current sheets spacing; t, g, h, h_w-dimensions along y; s-shift of upper and lower inductor]*

These methods permit to obtain a good approach to the exact solution for the regular part of the system and to study the influence of coil geometry, frequency and strip properties on coil efficiency and power factor. Analytical methods are particularly effective for multi-phase coil simulation, since they permit to study the power distribution due to the coil end-effects.[5] The influence of the slots on the coil parameters cannot be taken into account with these methods and has been studied by numerical 2D-FEM codes.

- *Numerical methods and calculations*

2D and 3D numerical codes have been used for studying both LFH and TFH systems.

2D-FEM codes allow to take into consideration the real geometry, e.g. yoke slots and coil conductors cross-section.

- A typical example of 2D LFH system simulation is given in figures 4-a and -b. It refers to a comparison of calculation results with experimental data available in the literature[7] for the heating of a rectangular slab (70x20 mm) in a LFH inductor at frequency 9825 Hz. Slab material is nonmagnetic stainless steel X5CrNi18/9.

The diagrams of figura 4-a give calculated and experimental results, while figura 4-b shows the temperature pattern in the slab cross-section of the slab at the end of the heating period.

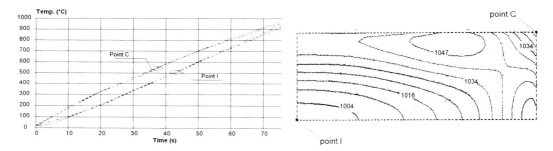

Fig. 4-a - *Comparison of calculated and experimental temperature values at points C and I of figure 4-b*

Fig. 4-b - *Temperature pattern for a quarter of the cross section of the steel slab at the end of the heating period*

Some difficulties arise, however, when simulating multi-phase systems or single-phase coils with parallel circuits. In fact, the used package requires the preliminary knowledge of all coil currents. When the exciting currents are not known a priori, an impedance matrix must be found to couple coil currents and applied voltages. Each element of the matrix corresponds to the mutual impedance of two coil circuits in the presence of the strip and a set of calculations must be performed. The matrix depends on the system geometry only and remains the same for different supplying conditions.

3D FEM codes require a big number of mesh elements;[6] their use creates some difficulties for big systems containing a set of slots. They represents, however, the unique accurate method for field calculation in the edges zones where the magnetic field is essentially 3D.

Taking into account a periodic coil structure and planes of symmetry (fig.2), relatively small parts of the induction system may be studied and a mesh of 20.000-40.000 elements can be usually sufficient for the problem solution. The results of simulations were checked with test data.[6]

- A special example of TFH calculation is constituted by the CAD of an inductor for gold strip heating with the following process specifications: *Material:* strip of 750‰ gold alloy, 45 mm width, 1.5 mm thickness; *Temperature:* 625±10 °C (final), 700 °C (max.); *Air-gap* (between upper and lower inductor): 25-35 mm (for insertion of a protective gas chamber); *Strip velocity:* 10 m/min.

Both TFHLC and TFHTC solutions have been analyzed. Due to the small strip width and the relatively big air-gap, conventional methods of power control for TFH were not applicable. High thermal material conductivity is favorable for temperature equalization. A frequency of 3 kHz has been chosen after preliminary evaluation. 3D simulation of coupled electromagnetic and thermal problems helped to find the most favorable dimensions of the coil for reaching good temperature uniformity, efficiency and power factor.

- **TFHLC** - Figure 5-a shows the system geometry: the inductor is subdivided into two separate sections with "hair-pin" windings and magnetic yokes for efficiency improvement. Fluxtrol A magnetodielectric material has been chosen for the yokes, due to its favorable magnetic and physical properties for this application. The windings connection allows different ampere-turns in the two sections, providing higher power density and heating acceleration under the first block in order to meet customer requirements for a limited overall inductor length.

Figure 5-b gives the corresponding transient temperature distribution in the gold sheet.

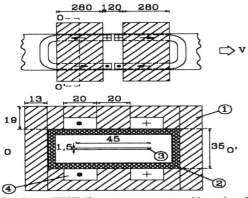
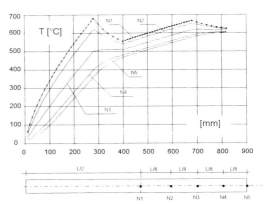

Fig. 5-a - *TFHLC system geometry [1- yoke; 2- refractory; 3-load: 750‰ gold strip; 4- inductor: 2 separate sections with different currents, f=3 kHz, I_1=1000 A, I_2=500 A]*

Fig.5-b - *TFHLC: transient temperature distributions at points $N_1,...,N_5$ of the strip cross-section [inductor: as in fig.5-a; strip velocity: 10 m/min]*

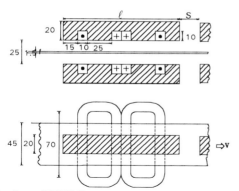

Fig.6-a - *TFHTC system geometry [inductor: 3 separate sections as the one in figure, f=3 kHz, l=120, I=2120 A, s=120 mm; load: 750‰ gold strip]*

Fig.6-b - *TFHTC: transient temperature distributions at points $N_1,...,N_5$ of the strip cross-section [inductor: 3 spaced sections as in fig. 6-a; stip velocity: 10 m/min]*

	Table I	
	TFHTC	TFHLC
Gap [mm]	35	25
I [A]	2120	1000/500
P_u [kW/m]	54.4-58.6	56.4/15.7
η	69 %	79.5 %
cos φ	0.20	0.32

- **TFHTC** - The inductor system consists of three equal sections, conveniently spaced in the movement direction, each corresponding to the schematic of figure 6-a, with "butterfly"-type windings. Figure 6-b gives, for this inductor design, the corresponding transient temperature distributions in different points of the strip cross-section. The main integral parameters (induced power per unit length P_u, efficiency η and power factor) for the two cases are given in table I.

Conclusions

A set of 2D and 3D codes for the electromagnetic and thermal CAD of induction heating systems for flat workpieces has been developed. The examples given show that transverse flux heating constitutes a flexible and efficient means which allows to obtain the required heating temperature distribution with different types of inductors supplied at medium frequency. The TFH inductors with either longitudinal or transverse conductors may be used: the first type gives higher efficiency while the second provides better temperature distribution.

References

1. S.LUPI et Al.: "Induction Heating/ Industrial Applications", *edition UIE*, Paris (France), 1992
2. J.M.DE HOE, J.LOVENS:"Continuous drying of coating on induction heated thin plates", UIE-Electrotech '92, *UIE-Electrotech '92, Proc. XIIth Congress Int. Union for Electroheat*, Montreal (Canada), 14-18 June, 1992, 140- 149.

3. N.ROSS et al.:"Transverse Flux Induction Heating of Steel Strip", *UIE-Electrotech '92, Proc. XIIth Congress Int. Union for Electroheat*, Montreal (Canada), 14-18 June, 1992, 110-119

4. B.NACKE, W.ANDREE, J.U.MOHRING, H.LEßMANN:"Induktive Querfelderwärmung - eine flexibile und effiziente erwärmungsmethode für metallische Flachprodukte im walzwerk", Elektrowärme Int. **51** (1993) B4, B156-B166

5. A.ALI, V.BUKANIN, F.DUGHIERO, S.LUPI, V.NEMKOV, P.SIEGA: "Simulation of multi-phase induction heating systems", *2nd Int. Conf. on Computation in Electromagnetics*, 12-14 April 1994, University of Nottingham (UK), 211-214

6. V.BUKANIN, F.DUGHIERO, S.LUPI, V.NEMKOV:"3D-FEM Simulation of Transverse-Flux Induction Heaters", CEFC 94, Aix les Bains (France), 5-7 July 1994, 306; *IEEE Trans. on Magnetics,* vol.31, n.3, May 1995, 2174-2177

7. C. CHABOUDEZ ET AL.: "Numerical modelling of induction heating of long workpieces", *IEEE Trans. on Magnetics*, vol.30, n.6, Nov. 1994

8. F.DUGHIERO, S.LUPI, P.SIEGA:"Analytical calculation of double-side planar travelling wave induction heating systems", *ISEF'95 - International Symposium on Electromagnetic Fields in Electrical Engineering*, Thessaloniki, 25-27 Sept. 1995, 311-314

Complementary Profile Heating

W.B. JACKSON
EA Technology, Capenhurst, Chester, CH1 6ES

ABSTRACT

Complementary Profile Heating is a new transverse flux induction technique that has been developed at EA Technology for the heating of flat metal strip and thin slab. Unlike existing methods it does not rely on a relative motion between the inductor and the workpiece but creates a required thermal profile by superimposing two or more profiles sequentially across a workpiece. The concept is a particularly powerful one, enabling the design of flexible inductors, working at up to 1 MHz on materials of variable width with dynamically changing thermal profiles.

The inductors can be operated with either the usual pulsating magnetic field or as travelling wave furnaces if circumstances demand. As a result of this thermal profiling flexibility, these devices are well suited to the needs of modern continuous strip processing lines where an on-line thermal profiling capability can be of crucial importance in obtaining specific material properties.

Complementary Profile Heating est une nouvelle technique d'induction de flux transversal qui a été mise au point par EA Technology pour le traitement thermique des feuillards et des brames aplaties.

A la différence des méthodes existantes, celle-ci ne repose pas sur un mouvement relatif entre la bobine d'inductance et la pièce à travailler, mais crée le profil thermique voulu en superposant successivement deux profils ou plus en travers de la pièce. Ce concept est particulièrement puissant puisqu'il permet de concevoir des bobines d'inductance flexibles, pouvant fonctionner jusqu'à 1 MHz sur des matériaux de largeur variable avec changement dynamique des profils thermiques. Les bobines d'inductance peuvent être utilisées soit avec le champ magnétique pulsé classique soit comme fours à ondes progressives si les circonstances l'exigent.

Du fait de cette flexibilité du profil thermique, ces dispositifs conviennent bien aux exigences des installations de recuit continu de feuillard où la capabilité du profil thermique en ligne peut être d'une importance capitale pour obtenir des propriétés spécifiques d'un matériau.

INTRODUCTION

It has long been recognised that Transverse Flux Induction Heating (TFIH) can be a particularly powerful and efficient method for heating continuous lengths of flat metal workpieces.[1] There is in principle no upper limit to the thicknesses that can be treated but in practice limitations may arise due to either the manner in which the technique is applied or, as with longitudinal heating, the increasingly nonuniform deposition of energy through the thickness of the workpiece necessitates some form of thermal soak cycle. For electrically thin materials it may be necessary to operate at frequencies above 0.5MHz but to all intents and purposes the method is applicable to all wide metal strip.

Given the fact that almost all rotating electrical machines are of a basically transverse flux design, it is surprising that the technique has not been exploited to a much greater degree in the field of induction heating. The main problem in utilising this familiar technique however has been that of harnessing the innately non-uniform heating distributions associated with transversely directed fluxes so as to produce sensibly controlled, even uniform, thermal profiles across the width of a workpiece. To date, almost all the work carried out in this field has utilised a spatial averaging concept to obtain a satisfactory level of thermal profile control. This makes use of the natural spatial averaging process, occurring as the heated material passes through the inductor, to create the required thermal profile out of an undesirable distribution of induced heating intensity.

Complementary Profile Heating (CPH) is a novel technique, applicable to either moving or statioanary workpieces, which overcomes this basic problem in an entirely different way. In this technique the required heating distribution is synthesised from an appropriate combination of two or more specific heating profiles. These profiles can be applied sequentially in time or contemporaneously and are arranged to be complementary to one another in a spatial sense so as to produce the requisite thermal profile. EA Technology have filed patent applications to cover this concept in the UK and abroad.[2]

Several different methods have been devised for applying the complementary components to workpieces, the elapsed cycle times varying from 0.15ms to 30ms. These all result in particularly small transient temperature differentials within the workpiece compared to the conventional TFIH methods using spatial averaging processes. This may have an important bearing on the amount of thermal distortion caused within strip material as a result of a rapid heating cycle.

The principal advantages of CPH over existing methods however relate more to its operational flexibility. Not only is it possible to cater for a wide range of material widths and thicknesses within one inductor but the thermal profiles produced can be varied in real time under fully loaded conditions, using fast switching methods.

THE PRINCIPLES OF CPH

As can be seen from the schematic cross sectional view through a typical CPH inductor shown in Figure 1, the inductor currents are directed into the paper, that is lengthways **along** the direction of strip travel rather than across the strip as in conventional TFIH heating.

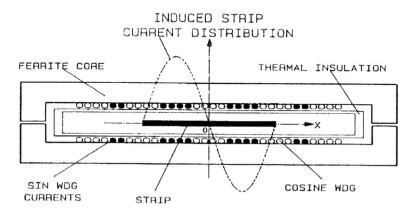

SCHEMATIC CROSS SECTION OF THE CPH INDUCTOR

Fig 1

This radically changes the nature of the eddy current distribution at the edges of the strip which now become flow lines along which the induced currents can flow quite easily rather than barriers across the path of the main current flow. The task of edge temperature control is thus considerably simplified.

A typical inductor has a conventional magnetic core which can be composed of either silicon iron lamination or ferrite blocks, depending on the frequency. The conductors may be either housed in slots or, as in the diagram, simply supported on the surface of the magnetic core. As indicated in Figure 1 these conductors are collected together into two groups, termed the SIN and the COS windings. The coil interconnections are carried out at the inlet and outlet ends of the furnace.

If the SIN winding is briefly energised, from say a medium frequency supply, the resulting induced current and power profiles across the strip width are as shown in Figures 1 and 2. Initially, it may be assumed that during this time the complementary COS windings are not energised. By using fast solid state switching techniques the SIN winding may be de-energised after only a few milliseconds and a similar power level applied instead to the COS winding. This produces a very different, but complementary, thermal profile generated across the width of the strip as shown in Figure 3. When the two sequential heating distributions have been applied for similar time periods, the net heating profile within the strip will then be the integral of the constituent SIN and COS profiles. This condition is depicted in Figure 4 and it can be seen that, transiently, the thermal profile is again uniform but a little hotter. Having established the required thermal profile across the workpiece it is now merely a question of repeating this process until the necessary temperature level is attained.

To obtain the constituent profiles of Figures 2 and 3 it is necessary to be able to control the currents in the various inductor windings depending on their spatial position relative to the edge of the strip. Details of how this can be achieved and in particular the crucial periodicity of the magnetic field as a function of both the width of the workpiece and the nature of the thermal profile are described in EA Technology's patent.[2] This is surprisingly easy to do in practice because the higher space harmonics of the inductor current distribution attenuate rapidly across the air gap. To demonstrate the principle, EA Technology have built and tested both

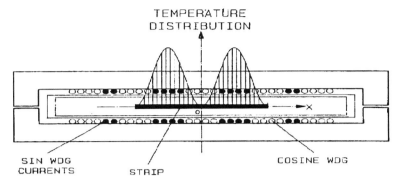

STRIP HEATING PROFILE FROM SIN CURRENTS

Fig 2

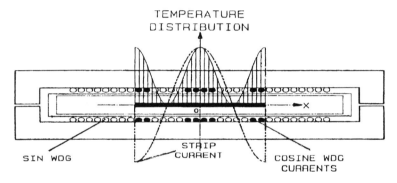

STRIP HEATING PROFILE FROM COS CURRENTS

Fig 3

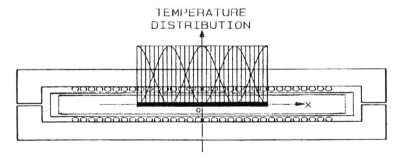

INSTANTANEOUS THERMAL PROFILE AFTER COMPLEMENTARY BURSTS

Fig 4

fixed and variable width CPH inductors of a 200kW rating. Test results for a simple uniform heating profile taken on 0.5m wide mild steel strip are shown in Figure 5.

The powerful CPH technique however is not merely restricted to the production of uniform profiles. Both the SIN and COS distributions discussed can be spatially modulated by **any** required profile. A mathematical discussion of this together with both travelling wave and multiple frequency options is presented in Appendix 1.

Since it is quite feasible to produce profiles that are asymmetric about the centre of the workpiece, this technique offers the possibility, at last, of a high power induction furnace capable of coping with workpieces that, as a result of their prior processing, have a dynamically varying temperature profile across their width and which simply need to be raised to a preferred processing temperature. This can be achieved using fast switching methods to control the current distribution across the inductor.

Considerable potential now exists for applying this technique to the heating of both thin slabs and strip as a result of the continuing developments in near-net-shape casting of seel. This has created a need for rapid in-line, profiled, heating of flat steel material in order to ensure that it has an optimum thermal profile at the start of the crucial hot rolling cycle.

APPENDIX 1   OPTIONS FOR POWER APPLICATION

Up to this point it has, for simplicity, been assumed that the COS and SIN windings were excited sequentially. This conveniently eliminates the possibility of there being any interaction between the two distinct current distributions. If the heating profile required across the workpiece is given by $\Theta(x)$ then these constituent current densities can be described as $J_s \propto \sqrt{\Theta(x)} . Sin(kx)$ for the SIN winding and $J_c \propto \sqrt{\Theta(x)} . Cos(kx)$ for the COS winding, where the pole pitch of the winding is π/k. Since these currents do not exist at the same time, the overall heating distribution is simply $H(x) \propto \Theta(x).Cos^2(x) + \Theta(x).Sin^2(x)$ which is obviously equal to $\Theta(x)$ right across the strip.

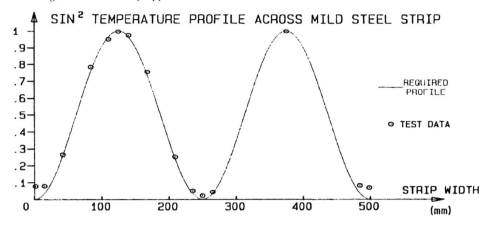
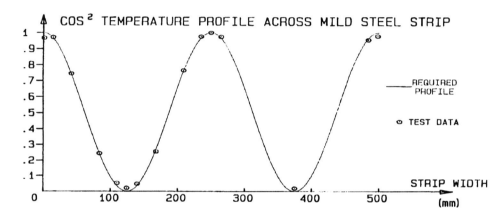
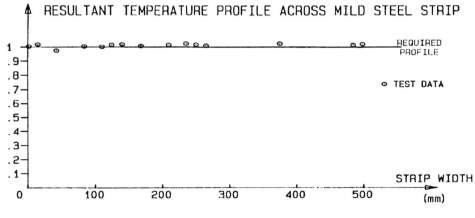

Fig 5

Alternatively however it is also possible to supply the COS and SIN windings from different static inverters operating at different frequencies. In this case the two sets of current will interfere with one another and the heating distribution becomes proportional to $\Theta(x).[Cos(kx).Cos(wt) + Sin(kx).Cos(vt)]^2$ where $w \neq v$. On time averaging this expression however only the $Cos^2(kx)$ and $Sin^2(kx)$ terms are non zero and again the $\Theta(x)$ profile is obtained. The feasibility of dual frequency schemes has already been investigated by Green and Williamson.[3]

There is yet a further possibility however when the two windings are supplied at the same frequency but with their time phases differing by 90°. In this case the constituent currents can be written as $J_c \propto \sqrt{\Theta(x)}.Cos(kx).Cos(wt)$ and $J_s \propto \sqrt{\Theta(x)}.Sin(kx).Cos(wt+\pi/2)$ and the overall heating profile becomes $H(x) \propto \Theta(x).[Cos(kx).Cos(wt) + Sin(kx).Cos(wt+\pi/2)]^2$.

Despite the complexity of this expression, when it is expanded out and time averaged over half a cycle the only non-vanishing terms are again those proportional to $Cos^2(kx)$ and $Sin^2(kx)$ and again the net profile is $\Theta(x)$.

It is instructive in this latter case to look carefully at the expression for J, the resultant induced current density in the workpiece. Since $J = J_c + J_s$ then from the above expressions this can now be written as

$J = \sqrt{\Theta(x)}.[Cos(kx).Cos(wt) + Sin(kx).Cos(wt+\pi/2)]$ or

$J = \sqrt{\Theta(x)}.[Cos(kx-wt)]$ which is the equation of a travelling wave whose amplitude varies as $\sqrt{\Theta(x)}$ as it crosses the strip.

Interestingly therefore the concept of complementary heating has serendipitously provided us with the solution to an old induction heating problem of how to obtain profiled heating in flat products using travelling wave inductors. The validity of this theory was **verified by tests on a 100kW, 50Hz, travelling wave plate heater.**

REFERENCES

1. W.B.Jackson: 'Transverse Flux Induction Heating of Flat Metal Products', 7th UIE Electroheat Conference, Warsaw, 1972.

2. International Patent Application, No. WO 93/11650, PCT/GB92/02212.

3. A.M. Green and A.C. Williamson: 'The Parallel Operation of Inverters Supplying Induction Heating Loads', Proc. EPE '95, 19-21 Sept 1995, Seville, p778- 783.

Development of a Direct Resistance Heating with Transformer Coupling

YUGO YAO. MASARU FUKUYAMA. FUMIAKI IKUTA. MORIHIKO TANINO.
Neturen Co. Ltd.   5893, Tamura, Hiratsuka, Kanagawa, 254 JAPAN
MASAMI ONODA. YOSIAKI HIROTA.
Nippon Steel Corp. Technical Develop. Bureau. 20-1, Shintomi, Futtsu, Chiba, 293 JAPAN
YOSIO SAITO.   TAKAO NAGASE.
Nippon Steel Corp. Hirohata   Works.   1, Fuji-cho, Hirohata, Himeji, Hyogo, 671-11 JAPAN

ABSTRACT

Although direct resistance heating has excellent features such as uniform heating, rapid heating for thinner metals and high electrical efficiency, sparking at the contact between a conductor roll and work keeps it from wider use in continuous process lines. In order to suppress sparking, numerous tests were done. In 1982, an induction heater in the continuous heat treatment line for wires was replaced to the direct resistance heating with transformer coupling. Since then replacement has progressed steadily with the number of the replaced machines to 17 at present. For steel strips, the 12,000kVA direct resistance heater with transformer coupling started to run as a heater of the galvanizing line at Nippon Steel Corp. Hirohata Works in 1990.

Résumé

Bien que le chauffage direct par résistance ait d'excellentes propriétés telles qu'un chauffage uniforme, un chauffage rapide pour métaux plus minces à grande efficience électrique, les étincellements qui se produisent au zones de contact entre le métal et les rouleaux de la ligne de production empêche une utilisation de plus large envergure dans ligne de fabrication à processus continu. Pour éliminer ce type d'étincellement de nombreux essais ont été effectuées. En 1982, un appareil de chauffage à induction pour le ligne de traitement continu des fils acier a été remplacé a le chauffage direct pur résistance. Depuis lors un remplacement progressif a eu lieu et le nombre d'appareils remplacées se monte à 17 à présent. Pour des bandes d'acier, des appareils de chauffages direct par résistance de 12000 kVA avec couplage de transformateur ont commencé à être mis en service en tant qu'appareils de chauffage pour la ligne de galvanisation aux chantiers de Nippon Steel à Hirohata en 1990.

THE PRINCIPLE OF THE DIRECT RESISTANCE HEATING WITH TRANSFORMER COUPLING

In practical use of direct resistance heating in process lines, the voltage of conductor rolls and bus bars must be very low which is free from electrical shock. Fig. 1 shows a schematic diagram of a direct resistance heater with transformer coupling (D.R.H.T.C.). The transformer core with the primary winding is set in a line. As a part of the secondary winding, the work goes through the window of the transformer core. A single short circuited secondary winding is formed by conductor rolls, a work, and bus bars. The electric resistance of the bus bar is designed negligibly small compared with that of the work. Therefore, it is obvious that almost all heat is generated in the work. The electromotive force (the secondary voltage V_2) is induced in the work surrounded by the core. The relationship among V_2, short circuit current I_2, and impedances of the secondary winding follows Kirchhoff's law.

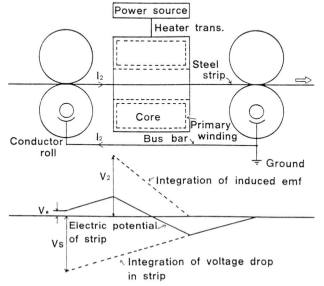

'Fig. 1' Schematic diagram of D.R.H.T.C. and electric potential distribution of strip

$$V_2 = I_2 Z_w + I_2 Z_b = V_w + V_b$$

where

Z_w impedance of the work

Z_b impedance of the bus bar

(The impedance of conductor rolls are neglected)

If a conductor roll is connected to the ground, the voltage of another roll (V_a) is equal to the voltage drop $I_2 Z_b$ (V_b). Z_b is extremely small compared with Z_w. Hence, the voltage of the conductor roll (V_a) is only several volts. This is one of the D.R.H.T.C.'s essential characteristics. By contrast, V_a of a conventional direct resistance heater must be V_w.

FEATURES OF D.R.H.T.C.

RAPID HEATING

Comparing with a conventional gas fueled furnace, D.R.H.T.C. can heat thin metals very rapidly. Fig. 2 shows the heating rate in accordance with the steel strip thickness for annealing at heating temperature of 750 ℃. The heating rate of gas fueled radiant tube type furnace is based on the actual operation data. The heating rate of D.R.H.T.C. is calculated in accordance with the current in unit width of the strip. The ratio of heating rate indicates the ratio of the heater length between a conventional gas fueled furnace and D.R.H.T.C.. By employing D.-R.H.T.C., it is possible to reduce the length of the heater about a fifteenth for thin metals. The compact heater enables to reduce the initial investment including building cost. With the accuracy of heating temperature, rapid heating can reduce the grain size, thus, to improve the material properties. Improvement in durability and ductility of middle-carbon steel wires have been confirmed by the actual operation.

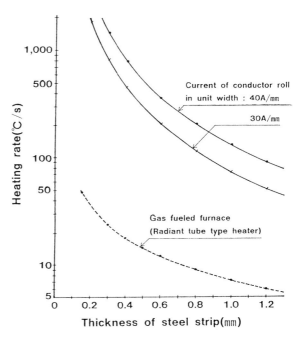

'Fig. 2' Heating rate in accordance with the steel strip thickness

HIGH EFFICIENCY

In case of applying D.R.H.T.C. to thin steel wires and strips, electrical heat efficiency converting from the delivered electric power to the temperature of the work is more than 90%.

EXCELLENT CONTROLLABILITY AND RESPONSE

As heat is generated in the work, heat inertia is zero. Therefore the heating temperature can be changed instantaneously in practical use. In terms of heating equipment, it is free from scheduling in sizes and heating temperatures. By employing D.R.H.T.C., heat treatment lines will be able to significantly reduce the restriction of scheduling. This results in a dramatic reduction in the inventory of metals. Thus, it is possible to save the production cost.

UNIFORM HEATING

If a suitable heating frequency in accordance with the material properties and sizes is chosen, the temperature of the cross section is uniform in practice.

CRUCIAL PROBLEMS AND SOLUTIONS

WIRES

In terms of wires, sparking at the exit side conductor roll was the crucial problem. One of the factors of sparking is cleanliness of the work surface. It is necessary to remove the electrical insulation materials such as rusts and deposits from the work surface before heating. For the conductor roll, there are a lot of factors to suppress sparking such as the material of the roll and the diameter of the roll, relative relations between the caliber of the roll slot and the wire diameter and the pressure between the roll and the wire. In order to identify the material characteristics for sparking, a lot of experimental tests by using different material rolls were conducted. Table-1 shows the test conditions. Fig. 3 shows the maximum non-sparking current of each material for 20 seconds operation.

'Table-1' Test conditions

Items	Conditions
Wire diameter	13 (mm)
Work	Steel
Heating temp.	900 (℃)
Roll dia.	350 (mm)
Environment	Air

Roll Material	Max. non-sparking current (10^3 A) 5 10 15
Copper alloy	————————————
Iron alloy	—————————
Cast iron	————
Carbon	————————————————

'Fig. 3' Relation between the materials and the maximum non-sparking current

The results show that there is little difference between materials on the maximum non-sparking current for short period operation. In case of the copper and iron alloy rolls, solid deposits on the roll surface accumulated with elapsed operation time. Because the resistivity of carbon is some 30 times grater than that of metals, the maximum current of carbon roll is restricted by its temperature caused by Joule heat under the continuous operation. For a kind of cast iron, the ware rate is not only slight but solid deposits are also rare. Fig. 4 shows the relation between the maximum non-sparking current and the diameter of the wires. Because carbon's Young modulus is about

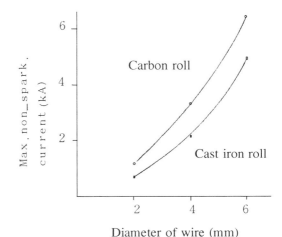

'Fig. 4' Relation between the diameter of the wire and the maximum non-sparking current

a twentieth that of steel, it makes a larger contact area. Therefore the maximum non-sparking current of the carbon roll is larger than that of the cast iron roll. But a higher wear rate of carbon roll attributes to a higher cost. At present suitable conductor rolls are determined by the size and the kind of metals of the work.

STRIPS

In addition to the problem of sparking, uniformity of temperature distribution in width had to be realized.

Sparking At The Entry Side Conductor Roll

By using a roll covered with rubber as a pinch roll and two sizes of conductor rolls, numerous experimental tests for low temperature steel strips with a pilot line were done. The data obtained from those trials were analyzed. Thereby the relationship among the maximum non-sparking current and the pressure to the conductor roll in unit width of the steel strip and the roll diameter can be described by equation (4). [1]

$$I_{max} = 46 P^{3/4} (D/700)^{0.3} \quad (A/mm) \quad (4)$$

where

I_{max} Maximum non-sparking current in unit width of the steel strip

P   Pressure to the conductor roll in unit width (kgf/mm)

D   Roll diameter (mm)

In the practical application, the roll size and the pressure to the roll are determined based on equation(4) including adequate surpluses.

Sparking At The Exit Side Conductor Roll

From a lot of experiments of conductor rolls for high temperature wires and low temperature strips, conductor rolls for high temperature strips were designed. Through numerous tests with pilot lines and accumulated analytical works, the designing method of the exit side conductor roll was established.

Temperature Deviation In Width

If a commercial frequency is used for the heater, strip edges tend to be over-heated as strip thickness become larger. Fig. 5 shows the distribution of the measured temperature and the squared current density in strip width calculated by the vector potential method.[2] Although the deviation of the calculated squared current density distribution is a little greater than that of the measured temperature distribution, it is possible to simulate the temperature distribution in strip width almost precisely. Choosing a suitable frequency, it is possible to restrain the temperature deviation by $\pm 5°C$.

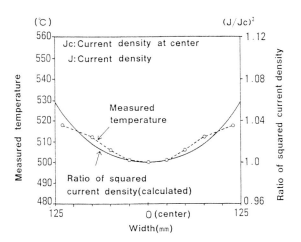

'Fig. 5' Temperature and squared current distribution in width

APPLICATIONS

FOR WIRES

An example of the maximum capacity heater of those 17 heaters applied for wires is as follows. Table-2 shows the specification of the heater and the operation data. The configuration of the heater is the same as the one showed in Fig. 1.

'Table-2' Specifications of the heater and the operation data

Item	Spec. & Data	Item	Spec. & Data
Work diameter	13 (mm)	Capacity	200 (kW)
Work	High speed steel	Secondary current	10 (kA)
	Stainless steel, etc	Frequency	60 (Hz)
Heating temp.	1200 (℃)	Heater length	2.5 (m)
Wire speed	120 (m/min)	Electrical efficiency	90 (%)
Installation (year)	1987	Volt. of conductor roll	4 (V)

FOR STRIPS

As a heater of the galvanizing line, D.R.H.T.C. has run without any problems since 1990. Table-3 shows the specifications of the heater and the operation data. Fig. 6 shows the construction of the secondary circuit. The molten zinc is used as an electric conductor instead of a conductor roll.

'Table-3' Specifications of the heater and the operation data

Item	Spec. & data	Item	Spec. & data
Thickness	1.2-6.0 (mm)	Capacity	12000 (kVA)
Width	570-1250 (mm)	Secondary current	90.5 (kA)
Work	Low carbon steel	Frequency	22-30 (Hz)
Heating temp.	450 (℃)	Heater length	17.7 (m)
Strip speed	40-100 (mpm)	Electrical efficiency	91 (%) average
Installation (year)	1990	Volt. of conductor roll	7.5 (V)

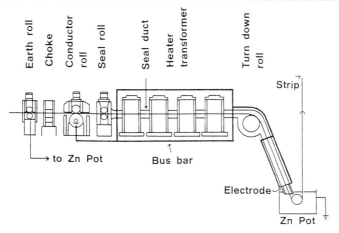

'Fig. 6' Configuration of the secondary circuit

Fig. 7 shows the temperature distribution in strip length. Fig8 shows the temperature distribution in strip width.

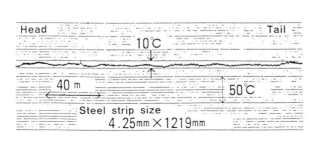

'Fig. 7' Temperature distribution in strip length

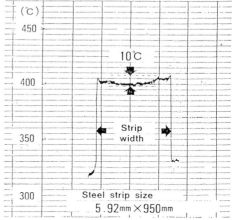

'Fig. 8' Temperature distribution in strip width

CONCLUSION

Employing D.R.H.T.C. for thin steel wires and strips, extremely compact heaters were realized. The rapid heater with uniform and accurate heating temperature enables to decrease the initial investment and to improve the properties of particular metals. In order to extend the application of D.R.H.T.C. the conductor roll at the exit side of the heater, where strip temperature is high, had to be developed. Although more trials are necessary, the designing method of the exit side conductor roll has been established.

REFERENCES

1. M. FUKUYAMA, M. HATTORI, Y. YAO, et. al.: ' An Application of A Direct Resistance Heating with Transformer Coupling to Steel Strip ' , *IEE JAPAN- IAS'91*, 1991, 177-182
2. T. NAKATA and N. TAKAHASHI: ' Finite Element Method in Electrical Engineering ' Morikita Publishing, Tokyo, 1988

Non-Deforming Induction Heating of Metal Discs

JERZY ZGRAJA

Technical University of Lodz, Department of Electroheat

Al.Politechniki 11, 90-924 Lodz, Poland

ABSTRACT

Nonuniform temperature distribution of induction heated steel discs may cause the disc deformation, especially its buckling. After short presentation of the theory of this phenomenon, some methods of avoiding the deformations are discussed. The results of computer calculations are presented. Also the influence of material parameters and of the geometry of the system are considered.

RESUMÉ

Héterogenè distribution de température dans un disque en acier, chauffé par induction peut resulter une deformation ou specialement flambage du disque. Aprés une discussion abbrevée de la théorie du phénomene on a presenté quelques méthodes pour eviter une deformation. Les resultats des calculs par ordinateur confirment la validité de la theorie. On a presente aussi l'influence des parametres de matiere et de la geometrie du systeme considere.

INTRODUCTION

Usually for plastic hot working, the steel discs are heated in electric or gas furnaces. Often the heated discs are of relatively large diameters as for example the discs used for the production of the container bottoms applied in chemical industry and in power engineering. The process of hot pressing concerns the discs of diameters of 0.5-4.0m and a thickness of several millimeters. Therefore large furnaces with high investment and operation costs, often not fully filled with charge, are used. Also the energy losses during transportation of the hot discs to forming machine are considerable. Alternative heating method before forming can be the induction heating. Well known advantages of this method permit to reduce the energy losses, to shorten the heating time and to increase the efficiency. It is particularly important if the heating could be realized on or near the forming machine. These clear advantages resulting from the application of induction heating aroused interest of factories producing vessels for the chemical industry and power engineering.

Nevertheless, the induction heating of plates with large diameter to thickness ratio encounters certain technological difficulties [1] connected with thermal deformation, resulting from non-uniform temperature distribution. The occurrence of thermal deformations of the disc was often the reason eliminating the application of induction heating for this purpose.

The heating of discs can be carried out in a technologically simple way by applying a plane spiral inductor. The problem of uniform heating of a disc, however, emerges here, since it concerns both the disc edge and its central part. A number of parameters influence the temperature distribution in a disc heated by a plane spiral inductor [2]. The appropriate design of the inductor enables the shaping of temperature field in the disc. However, applying a relatively simple design of the spiral inductor, there are some limitations resulting for example from the fact that the heating power in the inductor centre equals zero. Thus the question arises how large the temperature nonuniformities might be which still would not cause any undesired disc deformations. The intention of the author is to analyse this problem.

MATHEMATICAL MODEL AND ASSUMED SIMPLIFICATIONS

Essential in solving the problem of thermal deformations is the knowledge of temperature distribution in induction heated discs and the possible ways of improving the uniformity. From the mathematical point of view this problem involves the analysis of coupled electromagnetic and thermal fields in a two-dimensional, axially symmetrical system.

The mathematical analysis of coupled electromagnetic and temperature fields was carried out by solving Maxwell and Fourier-Kirchhoff equations. The electromagnetic field was assumed to be harmonic, which allowed employing a complex vector potential A for analysis. Hence, the analysis was reduced to solving mutually coupled equations:

$$j\omega\gamma \mathbf{A} + curl\,[(1/\mu)\,curl\mathbf{A}] = \mathbf{J} \qquad (1)$$

where:

ω - angular frequency, γ - electrical conductivity, μ - magnetic permeability,

$\mathbf{J}$ - current density,

and :

$$div[(-k)\,grad\,\vartheta\,] + \rho c \frac{\partial \vartheta}{\partial t} = p_v \qquad (2)$$

where:

k - thermal conductivity, ρ - density, c - specific heat, p_v - internal heat source intensity, with the existing boundary conditions taken into consideration.

Shaping of the temperature distribution, especially in the central part of the disc, can be achieved by using an inductor of variable turn pitch. In the paper, two-zone inductors with different turn pitch in each zone were analysed. The ratio of outer zone turn pitch of the inductor to that of the inner part was called a coefficient m. For computer analysis the rotational symmetry of the inductor-disc system was assumed. The mathematical model of "inductor-disc" system shown in Fig.1 was used.

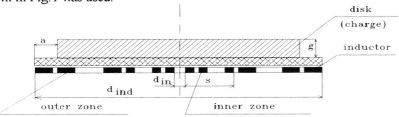

Fig.1. Computer model of "inductor-disc" system.

The analysis was limited to the medium frequency of 500–10000 Hz, most often used in induction through-heating. For better clarity of analysis, some calculations of heating of ferromagnetic discs were performed with an assumption of the magnetic permeability independent of magnetic field strength. All the materials used, were assumed to be isotropic.

The thermal deformations can be caused by the nonuniform temperature distribution across the disc thickness and also along the disc radius. In the presented analysis only the later ones are considered. The nonuniformity across the disc thickness can be neglected in the case of thin plates or also easily limited by applying the double-sided heating by two inductors.

Based on the above mentioned assumption, and also that the deformation is not large, the buckling of the disc can be analysed by solving the following three-dimensional equation:

$$(\mathbf{K} + \lambda_i\,\mathbf{K}_g)\,u_i = 0 \qquad (3)$$

where:

K - linear stiffness matrix, K_g - initial stress stiffness matrix, λ_i - ith multiplier to the load (temperature difference), u_i - ith displacement eigenvector or mode shape.

The numerical analysis of coupled electromagnetic and thermal field was based on computer program developed by author and on commercial program Flux2D. The buckling analysis was based on commercial ANSYS program.

RESULTS OF NUMERICAL ANALYSIS

From the technological point of view and also to avoid any deformations the uniform temperature distribution in the disc is most desirable. A plane spiral inductor of constant turn pitch cannot be applied during the whole process of heating. Radial distribution of surface temperature after 12 minutes heating of disc of austenitic steel of thickness g=10mm and diameter d=630mm by a typical plane inductor of a constant turn pitch, having the same outer diameter as the disk and inner diameter d_{in}=20mm, operating at the frequency f=8kHz is shown in Fig.2. This temperature distribution can be only calculated because in practice even during very slow induction heating, temperature deformation of disc is so large that induction heating cannot be applied.

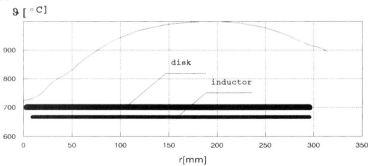

Fig.2. Radial distribution of surface temperature ϑ of the disc heated by a spiral inductor of constant turn pitch.

Radial temperature distribution, especially in the central part of the disc can be corrected by applying an inductor of two different turn pitch. Even a simple two-zone inductor (Fig.1) enables the correction of the distribution of heating power and thus also of temperature in the central part of the disc. In such a correction it is important to choose the correct value of the ratio m describing the outer to the inner zone turn pitch as well as the width s of the inner zone[2]

Applying a two-zone inductor for induction heating of an austenitic steel disc (d=630mm, g=10mm), the obtained results of radial distribution of volumetric power density p_v in the surface layer and of surface temperature ϑ of disc for s=70mm are presented in Fig.3.

The temperature distribution in the central and external edge parts of the disc depend on material parameters of the plate as well as on the supply frequency[3]. The heating power and

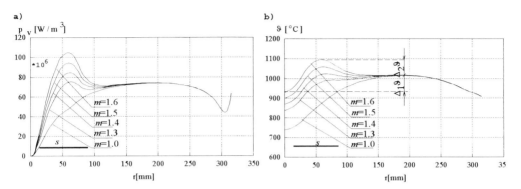

Fig.3. Radial distribution for a two-zone inductor with s=70mm :
a) of volumetric power density p_v; b) of surface temperature ϑ.

temperature distribution in the edge part of disc depend also markedly on the magnetic permeability of the steel[2,4]. The correction of temperature distribution in this part is easily accomplished by proper setting of the protrusion a of the inductor over the disc edge (Fig.1). Also the other methods of correction such as changing of inductor turn pitch, changing of the air gap between the inductor and the disc or modifying the thermal insulation can be applied. Fig. 4 shows the temperature distribution at the disk edge obtained for different values of protrusion a, during heating by similar power of disk (d=630mm, g=10mm) of austenitic steel (μ_r=1) and of ferromagnetic steel (μ_r=50) at the frequency f=8000Hz.

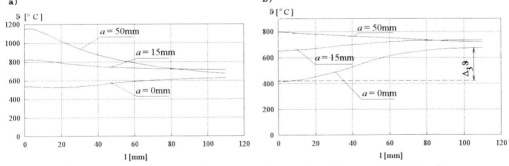

Fig.4. Temperature distribution ϑ as a function of the distance l from the disk edge, for different values a of the inductor protrusion:
a) austenitic steel, μ_r=1, b) ferromagnetic steel, μ_r=50

The optimum s value and of coefficient m (Fig.1) granting minimum $\Delta_1\vartheta$ value for $\Delta_2\vartheta \rightarrow 0$ (Fig.3) are not constant but depend upon the material and dimensions of the disc and also upon the heating frequency. So does the a - value (Fig.4) granting minimum $\Delta_3\vartheta$ deviation.

Analysing the curves in Fig.3 and Fig.4, three different zones can be distinguished in the heated disc. The first one, near the disc centre, is characterised by the marked temperature drop towards the disc axis, or applying the correction (large m -value) by a temperature maximum near the axis. The second zone is characterised by uniform temperature distribution and the third one, near the disc edge exhibits either a rise or a drop in temperature.

CENTRAL PART OF DISC

Analysing the temperature distribution in the central zone, it is assumed that there is a uniform temperature in the other two zones. Without any correction ($m=1$) there is a nearly linear temperature drop towards the disc axis. Applying the correction, for large m - values a rather trapezoidal temperature distribution is observed. For assumed characters of temperature distribution a three-dimensional analysis of disc buckling is performed. Its aim was to determine the most probable buckling modes and also the critical temperature difference $\Delta_{crit}\vartheta$ at which it takes place. There are three the most possible buckling modes as shown in Fig.5. That is: a) - bowl-shape buckling; b) - one side up and opposite one sagging down; c) - circumferential corrugation.

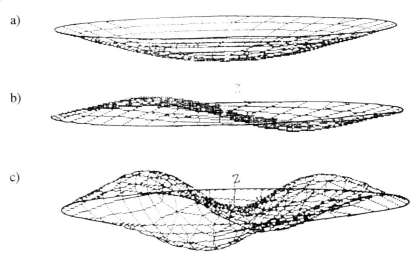

Fig.5. The most possible modes of disc buckling under the influence of radial temperature differences.

Fig.6 presents the influence of the disc thickness g and of the width s_1 of the temperature drop zone on the critical value of temperature difference $\Delta_{crit}\vartheta$ at which the buckling occurs. Easily observed is the strong influence of the thickness g and width s_1 on critical value of $\Delta_{crit}\vartheta$ at which the buckling occurs. In the some figure a really weak influence of the disc diameter d is seen. For steel disc, thinner than 10mm, heated by one-zone inductor, even at the uniform

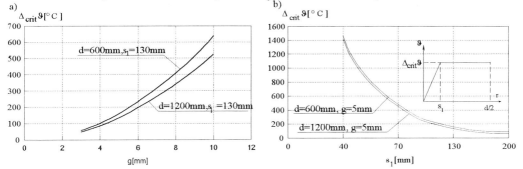

Fig.6. Critical temperature difference $\Delta_{crit}\vartheta$ as a function of disc thickness g and width s_1...

temperature distribution in the edge zone and across the disc thickness, there is a virtual possibility of buckling. It can be avoided by applying a two-zone inductor, which grants a decrease of $\Delta_1\vartheta$ values and also for well chosen $\Delta_2\vartheta/\Delta_1\vartheta$ ratio, leads to increase in $\Delta_{crit}\vartheta$.

EDGE PART OF DISC

As shown in Fig.4 approaching the disc edge the disc temperature may either increase or decrease, depending upon the inductor protrusion a as well as upon the material and geometry of the disc[2]. Let us assume as before that the temperature nonouniformity takes place only in the edge part of s_2 width and that it can be approximated as a linear function. Basing on these assumptions the buckling analysis of disc is performed, determining the most probable buckling modes as well as the critical $\Delta_{crit}\vartheta$ values at which the buckling occurs. Fig.8a presents for a

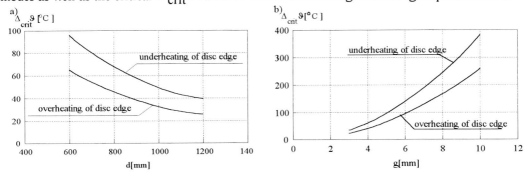

Fig.8. Dependence of the critical value of temperature difference $\Delta_{crit}\vartheta$ upon the disc diameter d (a) and upon the disc thickness g (b), for rising or falling temperature while approaching the disc edge.

steel disc of thickness $g=5$mm and of width of edge zone $s_2=80$mm the dependence the critical temperature difference $\Delta_{crit}\vartheta$ upon the disc diameter d and Fig.8b upon its thickness g.

As may be seen in Fig.8, the critical value of temperature difference $\Delta_{crit}\vartheta$ depends strongly upon the disc diameter and its thickness. These $\Delta_{crit}\vartheta$ value are lower in the case of overheating of disc edge. Quite different are also the buckling modes in case of overheating or underheating of disc edge. In the first case more probable is the form from Fig.3c, and in the second one that from Fig.3a.

CONCLUSIONS

In the case of induction heating of steel disc of large diameter to thickness ratio the probability of thermal deformation is high. It can occur even in the case of uniform temperature across the disc thickness when there is a nonuniform temperature distribution in centre or at its edge.

Especially dangerous is the case of edge overheating, because then the deformation can occur at smaller temperature differences. To prevent any deformation it is necessary to determine the critical values of temperature differences for central and edge parts. The applied inductor is to be designed in such a way as to prevent any overpassing of critical temperature differences.

REFERENCE

1. K. Januszkiewicz, J. Bereza and J. Zgraja: 'Induction Heating of Large Steel Disks: Computer Simulation and Experiments', Proc. of XIIth Congress of UIE, Montreal, Quebec, Canada, June 1992.1007-1015.
2. J. Zgraja, A. Skorek, V. Rajagopalan and M. Zaremba: 'Induction Heating of Metal Disks by the Plane Spiral Inductor', Proc. of International IEEE/IAS Conference of Industrial Automation and Control, Taipei, Taiwan, May 1995. 8-13.
3. J. Zgraja: 'Shaping of Temperature Field in Induction Heated Metal Plates', Proc.of 40 Internationales Wissenschaftliches Kolloquium, Ilmenau, Germany,September 1995. 82-87.
4. V. Nemkov, V. Demidovich, V. Ruchnev, O. Fishman: 'The use of end and edge effects for induction heating design', Proc. of XIIth Congress of UIE, Montreal, Quebec, Canada, June 1992. 180-187.

This work was supported by the National Research Committee (KBN), grant No. 8T10B05209

HIGH FLUX INDUCTION FOR THE FAST HEATING OF STEEL SEMI-PRODUCTS IN LINE WITH ROLLING

INDUCTEURS A HAUT FLUX POUR LE RECHAUFFAGE RAPIDE DES PRODUITS SIDERURGIQUES EN COURS DE LAMINAGE

G. PROST - J. HELLEGOUARC'H - CELES, LAUTENBACH, FRANCE.

J.C. BOURHIS - E.D.F. - D.E.R., LES RENARDIERES, FRANCE

G. GRIFFAY - IRSID, MAIZIERES-LES-METZ, FRANCE

ABSTRACT

In order to reduce equipment and operation costs of reheating process in mille. CELES has developed and qualified in collaboration with E.D.F. ans IRSID a new technology of high flux induction decalled CELINE. It is designed to reach more easily a good temperature rise homogeneity and to be installed on line there where it was previously impossible to put conventionnal technologies.

With a new manufacturing technology including a composite conductor, this inductor is characterical by :

* High specific power : 4 MW m^{-2}

* High efficiency when with a wide mechanical gap between the inductor and the material

* High reliability due to its mechanical internal protection and thermal screens.

A 500 KW prototype has been manufactured. The results of the performed tests are presented confirming the above objectives.

This technology is now on operation on industrial sites

RESUME

Afin de réduire les coûts d'investissement et de fonctionnement du réchauffage des produits sidérurgiques en cours de laminage, CELES a développé et testé en collaboration avec IRSID et E.D.F. un nouveau type d'inducteur dit à haut flux appelé CELINE.

Il est conçu pour atteindre rapidement une température homogène en étant installé dans un espace réduit, non accessible aux inducteurs conventionnels.

Grâce à la mise en oeuvre d'un nouveau conducteur, il est possible d'atteindre des puissances spécifiques élevées (4 MW m^{-2}) avec un très bon rendement en dépit d'un jeu mécanique important entre l'inducteur et la charge.

Des écrans thermiques et des protections mécaniques intérieures garantissent la fiabilité de l'inducteur.

Un prototype d'une puissance de 500 KW a été fabriqué et testé. Les résultats obtenus confirment ces performances.

Des équipements construits selon cette technique sont en fonctionnement industriel.

INTRODUCTION

Induction is the physical phenomenon combining Lenz law and the Joule effect. Therefore it presents many advantages for the reheating of metals.

- creation of the heat inside the material to be reheated,
- fast start,
- low inertia,
- easy control.

However, up to-day, induction has only seldom been used to reheat metallic products with high production rates because of the low heat transfer rate to metals in amagnetic phase.

The expertise of CELES for high intensity magnetic fields created by low loss coils has allowed the creation of a new technology for coils featuring :

- improved tempeature homogeneity as a result of lower working frequency,
- higher transfered power levels and efficiencies compared to conventional solutions,
- injection of high power within a short space in-line with a rolling mill.

OBJECTIVES

The initial goal was reheating billets before a roughing mill by injection of a power density of $4\ MW\ m^{-2}$ of surface. In a second phase, this objective has been spread to all types of cross sections of long and flat products.

Following problems were to be solved :

- creation of an intense enough magnetic field in order to generate the aimed power density,
- design of a coil with reduced losses in order to minimize the energy consumption,
- adaptation of the equipment to the mill environment which means :
    * good protection against mechanical shock,
    * sound thermal insulation,
    * good resistance to humid atmosphere,
    * evacuation of scale.

- minimization of the lost magnetic flux outside the inductor in order to avoid parasitic reheating of structure and rollers.

INCREASE OF POWER DENSITY

The power transferred by induction can be written as follows :

$$P = KH^2 \sqrt{f}$$

In order to increase the power density increasing the magnetic field looks more efficient than increasing the frequency (which is unfavorable to homogeneous reheating)

The magnetic flux created by a solenoid can be written as follows :

$$H = k' \frac{NI}{l}$$

Increasing the magnetic field will result either from increasing the inducing current or the number of turns. Increasing the inducing current causes an important elevation of the losses in the coil (which vary as the square of the current).

A physical limit (evacuation of the losses) and an economical limit (efficiency) are, in this way, reached quickly. Increasing the number of turns leads to multiple layers of turns which enable to hope a reduction of the losses.

- for a one-layer coil :

$$\eta 1 = \frac{Pch1}{Pch1 + Pind1} \qquad Pch1 = \alpha N^2 I^2 \qquad Pind1 = \beta I^2$$

- for a two-layer coil transporting the same current :

$$Pch2 = 4 \, Pch1 \qquad Pind2 = 2 \, Pind1$$

$$\eta 2 = \frac{Pch2}{Pch2 + Pind2} = \frac{2}{1 + \frac{1}{\eta 1}}$$

- for a n-layer coil :

$$\eta n = \frac{n}{(n-1) + \frac{1}{\eta 1}}$$

With :
- Pch1 = power transferred to the load with a 1-layer coil,
- Pind1 = electrical losses in the 1-layer coil.

The result is that the efficiency increases when the number of layers increases. If the efficiency of the 1-layer coil is 60 %, the progression is as follows :

Number of layers	Efficiency
1	60 %
2	75 %
4	85 %

This demonstration is of course only valid if the coil is achieved in such a way as to ensure that the inner layers are not reheated under the influence of the field generated by the outer layers. This led us to develop an adapted conductor which we named « composite » conductor. It is achieved by assembling around a water-cooled central core, insulated copper wires. The section of these wires is small enough to avoid induced currents. (fig.1)

In a cross-section of 10 x 10 mm², this conductor can carry an intensity of 1000 A at 1000 Hz in a magnetic field superior to 300000 At/m. A patent has been filed for this solution.

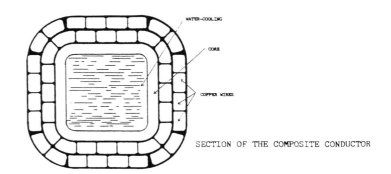

SECTION OF THE COMPOSITE CONDUCTOR

TECHNOLOGICAL SOLUTIONS

The coil is multilayer wiring using our « composite » conductor. It is constituted by an assembly of double pan-cake coils. Each basic coil is insulated by impregnation under vacuum und pressure of an epoxy resin. This technique allows the coil to work under water thus in a humid environment. (fig.2)

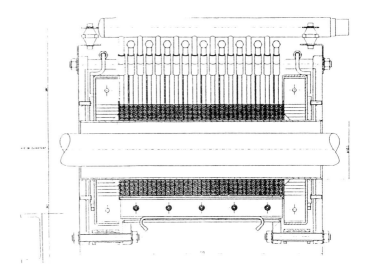

The coil electrical insulation is protected from the radiation of the load (temperatures up to 1250°C) and from scale by a heat-shield which has the feature of being transparent to the magnetic field. The ceramic heat-shields are water-cooled in order to withstand the temperature gradient on a reduced thickness. This system also has been filed for patent.

The mechanical protections (« skis ») are in water-cooled amagnetic steel. They protect the heat shield and the wiring. They are essential to the long life time of the inductor.

Magnetic yokes capture the flux outside the coil and allow the use of conventionnal rollers and the setting of the coil in the immediate vicinity of the rollers.

PERFORMANCES

The performances have been tested by E.D.F. and IRSID on a prototype built by CELES.

The features of the prototype are as follows :

- inner diameter of the coil   180 mm
- opening diameter   120 mm
- length of the coil   500 mm
- overall length of the inductor   700 mm
- number of pancakes   18
- number of layers   4

The inductor is supplied with middle frequency current by an inverter (500 kW - 600 V MF -1000 Hz).

EFFICIENCY OF THE INDUCTOR

The tests have been performed on several amagnetic steel cylindrical billets. The efficiency of the coil is the ratio between all the losses of the coil (measured by calorimetry) and the middle frequency power delivered to the inductor.

Diameter in mm	Efficiency in %	Conventionnal coil - Efficiency in %
95	85	63
75	78	56
60	66	48
50	49	33

These tests show that the efficiency is higher than the one of conventional coils and that the efficiency decreases slower when the load diameter decreases.

For a copper billet :

Diameter in mm	Efficiency in %	Conventionnal coil - Efficiency in %
80	60	40

TEMPERATURE HOMOGENEITY

Fig. 3 shows the evolution of the temperature from ambiant for a 95 mm diameter billet in stainless steel. After the heating stops, a soaking occurs in an insulated tunnel.

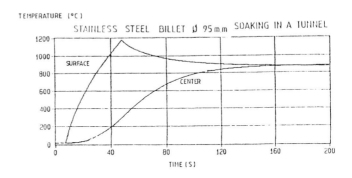

Fig. 4 simulates the reheating before rolling after coming out from a furnace set to a reduced temperature. Soaking occurs without thermal protection. When both curves cross (30 s after end of heating), temperature homogeneity in one section is ± 20°C.

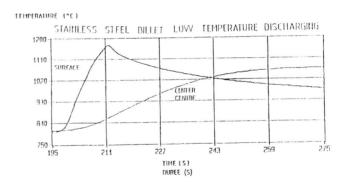

Fig 5 shows the evolution of the temperature of the center of the face and of the corner for a 70 mm square billet.

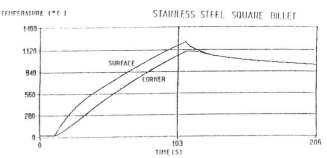

CONCLUSION OF TESTING

The tests have validated our technological choices and confirm that the CELINE technology enables :
- to transfer to amagnetic steel surface power densities of 4 MW m^{-2},
- to transfer these power densities with coil efficiencies higher than 80 % on well coupled plain loads.

EXAMPLES OF USE

OBJECTIVE Reduction of the steel billet reheating furnace temperature setting

Increasing the temperature of the billet in-line with the rougher or the finishing mill

INTEREST Scale reduction

Decarburization reduction

Production boost if needed

Process improvement

PERFORMANCES

Billet diameter	125 mm
Production	55 t/h
Injected power	1 MW
Temperature increase	100 °C
Coil overall length	1300 mm.

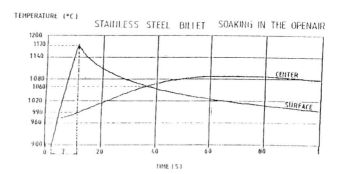

THIN SLAB REHEATING

OBJECTIVE Increasing the thin slab temperature after continuous casting

Reheating homogeneously

INTEREST Scale reduction

Reduction of reheating equipment length thus radiation losses

Process improvement.

PERFORMANCES

Width	1300 mm
Thickness	30 mm
Speed	0,167 m/s
Temperature increase	145 °C
Injected power	4,6 MW
Inductor dimensions	

CONCLUSIONS

The tests have confirmed that CELINE is well adapted for reheating amagnetic metals with efficiencies higher than 80 % and transferring power densities higher than 4 MW/m². The thermal and mechanical shields are designed to ensure reduced maintainance costs under most difficult working conditions.

CELINE is used for industrial applications:
- round billets,
- square billets,

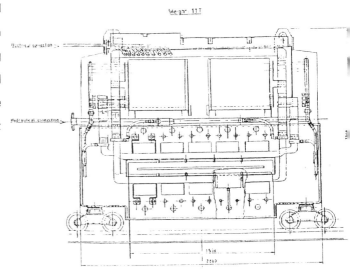

It may be used for there slabs and rough bars (Fig. 7)

CELINE is featured for:the reheating of,
- all steel grades in amagnetic phase
- copper and copper alloys,
- aluminium and aluminium alloys, etc...

CELINE proposes an interesting solution for :
- reheating before the rougher mill or in-between finishing mills,
- boosting an existing furnace,
- optimizing the power of new furnace.

The simulation of non-linear electromagnetic heating and deformation fields in the induction heating system.

A. VASILIEV, V. ZAREVSKY, O. IABLONSKAJ

This report discribes complex design and research methods of inductive heat unit. Consideration includes all parts of plant, such as power source and inductor-workpiece system. Resulted structure of material received were also examined as well as valuation of strain stresses distribution . Simulation methods of common used power sources, such as vacuum tube invertor, thiristor and transistor ivertors in the pulse and continuous operation modes are also subjects of this report. Joint decision of electrical, thermal and deformation tasks are included as phisical base of problem. The main problem of report is inductiv heating process with high power -weight ratio(10-15kwt·cm^{-2}). This technology viewed as a way to get perfect metal structure and mechanical propertis. Pulse operation mode of htating unit (with different modulation factors and pulsefrequency) were reseached from the point of getting new material structure.

Le présent exposé vise à mettre au point les méthodes des calculs et d'étude des installations pour le chauffage d'induction y compris la source d'alimentation, le réseau et le système de l'inducteur et de pièce à travailler, ainsi que l'évaluation de la structure des matières obtenues avec la distribution des contraintes résiduelles. Les méthodes de la simulation des sources d'alimentation les bien repandues (le générateur à lampes, convertisseurs aux thyristors et aux transistors) en régimes impulsionelles et continues sont décrites. La résolution jointe des problèmes électriques, thermiques et ceux de la déformation pour le système de l'inducteur et de pièce à travailler est donné. L'étude de la chauffage inductive avec la puissance spécifiqelévée (10-15kwt·cm^{-2}) qui change le caractère du traitement thermique et qui donne la possibilité d'obtenir les structures des métaux plus perfectes avec les caracteristiques mécaniques ameliorées est d'importance pariculière. L'application des régimes impulsionelles des installations avec des profondeurs de la modulation et des fréquences des impulsions differents est mise au point ce qui donne la possibilité d'obtenir les structures des matières nouvelles.

Processes with the high energy concentration in a small volumes are very meaningful in modern industry. Induction heating is the technology process of this class. In Russia inductive and plasma heating are frequntly using technologies. Previuos years were characterized by the power-weight ratio about 1- 1.5 KWt·cm^{-2}. Modern materials and calculating methods moved this limit on the new level of the 10-15KWt·cm^{-2}. This new power-weight ratio had opened new possibilities in the material treating technology.

High-frequency power modulation being used for the purposes of getting new properties of materials, is very actual. Two aspects may be considered in the context of this technology.

High modulation level, when process may be characterized by the pulse performance, allows precise energy portioning. It is exclusively important, particularly, in the case of fast processes.

This intensive technology has different variants. Some problems in technology design must be considered in detail.

Surface hardening with the high power-weight ratio (to 15 KWt·cm^{-2}) is the case. Till now process were considered as a electro-thermal task. Following this model causes co-joint solution for the thermal and electomagnetic fields. First has a Fourier equation description, second has a description of Maxwell equations. Due to dealing (in most cases) with the bodies of axis symmetry, task may be simplified to the 2 - dimensional case. Solution of this electomagnetic task provides distribution of heat sources as well as the **electro-thermal** one; latter deliveres temperature field distribution. Circuitry types used were thiristor, transistor and vacuum tube invertors.

In the St. Petersburg electrotechnical university class of 2- dimensional tasks on the non-magnetic and ferro-magnetic matirials was completely explored.

At the same time design methods of power sources also has been developed. In most cases task was reduced to the solution of differential equations system (for piecewise linear regions) and to the determination of stationary state by means of transient. At last time were found solutions for some circuit schemes as a stationary task with the boundary value. Taking given forward period value of process is implied. Full physical picture may be obtained by joining the equation solutions, which discribes power sources, with the solutions of electro-thermal task.

The separate case appears when techological piece has a region of molten metal with unknown border of solid phase. In some cases the task appears difficult (micro-wire casting, high-frequency tube welding with smelted droplet, and so on). For tube welding technology one of possible decisions consistes of the low-frequancy modulation in power source. This decision gets high quality joint of workpieces. Such kind of joint contains no fins. Very closely to the same tasks stands problem of operate mode estimation for plasma burner, directed to the optimization of cut quality and width.

The strain stresses analysis during heat treatment have not been done till nowdays.

Solving thermal tasks, we are dealing with greater values of time constants than exists in the electromagnetic part of heat treatment equipment. It makes possible to solve these tasks separately.

Usually, electromagnetic processes are considered as quasi-stationary sustained ones. Transient does not considered in this model in spite of that is not completely correct in the case of thiristor's and transistor's power sourcers usage. During turning on such kind of power sourcers, settling time may take more than some periods of current, and electric resistance may be essentially different than on the sustained phase of process.

As an example may be considered half-space $(x > 0)$ with the constant parameteres ρ=const and μ=const, which is affected by the current, i.e., field H, or voltage, i.e., field E.

For plane wave

$$\frac{\partial H}{\partial x} = -\gamma E, \qquad \frac{\partial E}{\partial x} = -\mu_0 \frac{\partial H}{\partial t}. \qquad (1)$$

In the case of sinusoidal magnetic field, boundary conditions on the sirface and on the infinity has a next view:

$$H(0,t) = H \sin(\omega t + \alpha), \qquad t \geq 0; \qquad (2)$$

$$H(0,t) = 0, \qquad t < 0; \qquad (3)$$

$$H(x,t) = 0, \qquad x \to \infty. \qquad (4)$$

In operational form

$$\frac{d^2 H}{dx^2} = \gamma \mu_0 S(H). \qquad (5)$$

With the sinusoidal signal

$$H = H_0 \sin(\omega t + \alpha) = Jm[H_0 e^{j(\omega t + \alpha)}]. \qquad (6)$$

Boundary conditions in operational form are

$$H_{x=0} = \text{Jm}[H_0(S-j\omega)^{-1}e^{j\alpha}], \qquad (7)$$

$$H_{x\to\infty} = 0. \qquad (8)$$

Equation solving

$$H = \text{Jm}[H_0(S-j\omega)^{-1}e^{j\alpha}\exp(-x(\mu_0\gamma S)^{0,5})] \qquad (9)$$

for x=0 (10)

$$E_0 = -\gamma^{-1}\frac{dH}{dx} = \text{Jm}[\gamma^{-1}H_0 e^{j\alpha}\sqrt{\mu_0\gamma}(S-j\omega)^{-1}\sqrt{S}]. \qquad (11)$$

And pertaining to time

$$E_0(t) = \text{Jm}[\gamma^{-1}H_0 e^{j\alpha}\sqrt{\mu_0\gamma}((\pi t)^{0,5} + j\omega\, e^{j\omega}\,\text{erf}\sqrt{j\omega t})]. \qquad (12)$$

Where error function

$$\text{erf}(j\omega t)^{0,5} = (2j)^{0,5}[C(\omega t) - jS(\omega t)],$$

and $C(\omega t)$, $S(\omega t)$ - Fresnel integral.

Finally for $E_0(t)$

$$E_0(t) = H_0(\mu\omega\gamma^{-1})^{0,5}((\pi\omega t)^{-0,5}\sin\alpha + M\sin(\omega t + 0,5\pi - \text{arctg}(SC^{-1}))). \qquad (13)$$

Another view of final expressions take place due to the first case refer to the current source, but second is the case of voltage one. And the decision with the action of

$$E(t) = E_0 \sin(\omega t + \alpha), \qquad (14)$$

$$E = \text{Jm}[E_0 e^{j\alpha}(S-j\omega)^{-1}\exp(-x(\mu_0\gamma S)^{0,5}), \qquad (15)$$

$$H_0(t) = E_0(\mu_0\omega\gamma^{-1})^{-0,5} M\sin(\omega t + \alpha - \text{arctg}(SC^{-1})). \qquad (16)$$

. Because $t \to \infty$, Fresnel integrals $C(\omega t)$ и $S(\omega t)$ tend to value 0.5, the phase angle approaches to the value of $0,25\pi$

$$\text{arctg}(S(\omega t)C^{-1}(\omega t)) \to 0,25\pi, \qquad t \to \infty.$$

As may be seen, at the moment t =0 half-space response essentually differs from the steady-state phase.

$$\text{arctg}(S(\omega t)C^{-1}(\omega t)) \to 0, \qquad t \to 0,$$

half-space response in the case of $H\sin(\omega t)$ is clearly inductive, but in the case of $E\sin(\omega t)$, response will be clearly active. Absolute response value

$$M = (2[C^2(\omega t) + S^2(\omega t)])^{0,5} \qquad \text{approaches to the one,}$$

slowly deviating from one value only when $\omega t = 20\text{-}25$, i.e., after 3-4 periods.

The case of input disturbance $E \sin(\omega t + \alpha)$ is really interesting. Table containes the difference Δ in responses between transient and sustained stages ($\alpha = 0$, $\alpha = \pi/4$).

Table

$t \cdot T^{-1}$	0.1	0.2	0.3	0.4	0.5	0.6	0.7	0.8	1.0
$\Delta(\alpha=0)$	0.517	0.420	0.354	0.322	0.303	0.285	0.267	0.247	0.220
$\Delta(\alpha=0.25\pi)$	0.106	0.100	0.096	0.081	0.061	0.052	0.047	0.034	0.024

As may be seen, in $\alpha = 0$, response on the end of first half-period yet has a differens about 20% from sustained phase.

In $\alpha = 0, 25\pi$ transient component fades faster and approaches the 5% level in $t = 0,6$; thus, transient time in first case is remarkably greater, than in $R = \omega L$ ($\alpha = 0, 25\pi$).

Sustained mode computation is based on the iteration calculations and reduced by the step-by-step approaching to the true decision of electrical and thermal tasks.

First must be defined input data, including geometric sizes, frequency, approximated temperature distribution. Then should be defined constant of material, where the currents are induced. After completing these steps electrical task will be solved. The result is calculated distribution of heat sources. Next step is the calculation of temperature distribution and comparing with the predifined (approximated) distribution in begining of calculations. Frequently used methods for solving electrical task are integral equations methods and for thermal task the net ones.

However, if for heating non-magnetic workpieces integral equations method is very suitable, for ferro-magnetics workpieces its usage too complicated. It is the reason for solving electrical task by the difference methods and calculating the field outside of workpiece using integral equations methods. The way is co-joint solving of equations, which discribes the processes inside and outside the workpiece.

Results of research are computer programs for calculating elctromagnetic and thermal fields by means the 1-dimensional difference method as well as programs for co-joint calculations physical fields parameters.

For electrical task were used integral equations methods as well as net ones. For thermal task the only net methods were used.

For field calculations in the workpieces with high complicated shapes 3-dimensional and quasi-3-dimensional methods were used. Mathematically they are identical to the described above.

Simulating processes inside metall workpieces during inductive heating must be considered as closely connected with the way of energy feeding inside the workpiece. Energy feeding viewed as depending on power source operation mode. Power sources may use thiristor's, transistor's or vacuum tube's circuit technology. Such kind approach in the principles of simulating is a good way for power source circuitry design as well as for conrol algorithm development (for whole heating process). The task becomes more complicated due to non-linear load influence. Besides that, other non-linear tasks should be added. They are connected with the non-linear switching processes and non-linear properties of active devices and components like transformers and chokes with magnetic cores. State-space approach is a base of widespread simulating methods. On the basis of simulation stands system of equations, which discribes processes in circuit components during between-pulse commutation interval. Duration of this interval usually unknown and causes additional non-linearity.

For composing space-state equation's systems, topological and component system of equations are used. Topological equations has descriptions on the base of Kirchhoff rule for current and voltage. Component equations describes mathematical relations between current and voltage inside the circuit branch. Then system of equation may simplified. Next step is extracting the ordinary differential equations in Cauchy form. Solution of latter equations by the numerical methods is a way to calculate elctromagnetic transient (process evolution in time). Following the same path, algebraic equation system may be obtained. It gets connectivity with another variables.

The most complicated problem is switching point calculation in self-excitation circuitry schemes based on the thiristor or transistor circuitry. It is nesessary to describe switching conditions on every step, i.e., external input signals, current and voltage values on the device terminals [1,2].

As a result of power sourses simulation, circuitry optimizations for power sources had been done. Influence on the feeding lines were also calculated, as well as dynamic characteristics, emergency mode operation, peak power (transfering into the load). Pulse behavioure of inductive heating invertor may be demonstrated on the example of double bridge invertor with frequency doubling and reverse diods. The way of good power control is frequency and phase regulation [3].

Invertor performance with different load impedances demonstrates trasferring harmonics power into the feeding lines. The essential fact is the multiplying harmonic amplitude due to resonance coincidence in the invertor circuits and external lines. Good way to reduce harmonic influence provides precisely designed serial resonance filter, connected to the invertor output.

Designed research and optimization methods for power sources of electro-technological equipment brought interesting results on the field of thermal treatment technology, including hardening, welding, cutting, push broaching.

On the research base of non-stationary thermal fields, straines and stresses stands 3-dimensional decision of time-dependent electromagnetic thermal task (including Stephan task). Decision for dynamic task of strains and stresses, as well as for smelting and vaporing are also part of common decision for non-stationary thermal fields in the condition of spreading energy stream. Treating material according this approach considered as having no-linear thermal as well as elastical and plastical characteristics. Mechanical properties modification as a result of structure change had been taking in view, using special curves. Latter were specially processed. These curves mirrores volume change depending on structure conversion. Relaxation curves also was processed and used. Finite element method is commonly used for solving thermal task. Numerical dicision includes calculating phisical properties of every element and placing results in memory. On next step this data are used for solving deformation task (modulus of elasticity, lower yield point, linear expansion coefficient).

Structure conversion calculation realizes by means of computing free metal volume modifications and temperature lower yield point variations. Curves mirroring volume change depending on structure conversion may be used as well as relaxation curves [4]. First type curves may be approximated by the broken line.

Building lines y_3 and y_4 may be explained from the Figure 2. It bases on the assumption that line y_1, having the coefficient α_1, corresponds with perlite-ferrit heating metal. This line crosses line y_2, which has the coefficient α_2, conforming to the austenite in the cooling stage with the temperature T_s. Taking in view that values α_1 and α_2 are given, building lines y_3 and y_4 may be done.

$$\alpha_3 = (\alpha_1(T_s - T_c) - \alpha_2(T_s - T_b))(T_b - T_c)^{-1} \qquad (17)$$

$$\alpha_4 = (\alpha_1(T_s - A_{c1}) - \alpha_2(T_s - A_{c3}))(A_{c3} - A_{c1})^{-1} \qquad (18)$$

After that there is a possibility to correct curves, mirroring volume change depending on structure conversion, according to the given forward heating and cooling speed value.

After that there is a possibility to correct curves, mirroring volume change depending on structure conversion, according to the given forward heating and cooling speed value. Relaxation curve may be corrected according to the initial values of lower yield point on the cooling stage. A way to compute relaxation curve is thermo-kinetic diagrams usage. Because cooling speed was given, steel hardness after cooling. After that lower yield point also may be found. Using this method, particularly, surface inductive hardening technology in the air enviroment was designed. This technology suitable for hardening such workpieces as gears with big module and a kind of long-size parts. This method provides good durability and preferable distribution of remanent stresses. It bases on the clear phisical principles that looks simple. Form distortion during surface hardening undoubtedly follows the proprtion law depending on value of volume been heated. Decreasing of this volume may be reached by the reduction of hardening layer. Connection with the heating speed also take place. Another clear fact take place in concern with decreasing of hardening layer. Thin hardenig layer may be produced on the firm base. Getting firm workpiece base is possible by means of selection suitable steel composition or by the previously done volume heat treatment of workpiece. Surface hardening method named "from within" inevitably brings a remanent compressing strains on the skin layer of workpiece. Explanation bases on the fact, that martensite first appeares inside layer and then goes into the surface. A way to correct remanent strains degree is previous volume heat treatment.

For heat cutting processes thermal model also had been designed. It is suitable for laser, plasma-arc and other types of cutting. Analyzing quality (roughness of cutting sirface, perpendicularity, thickness of flasing layer, remanent strain and stresses) as a function of mode operation of cutting and power source appears important applying to this model.

Methods of heating and local heat treatment (welding joints of high pressure thick-walled vessels) were patented.

1. Васильев А.С. Статические преобразователи частоты для индукционого нагрева. М.: Энергия, 1974.
2. Васильев А.С., Гуревич С.Г., Иоффе Ю.С. Источники питания электротермических установок. М.: Энергоатомиздат, 1985.
3. Васильев А.С., Дзлиев С.В. Методы машинного проектирования преобразователей электрической энергии для электротехнологий. С.-Петербург.: Энергоатомиздат, 1993.
4. Гатовский К.М., Марков С.П., Шемелов С.А. Оценка напряжённого состояния сварных соединений сталей 09Г2 и 10ХСНД с учётом структурных превращений // Автоматическая сварка. № 4, 1980.

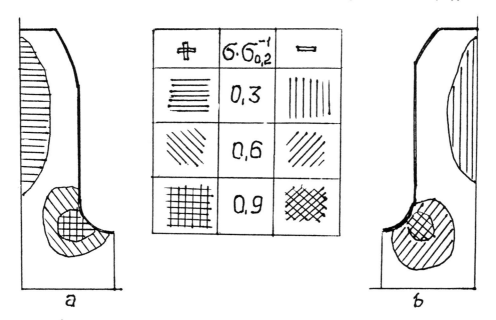

Fig.1. Remnant straines after the surface hardening calculaed without reference to structure transformations (a) and with reference (b)

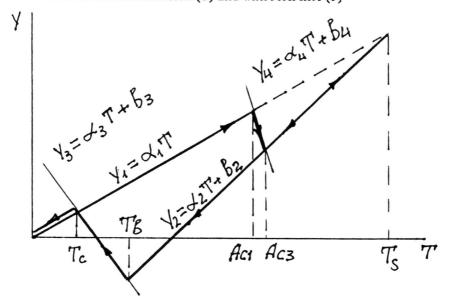

Fig.2. Approximation of curves mirroing volume and change depending on structure conversion

The Design and Performance of Multi-Megawatt Variable Width TFX Induction Heaters for the Ultra Rapid Heat Treatment of Metal Strip

ROGER GIBSON

EA Technology, Capenhurst, Chester, CH1 6ES, UK.

ABSTRACT

A high power transverse flux induction heater design developed at EA Technology, licensed to Davy International, and marketed under the registered name TFX, has been in successful commercial operation for a number of years at powers up to 3MW. The further development of this reliable design to give a fully continuously variable width capability over at least a two to one strip width range is described. Operating economics, efficiency, power intensity and electromagnetic forces are discussed. The increased operational flexibility of this design allows a very wide range of strip materials, including magnetic strip, to be heated in a common inductor, using commercially available variable frequency power supplies. This extends the use of TFX induction heating to applications such as coating, where steel and aluminium strip may be intermixed in a single line.

Un inducteur de grande puissance à flux transversal mis au point par EA Technology, sous licence auprès de Davy International, et commercialisé sous la marque déposée TFX, connaît depuis un certain nombre d'années la réussite commerciale à des puissances allant jusqu'à 3MW. L'évolution de cette conception fiable pour obtenir une capacité de largeur totalement variable en continu sur une gamme de largeur de bande dans un rapport de deux à un a été décrite. L'économie d'exploitation, l'efficacité, l'intensité de puissance et les forces électromagnétiques sont discutées. Cette conception d'une plus grande souplesse d'utilisation opérationnelle permet de chauffer une gamme très étendue de bandes, y compris bande magnétique, dans un inducteur commun, en utilisant des alimentations électriques à fréquence variable du commerce. L'inducteur TFX peut ainsi être utilisé pour des applications telles que le revêtement où des bandes d'acier et d'aluminium peuvent être mêlées sur une seule ligne.

INTRODUCTION

The principles of the induction heating of metal workpieces have been known for over a hundred years. Although it has long been recognised that efficient high power intensity heating of thin metal strip requires a magnetic field perpendicular to the strip to induce currents in the plane of the strip, a practical application of this transverse flux principle has been hampered for many decades by difficulties in producing a sufficiently uniform temperature rise in strip passing through the heater. Noise, vibration and unacceptable strip forces have also caused problems. However, a design developed at EA Technology, licensed to Davy International and marketed under the registered name TFX, has been in successful commercial operation for a number of years at powers up to 3MW for the annealing and other heat treatments of aluminium strip. Smaller units are in operation for coating operations and brass annealing. This technology has also been successfully applied to the annealing of magnetic steel strip.

NARROW WIDTH RANGE TFX STRIP HEATERS

TFX strip heaters used in production installations hitherto have a relatively narrow strip width range of ±30mm to ±50mm about a nominal strip width. Fig. 1 shows a typical design.

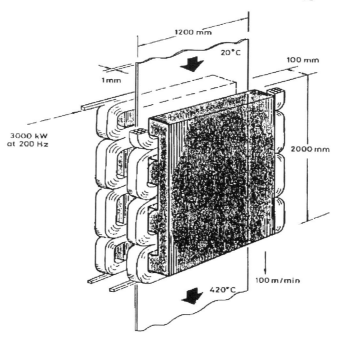

Fig. 1. A 3MW narrow width range TFX aluminium strip heater.

When researching the design of such an inductor, the EA Technology computer model SCEDDY was used to calculate the shape of a transverse magnetic field that would produce a uniform temperature profile. A suitable transverse magnetic flux profile is shown in Fig. 2, where strip motion through the inductor is from bottom left to top right.

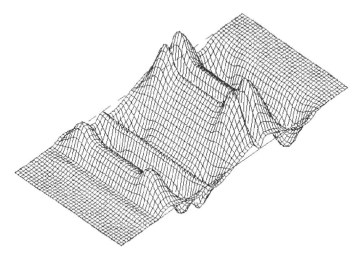

Fig. 2. Transverse magnetic flux profile in a 3 pole pitch TFX inductor.

Current paths induced in the plane of the strip by this field are shown in Fig. 3. The inductor core outline drawn next to the strip shows the positional relationship between strip currents and the inductor winding. Strip motion is from left to right. From the closer contours at the left, or entry end of the inductor, it can be seen that a higher current density flows here, due to the less resistive colder strip. A close observation of the flux density profile in Fig. 2 shows that there is a lower flux density at this point due to the increased demagnetising effect of the higher strip current.

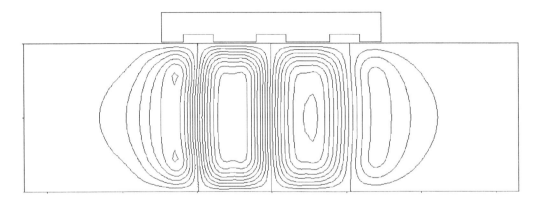

Fig. 3. Current paths induced in the strip by the flux profile in Fig. 2.

The consequent apparently very non-uniform power density distribution produced by these currents is shown in Fig. 4, where it can be seen that the higher strip resistivity towards the right, or strip exit end of the inductor compensates for the lower current at this point to give a relatively high power intensity.

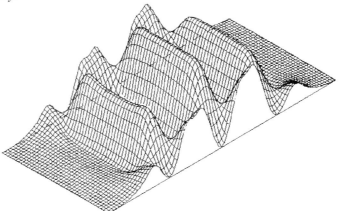

Fig. 4. Power intensity profile produced by the currents in Fig. 3.

If the strip moves through the heater at a uniform speed, it is evident that the central portion, where the current flows in parallel paths across the strip will be uniformly heated. At the edges of the strip, however, the currents have to turn to return back under the next inductor slot, locally producing very non-uniform heating. A careful examination of the power intensity profile shows that although the peaks and troughs are of different heights, and spatially displaced from those at the strip centre, the net amount of heat induced across the strip edge can be made to integrate to the same as that at the strip centre, as shown in the resultant temperature rise shown in Fig. 5.

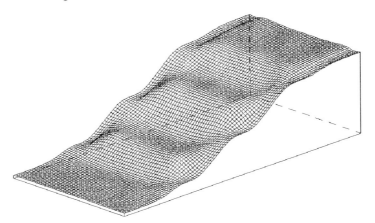

Fig. 5. Temperature rise in the strip as it passes through the inductor

It can be seen that although the temperature across the strip is not everywhere uniform within the inductor, the carefully shaped distribution of strip currents ensures a uniform temperature profile at the inductor exit. It is apparent that a single pole pitch inductor would produce a very non-uniform temperature distribution within the inductor itself, so a sufficient number of inductor poles is required to reduce the this lack of temperature uniformity to a level below which differential thermal expansion could cause strip distortion. Although strip thermal conduction is fully catered for in the SCEDDY model, strip transit through the inductor takes only a few seconds, leaving no time for thermal diffusion across the strip to help equalise the temperature uniformity. Each portion of strip needs to receive its equal allocation of induced power.

The feature that ensures the success of this design is the determination of a flux pattern to shape the currents to give a uniform temperature rise, making due allowance for changing strip resistivity caused by the temperature rise. This theoretical flux shape has also to be practically achievable in a reliable TFX inductor. An examination of the optimum flux shape (Fig. 2) shows that it is relatively uniform across the central portion of the inductor, rises towards the edge of the inductor, and then falls to a small value right at the strip edge. In practice this is achieved by using carefully shaped magnetic flux concentrators inboard of the edge of the strip to provide the flux humps, and carefully matching the inductor width to the strip width to produce the low flux level at the edge of the strip. It is this need to match the inductor to the strip width that limited the strip width range of early TFX inductors.

Although extensive successful operational experience has been achieved with narrow width range inductors to produce high added value product with continuous shift working, the excessive capital cost of a magazine of such inductors prompted an investigation into a fully continuously variable width design.

THEORETICAL DESIGN OF VARIABLE WIDTH TFX STRIP HEATERS

Using the SCEDDY computer model described above to study various options for a continuously variable width design, it was soon realised that it was essential to retain the basic magnetic flux distribution shown in Fig. 2. The problem thus posed is to design an effective variably positioned notional inductor edge. Experience with the narrow width range design has shown that for long term reliable operation at multi-megawatt power levels, it is essential to use rigidly constrained water cooled hollow copper inductor windings. The computer model showed that if this full width rigid winding was to be retained, the removal of practicably sized portions of core behind this winding had an insufficiently controlled effect on the resultant magnetic field distribution, as well as posing significant mechanical design problems.

The only effective way found to produce the required carefully contoured flux shape was the use of additional copper current carrying components. These are combined with the carefully shaped flux hump concentrators to produce an enhanced flux modification device. These are mounted on the face of the inductor, and travel in from the inductor edge to suit the strip width being heated. The width range possible in a single variable width inductor is theoretically limited to a minimum when the flat central portion of the flux shape disappears as the two flux humps meet at the centre of the inductor. For a 2m wide strip inductor, this can be as narrow as 0.5m strip width. However, for practical usage, a 2 to 1 turn down ratio is considered a more realistic design compromise. The design study undertaken shows that the high power intensities of around $1.5 MWm^{-3}$ achieved in the narrow strip width inductor can be retained. This limit is a practical guideline, governed by a range of design considerations, including winding voltage and current limits, flux density values, and strip forces.

PRACTICAL DEMONSTRATION OF VARIABLE WIDTH TFX STRIP HEATER DESIGN

Operational experience with the narrow width range design has shown that it was essential not only to prove the temperature uniformity of the proposed variable width solution, but to ensure that a mechanically reliable design could be manufactured. To demonstrate the effectiveness of the method, a programme was undertaken at EA Technology, funded by five UK electricity companies, Eastern Electricity plc, East Midlands Electricity plc, Midlands Electricity plc, National Power plc and Yorkshire Electricity Group plc. In this project, a 1MW, nominal 1200mm to 1250mm wide narrow width range inductor was fitted with the proposed new enhanced flux modification devices, and tested in the EA Technology strip process line. Two tonne coils of both steel and aluminium were heated over a wide range of operating conditions. Temperature profiles at the 2% target level were achieved over a 2 to 1 strip width range.

Design reliability was then demonstrated by replacing the initial experimental enhanced flux modification devices with rugged units designed and manufactured by Davy International to their full stringent production equipment standards. The temperature profile test results were successfully repeated. Monitoring of component vibration with an array of accelerometers was also undertaken to ensure that all components would operate reliably over the design life. This design is sufficiently rugged that, although not normally required, strip width variation can be followed at all power levels. Instantaneous change to a new width can be made as the strip join passes through the inductor.

A particular feature of these demonstrations is that both aluminium and magnetic steel strip were heated equally effectively in the one inductor. The only operational difference was a frequency change to match the heater to the contrasting strip resistivities. The frequency range required is well within the scope of commercially available variable frequency power supplies and associated matching transformers and tuning capacitors, and is below the critical 10kHz F.C.C. regulations limit. Although of little advantage to metal strip producers, this feature is of particular interest to end users such as strip coaters, who often process both aluminium and steel strip in a single line.

Heating flexibility is so wide that a single inductor could treat the unlikely combination of a thin narrow coil of steel heated to 1200°C, with a short cold tail stitched to a thick wide aluminium coil for partial annealing at 300°C. For the second or so that the stitch passed through the inductor, power would be switched off while the enhanced flux modification devices moved to the new position, and the power supply and associated capacitors and matching transformer switched to match the new heating condition. This extreme example shows that there is no necessity to carefully sequence consecutive coils as is required in fuel fired continuous annealing lines.

Another objective of these tests was the investigation of the behaviour of wide magnetic strip under the influence of the magnetic field in the inductor air gap. Earlier studies with the computer model had shown that, due to the high magnetic saturation caused by magnetic flux fringing through the length of the strip, magnetic forces pulling the strip towards the inductor face would be sufficiently small so as to be easily controlled by normal strip tension levels. However, as computer models cannot allow for every twisted shape that strip can take up in the inductor air gap, a further series of tests was undertaken in the strip process line to confirm that magnetic strip attraction to the inductor faces could be successfully countered by normal strip tensions.

As well as the design of effective flux shaping measures, the computer model is also used to determine other operational parameters such as inductor cooling requirements, water flows and pressure drops, and tuning capacitor and output transformer settings. The tests demonstrated the effectiveness of these computer model based design methods.

ECONOMIC FACTORS IN VARIABLE WIDTH TFX INDUCTOR OPERATION

The ability to use just one continuously variable width inductor instead of a magazine of narrow width range inductors has significantly reduced the capital cost of a TFX heat treatment line, both by the reduction in the number of inductors required, and by the elimination of the complex magazine and docking mechanisms supplying services to the interchangeable inductors.

However, the inductor winding losses do not reduce proportionately to the lower power required to heat narrow strip, leading to a small loss of efficiency when operating below maximum strip width. There is also an inevitable small energy loss in the current carrying components of the enhanced flux modification devices. Overall efficiency is still very high however. Any additional energy cost of variable width operation can easily be offset against reduced capital costs and increased productivity by elimination of inductor change times.

FUTURE DEVELOPMENTS

Academic studies into the novel metallurgy possible using ultra rapid heating and quenching rates of several hundred $°Cs^{-1}$ have been undertaken world wide. Now that an economically viable variable width inductor design has been proven, attention is now being turned to the new and improved metallurgical processes and products made possible by the use of TFX strip heating to provide these ultra rapid heating rates. Development projects currently being undertaken at EA Technology are demonstrating significantly improved product properties for low carbon steels, austenitic and ferritic stainless steels. Other metals and alloys are also under investigation, and are showing promising results.

The future is indeed exciting for TFX strip heating, not only for its ability to provide a compact efficient clean and reliable heating facility, but also in the wide range of new products made possible by the increased ultra rapid heating rates being achieved using this technique.

Surface Treatment

Contents

Les plasma-froids microondes en écoulement: vecteures d'énergie électrique pour le developpement de processes surfaciques performants 1
Dessaux, O., Goudmand, P. (Université des Sciences et Technologiques de Lille) FRANCE

Bulk mode vacuum coating – The industrial concept 9
Cattentot, J.M. (Surface Engineering), Goudouneix, H. (Novelec) FRANCE

Electron beam curing: the future of coated steel production? 15
Williams, B. (British Steel Strip Products) UK

Changeover from an electrodeposition technique based on low current density to the utilization of high current densities for copper coatings 23
Volvert, A. (Bundy S.A.) BELGIUM

New magnetodielectric materials for induction heating applications 31
Ruffini, R.S., Nemkov, V.S. (Centre for Induction Technology) USA

Thermal stresses in the frequency induction pulse-hardening of non-symmetrical workpieces 39
Blinov, Y., Kachanov, B., Kogan, B., Fedorova, V. (Electrotechnical University of St Petersburg) RUSSIA

Effect of temperature on vibration and magnetostriction in mild steel heated by induction 47
Baker, R.T., Oliver, T.N. (Aston University) UK

Effective micro-metallurgical induction heating process for surface macro-reinforcing of mechanised parts 55
Vologdin, V. (FREAL Ltd), Ganjuchenko, V.M. (Electrotechnical University of St Petersburg), Mamykin, S.M. (SPLAV Ltd) RUSSIA

The use of pulsed radiant flux with variable time structure for laser technologies 63
Ganjuchenko, V.M, Vologdina, S.G. (Electrotechnical University of St Petersburg) RUSSIA, Kostrubiec, F. (Technical University of Lodz) POLAND

Resource saving technologies of thermal treatment with high frequency induction heating in infra-structure of railway transport 71
Bezmenov, F.V., Vologdin, V.V., Filippov, K.P., Chervinsky, V.I. (VNIITVCh) RUSSIA

Etude et optimisation de l'installation de coupage des métaux par la torche à plasma d'air 81
Pogora, V., Protsouc, I., Stantchou, T. (Université Technique de Moldova) MOLDOVA

Les Plasmas Froids Microondes en Écoulement Vecteurs d'Énergie Électrique pour le Développement de Processes Surfaciques Performants.

ODILE DESSAUX ET PIERRE GOUDMAND

L.P.C.E.P./L.E.F.E.M.O. - EA MRES N°1761 - UNIVERSITÉ DES SCIENCES ET TECHNOLOGIES DE LILLE
F 59655 VILLENEUVE D'ASCQ CEDEX (FRANCE)

Flowing post-discharge cold plasmas are obtained by microwave, or high frequencies discharges in flowing gas ; they make up an non ionic reactive medium.

If dinitrogen is used as gas, post-decharge plasma (C.R.N.P.) are obtained with an important volumic extension and with a low level of viscosity. C.R.N.P. is on efficient energy vector and his hight chemical reactivity is originally :
- on the one hand : the ability to obtain intensive modifications on the surfaces of plastic materials with a good durability of the adhesive properties,
- on the other hand specific reactions to obtain thin films with unusual properties as high level dielectric polymers, hard coating on conductive metallic films with a good adhesivity on polymers.
Industrials networks developing these possibilities are now working in some french compagnies.

Les plasmas froids en post-décharge et en écoulement sont obtenus à partir de décharges microondes ou hautes fréquences dans un gaz ; ils constituent un milieu réactif non ionique. Lorsque le gaz plasmagène est le diazote, le plasma post-décharge est réalisé sous un volume important avec une bonne homogénéité et une viscosité faible. Son caractère de vecteurs d'énergie et sa réactivité chimique lui confère, d'une part la possibilité de modifier des surfaces plastiques d'une manière importante et durable quant aux propriétés d'adhésion, d'autre part est à l'origine de réactions spécifiques pour créer des couches minces de propriétés particulières qui sont à l'origine de polymères aux caractéristiques diélectriques remarquables, de dépôts durs ou métalliques adhérant sur polymère.
Des réacteurs industriels développant ces possibilités fonctionnent industriellement.

INTRODUCTION

Le génie des procédés de modifications des propriétés de surface induit des opérations industrielles de niveau technologique avancé ; leur quasi totalité met en oeuvre l'énergie électrique. Parmi eux, les plasmas froids de décharge ont pris une place importante ; ils sont obtenus à partir de l'injection de l'énergie électrique dans un système gazeux et leur réactivité tient à des électrons rapides, des ions, des photons énergétiques ainsi qu'à des atomes et des molécules excitées.

L'interaction plasma-surface conduit à plusieurs classes d'applications :

* Modification des propriétés surfaciques, en particulier l'énergie libre de surface, la fonctionnalisation physicochimique pour orienter des propriétés d'adhésion, de structure ou de mécanique superficielle.

* Dépôts assistés par plasma froid d'une couche mince à propriétés spécifiques thermodynamiques, électrique ou électromagnétique ou encore présentant un effet barrière

* Gravures assistées par plasmas froids. Depuis quelques années un procédé original consiste à mettre en oeuvre des plasmas froids en post-décharge et en écoulement ; ils sont obtenus à partir d'espèces non ioniques extraites d'une décharge électrique. Cette dernière est préférentiellement excitée, sans électrode, par un champ électromagnétique haute fréquence ou très haute fréquence (microonde) qui permet l'obtention de décharges en non équilibre thermodynamique jusqu'à des pressions d'une centaine hecto pascals.

Dans ce cadre notre laboratoire a développé une méthodologie - qui lui est originale - de plasma en post-décharge d'azote - P.F.D.A.. Nous présentons les particularités de cette méthodologie en la remplaçant dans la perspective générale de l'utilisation industrielle des plasmas froids.

LES PLASMAS FROIDS MICROONDE ET HAUTE FRÉQUENCE

I - PLASMAS DE DÉCHARGE

Un plasma est un gaz ionisé, électroniquement neutre. En s'appuyant sur un modèle très simple on peut dégager une classification des plasmas fondée sur deux facteurs :

- leur concentration en électrons
- les propriété énergétiques des particules constitutives du plasma : les plasmas froids industriels sont faiblement ionisés, les particules qui les composent sont caractérisées par une température de translation définie dans la relation :

$$\varepsilon_C = \frac{3}{2} kT \quad \text{avec} \quad \begin{cases} \varepsilon_C = \text{énergie cinétique de la particule} \\ k = \text{constante de} \\ T = \text{température exprimée de Kelvin} \end{cases}$$

Un autre paramètre important des plasmas est la fonction de distribution de l'énergie de translation des électrons ; dans le cas particulier de l'équilibre thermodynamique elle est donnée par la loi bien connue de distribution des vitesses de MAXWELL.

La physicochimie des plasmas est régie par des collisions entre particules :

* Collisions élastiques où il y a conservation de l'énergie cinétique et dont le rôle est d'assurer une équipartition de l'énergie tendant à évoluer vers l'équilibre thermodynamique.

* Collisions inélastiques pour lesquelles une partie de l'énergie est utilisée pour entraîner des processus de transformation physicochimiques. Parmi ces collisions, celles intervenant entre électrons et particules neutres (atomes ou molécules) de fréquence v_{ne} sont les plus nombreuses ; elles régissent la physicochimie du milieu en créant les entités réactives des plasmas et en assurent l'entretien par les électrons et les ions qui en résultent.

La différence de masse entre électrons et particules conduit à une première classification des plasmas :

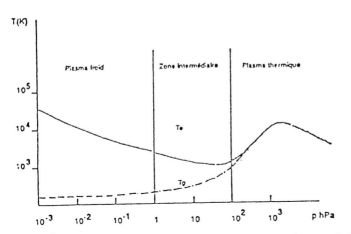

Figure 1. Température de translation dans un plasma froid
Te : température d'électrons
Tg : température d'ions et particules neutres

La figure 1 donne en fonction de la pression, c'est-à-dire de v_{ne}, l'évolution de la température de translation des électrons et des particules atomiques ou moléculaires. On met ainsi en relief, pour des pressions inférieures à 10^2 hPa, des domaines caractérisés par un important non équilibre thermodynamique. Ainsi pour des pressions inférieures à 2 hPa, la température de translation des électrons - de l'ordre de 20.000 K à 100.000 K - reste très élevée devant celle des autres particules ; ce plasma est une source d'électrons rapides et de photons énergétiques. Pour des pressions plus élevées - entre 3 et 100 hPa, le non-équilibre thermodynamique reste important. Ces plasmas intermédiaires sont des sources très riches en espèces chimiques réactives et en molécules excitées, vecteur d'énergie ; ils sont des précurseurs de choix pour des plasmas en post-décharge et en écoulement.

Tableau 1

p hPa	fréquence 13,50 MHz ν_{en}/ω	fréquence 2450 MHz ν_{en}/ω
10	0,8	4.10-3
50	2,3	0,01
500	11	0,06

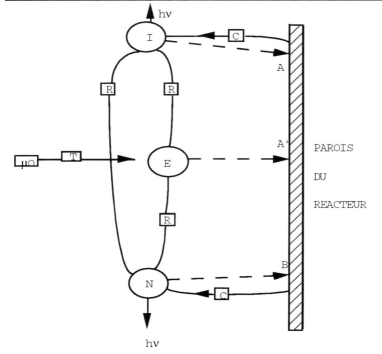

μo	Source d'énergie électrique microonde
I	Gaz d'ion lourd
E	Gaz d'électron
N	Gaz de particules neutres (atomes, molécules, radicaux libres)
A et A'	Recombinaison d'électrons et d'ions. Electrisation de la paroi
B	Recombinaison de radicaux libres
R	Collisions élastiques et inélastiques
T	Chauffage par accélération des électrons
C	Transfert de chaleur convectif
➡ hv	Emission de photons (ultraviolet et visible)

L'excitation d'une décharge électrique est généralement obtenue sans électrode par des dispositifs mettant en jeu des hautes fréquences - 100 kHz ; 13,56 MHz ; 27,12 MHz - ou des microondes - 433 MHz ; 896-915 MHz ; 2450 MHz. Le caractère spécifique de cette excitation tient à la différence de masse entre les électrons et les ions "lourds" atomiques ou moléculaires : les électrons sont accélérés sélectivement, tandis que du fait de leur inertie les ions "lourds" restent quasi immobiles. La fréquence rapide impose aux trajectoires des électrons vers les parois du réacteur de s'inverser rapidement ; la conséquence en est de limiter la diffusion de ces électrons vers les parois du réacteur et parallèlement, par l'augmentation de la fréquence des collisions inélastiques électrons-molécules d'enrichir le plasma en espèces chimiquement actives et d'accentuer le non-équilibre thermodynamique. D'une manière plus quantitative, le rapport v_{en}/ω - fréquence des collisions sur fréquence angulaire électromagnétique - joue un rôle important sur les propriétés des plasmas : pour que le plasma comporte les propriétés de non équilibre, il faut que ce rapport $v_{en}/\omega < 1$; la distribution des électrons se trouve alors déplacée vers les énergies élevées par rapport à la distribution de MAXWELL : ce rapport croît rapidement avec la pression de décroît avec la fréquence. Le tableau 1 ci-dontre permet de fixer les idées sur l'évolution de ce rapport pour un même gaz plasmagène pour deux modèles d'excitation.

La figure 2 schématise le modèle de décharge ; elle représente les divers processus significatifs au sein du plasma en mettant en relief le rôle important des parois du réacteur dans la thermalisation vers l'équilibre par convection naturelle du plasma;

Un aspect fondamental pour l'utilisation de plasmas pour l'industrie est la réalisation de coupleurs performants transférant avec un excellent rendement des énergies électromagnétiques de plus en plus importantes. La spécificité des plasmas en post-décharge implique que ces coupleurs présentent une efficacité optimale pour des pressions relativement élevées de l'ordre de 4 à 20 hPa. Le L.P.C.E.P./LEFEMO est à l'origine d'une nouvelle génération de cavités cylindriques microondes fonctionnant à 2450 MHz, 915 MHz et 433 MHz[1] ainsi que des cavités de volume important du type hélicon pour l'obtention de plasma de décharge haute fréquence à 13,56 MHz.[2]

II - PLASMA EN POST-DECHARGE ET EN ECOULEMENT.

La notion de plasmas froids en écoulement mérite d'être précisée car elle recouvre plusieurs situations relatives aux écoulements en aval de la zone de décharge. La figure 3 rend compte de différents types de plasma dans un écoulement.[3]

Un plasma en post-décharge non ionique a pour espèces réactives des atomes radicaux libres, des espèces moléculaires - pratiquement toujours diatomiques - électroniquement et vibrationnellement excitées, celles de longue durée de relaxation radiative ou collisionnelle offrant un intérêt tout particulier. Les collisions inélastiques entre particules non ioniques confèrent à ce milieu non ionisé une composition qui diffère de celle attendue après la seule détente des espèces énergétiques de la décharge. Ce milieu est exempt d'électrons ou de photons énergétiques (UV) susceptibles de rompre des liaisons chimiques. Sa non-ionisation implique une viscosité faible lui conférant une bonne capacité d'action dans des situations géométriques variées.

Les plasmas post-décharge sa caractérisent par un important non équilibre thermodynamique qui se traduit par une sélectivité dans la répartition, sur certains degrés de liberté, de l'énergie dans les atomes et surtout les molécules, qui gardent des températures de translation peu

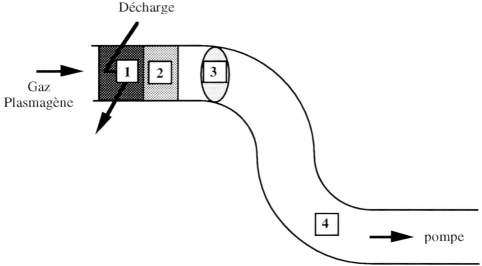

Figure 3. [1] plasma de décharge (*ionique*) - [2] plasma de transition à gradient de concentration P.T.G.C. (*ionique*) - [3] plasma secondaire dont une réionisation est à l'origine. Exemple post-luminescence rose de l'azote (*ionique*) - [4] post-décharge en écoulement ou plasma froid différé (*non ionique*).

élevée. Un plasma post-décharge est un vecteur d'énergies bien identifiées et de température de translation faible. Cette caractéristique lui confère des propriétés réactives remarquables et permet d'obtenir à la température ambiante des actions physicochimiques normalement atteintes à des températures très élevées. Sa seule source d'énergie est le plasma de décharge, le passage en post-décharge confère à cette énergie des formes spécifiques non thermiques.

La nature des parois du réacteur et des canalisations a, dans ce dernier cas, une importance essentielle - contrairement au cas des plasmas de décharge - dans la mesure où le PFD n'est pas l'objet d'un apport extérieur d'énergie ; ces parois doivent être choisies de manière à minimiser les désactivations surfaciques : les matériaux adéquats sont le quartz, les verres et les émaux, l'alumine (donc l'aluminium), certains matériaux plastiques ou des polymères d'hydrocarbures fluorés.

Ces plasmas sont préférentiellement obtenus à partir de décharges microondes qui permettent un important non équilibre thermodynamique jusqu'à des pressions de plusieurs hPa à l'origine d'une concentration remarquable en espèces actives (plasmas secondaires) et sont donc de bons précurseurs du PFD. Toutefois les hautes fréquences par le biais de coupleurs hélicons donnent des sources de bonne qualité.

La mécanistique du fonctionnement d'un PFD dépend simultanément des propriétés cinétiques des espèces actives du milieu, de la capacité des parois du réacteur à la non désactivation des espèces actives du régime d'écoulement et du volume du système gazeux. L'ensemble de ces conditions permet de déterminer une plage de fonctionnement optimal - quant à la pression de travail - pour l'utilisation du plasma.

Ia - Plasmas en post-décharge de dioxygène et de diazote

On rencontre essentiellement deux gaz plasmagènes qui sont à l'origine de réacteurs industriels : le dioxygène et le diazote :
les espèces actives dans les plasmas post-décharge d'oxygène sont l'oxygène atomique $O(^3P)$ ainsi que deux états électroniques électroniquement excités $O(^1D)$ et $O(^1S)$ accompagnés par des espèces moléculaires métastables $O_2(^1\Delta)$ et $O_2(^1\Sigma)$. Les durées de vie de $O(^3P)$ et $O_2(^1\Delta)$ sont importantes, de l'ordre de plusieurs minutes.

Dans l'azote les espèces actives sont l'atome d'azote dans son état fondamental $N(^4S)$ d'une durée de vie de 10 secondes et le diazote énergétiquement excité $N_2(^1\Sigma_g^+)_v$ molécule à l'état fondamental vibrationnellement peuplé ou à l'état triplet $N_2(^3\Sigma_u^-)$. Des atomes électroniquement excités $N(^2D)$ ou $N(^2P)$ en concentration faible interviennent, le premier avec une durée de vie de 26 heures.

Par ailleurs, le taux de dissociation moyen de N_2 dans une décharge microonde est de l'ordre de 5% ; il est cinq ou six fois plus élevé dans le cas de O_2. Sous ces conditions, les plasmas post-décharge d'oxygène apparaissent à priori plus intéressant pour la réalisation de réacteur pour la fonctionnalisation de surface que les plasmas correspondant d'azote.

En réalité, la situation se présente d'une manière totalement différente : le processus de perte de $O(^3P)$ par collision est très important et les plasmas de post-décharge d'oxygène sont très inhomogènes. Dans le cas des plasmas d'azote, d'une part le processus de perte de $N(^4S)$ par collision est lent ; par ailleurs il a été montré récemment que les espèces actives du PFDA -$N_2(^4S)$ et $N_2(^1\Sigma^+)_v$- sont créées en permanence dans la post-décharge par un mécanisme complexe de transfert d'énergie intermoléculaire. Il faut aussi souligner l'efficacité et la durabilité très importante de ces traitements par PFDA vis à vis de traitements correspondants à partir du dioxygène.[4]

CONCLUSIONS

Les plasmas froids en écoulement et en post-décharge d'azote ont été développés au LPCEP/LEFEMO dans deux directions : la conception de gros réacteurs industriels (2-4 m^3) de traitement de surface afin de leur conférer des propriétés d'adhésion. Ces appareils sont réalisés à partir d'une collaboration contractuelle entre le laboratoire et la Société ATEA FRAMATOME et sont utilisés chez des équipementiers automobiles pour la peinture et le collage sur des pièces en polypropylène de tailles importantes, telles des tableaux de bord ou boucliers. Les traitements se sont avérés d'une efficacité inattendue pour le traitement des composés plastiques poreux.[5]

Ce même PFDA est utilisé pour obtenir des couches minces conductrices à température ambiante avec des propriétés spécifiques (faible énergie surfacique, diélectrique, mécanique...) inattendues, différentes de celles obtenues avec des plasmas de décharge. La vitesse de dépôt est de 10 à 100 fois plus importante qu'en plasma de décharge ou proche post-décharge. Ces dépôts ont, suivant le cas, des constantes diélectriques (ε_r 20-30) pour de faibles facteurs de perte ou de dureté remarquable. Ainsi récemment un dépôt dur 9,5 GHa pour un faible module de Young 23 GPa doit conduire à des applications nombreuses et innovantes.

Références

1. C. DUPRET, O. DESSAUX et P. GOUDMAND : BF 95 04802

   C. DUPRET, O. DESSAUX, P. GOUDMAND : 9ème colloque international sur les procédés plasmas. Antibes 1993

   C. DUPRET, P. SUPIOT, O. DESSAUX, P. GOUDMAND : Rev. Scient. Inst. 65, 3439, 1994

2.. DUPRET, B. MUTEL, O. DESSAUX et P. GOUDMAND : BF 95 06301

3. O. DESSAUX et P. GOUDMAND : Congrès international microonde et haute fréquence CFE vol.II, Nice 1991

4. B. MUTEL, Thèse de Doctorat d'Etat n°704, Lille, 1986

5. B. MUTEL, C. HOYEZ, O. DESSAUX, P. GOUDMAND, L. GENGEMBRE et J. GRIMBLOT : 12 ISPC Minneapolis 21-25/8/1995

Bulk mode vacuum coating - The industrial concept

J.M. CATTENOT[1], H. GOUDOUNEIX [2]

1- SURFACE ENGINEERING - 260 route de Pontarlier - 39300 Champagnole, France
Tel. (33) 84 52 61 66 - Fax (33) 84 52 61 67
2- NOVELECT - EDF - 5 Chemein du Fort Benoit - 25000 Besançon, France

Abstract : Coating small parts under vacuum with a competitive price is almost incompatible with the utilization of holder. For this purpose an industrial barrel coater was developed. The system is described in detail and some of the potential applications are presented.

Résumé : Le revêtement sous vide de petites pièces à un prix compétitif est pratiquement incompatible avec l"utilisation de supports de pièces. Un procédé industriel de dépôt sous vide au tonneau a été développé pour cette application. Le système est décrit en détail et quelques applications potentielles sont présentées.

Introduction : Most of the vacuum coating machines on the market now, required the use of a substrate holder. The operation to put parts on a holder is time consuming and is incompatible with the full coverage of the part. The solution is to coat parts in a barrel. This concept, already used to coat resistor or aircraft fastener, is improved by the adjunction of load locks and RF bias capability to create a more universal equipment, with extended capabilities.

DESIGN OF THE EQUIPMENT

- **Competing technologies** : Several types of equipment was developed in the path using mainly evaporation technology to coat parts in bulk mode. The basis of the coater is using a mesh barrel and an evaporation source underneath [1]. The parts are tumbled during the coating inside the barrel. The coating coming from outside of the barrel, is gradually closing the holes of the mesh. The barrel needs to be cleaned off often, which is difficult to match with a load lock system supposed to work on a continuous basis. A second problem is relative to the minimum size of parts that could be coated, they have to be bigger than the holes of the mesh. Finally this type of coater can be modified to work in ion plating mode with a polarization on or in the barrel[2]. This technology require to work at high pressure and the result on the layer is often a columnar structure needing a post treatment like bead blasting.

Only a few systems are using magnetron sputtering, the main advantage of this kind of arrangement is the transfer of sputtering material from the cathode to the parts. The magnetron is mounted inside the barrel and the parts are mixed in the coating area [3-4]

- **Architecture** : The "FACET" system developed is one of the second generation. It is using a magnetron cathode mounted inside the barrel in sputter down mode. The coating chamber is loaded through an input load lock, and unloaded through an output load lock. The parts are traveling from air to vacuum to air by gravity. The bottom of the barrel is articulated to be open and to unload the parts (fig 1). The parts are mixed in a barrel on an inclined axis, this configuration helps a lot for an homogenous mix of the parts. The rotation movement of the barrel is obtain by a shaft using a feedthrough bellows sealed. This shaft is also moving the bottom of the barrel in translation and is powering the barrel during treatment.

The chamber is turbo-pumped and the load locks are pumped by a mechanical roughing pump. The pressure in the chamber is measured by a 4 decades capacitive gauge, for process and a Penning gauge for base pressure. In the load locks the pressure is measured by Pirani gauge.

The gas is controlled by 4 electronic massflow controllers.

The magnetron cathode is powered by a DC pulsed generator. The barrel can be polarized (biased) with an 13.56MHz RF generator through an automatic matching network.

The machine is totally controlled with a PLC, the dialog, process recipe programming and maintenance is done with a PC.

- **Capabilities and advantages** : The machine was design to be integrated in a coating line. The entry lock can be loaded by an automatic feeding system. One liter of parts is coated per run, and the load locks are pumped or vented during the deposition to minimize time. Depending of the size and the material to be coated, a load lock is pumped in less than 2 mn, generally in 30 s. With the used of the load lock the coating chamber is always under high vacuum, this condition is necessary for stable and reproducible reactive sputtering.

By the use of a solid wall barrel, and cups to contain the parts in the load locks, parts as small as $0.5mm^3$ can be treated without any problem.

The magnetron cathode is powered by a DC pulsed power supply, the period of the pulse is adjustable and the size is in the range of 10 to 50 µs. A positive pulse of a short time is used to neutralized the negative charge at the surface of the target and partially responsible of arcing. This function present a main advantage on target of low electrical conduction that required normally RF generator. In comparison the efficiency is almost the one of DC with the advantages of RF [5].

The barrel is polarized with a RF 13.56 MHz powered. This device allowed an efficient etching of the parts before coating. RF can etch either conductive or non conductive materials like oxides. RF can also be used during sputtering to bias the parts. In this case it resputter the coating material having a low adhesion, an it facilitate the penetration of the coating material in recess.

RF polarization allows the system to work in PECVD mode and by combining sputtering mode and PECVD its possible to first coat the parts with an adhesion layer and then to continue in plasma mode. An other aspect of the RF polarization is the possibility to plasma treat plastic before coating it, with reactive gases. After plasma treatment the parts can be unloaded for an other operation (painting...) or either coated by sputtering or by PECVD.

The gas are controlled by massflow controllers, to obtain reproducible condition from run to run especially on reactive sputtering applications.

The machine was engineered for production and access for maintenance to the different subassemblies is maximized. A mechanism is mounted on the front of the coating chamber to facilitate the opening of the front plate. This allows to access to the target and to change it in less than 5mn. To decrease maintenance cost, pumping accessories and valves including the transfer valves are standard of the shelf items.

APPLICATIONS

-Plasma treatment on polycarbonate : One of the first application of the equipment was a simple plasma treatment of small polycarbonate parts used as decorative item in an assembly that needs to be painted. A plasma of oxygen at 1 Pa (0.75 mTorr) for 1mn gave the required efficiency. The efficiency can but evaluated by the test of the wettability of the surface [6]. The plasma treatment of polycarbonate parts is relatively easy, but some plastics are out-gasing a lot. In case of heavy gas load generated by the out-gassing of the parts, a large pumping speed and a low residual vacuum is necessary. If not the parts are plasma treat in an atmosphere of reactive gas and of the gas coming from the parts [7]. This is the situation with systems using only a mechanical roughing pump as process pump, the result is often a bad efficiency and a poor reproducibility.

- **Copper coating of plastic connectors** : Non conductive connectors needs some time to be coated with a conductive layer to protect the electronic against EMI and RFI [8]. To apply a thick copper layer on plastic connectors by a galvanic growth it is possible to apply first a thin layer of copper by sputtering to create a conductive surface and then to grow copper chemically. This was the solution experimented on PBT and PCL plastic connectors. The connectors was first plasma treated by different gases O_2, N_2, Ar. Then the parts was coated with a standard process of 0.5 to 0.7 μm of copper. The adhesion on the plastics was tested by the adhesive tape "electrical tape 3M n°5". The adhesion of the copper parts coated without plasma treatment first was very poor, the layer was peeling off with the tape. In contrary the three gases gave equal results. The observation under the microscope (magnification 200) of the number of holes created by the tape did not shown any difference, no measurable hole was founded. To simplify the process an Ar plasma treatment was selected. The parts was then successfully coated with acid copper or chemical nickel.

- **Titanium nitride and titanium carbonitride on metal** : As decorative layer TiN or TiCN are often used. On eye glass frame the use of titanium as base material is becoming very common. The problem of titanium is the absence of galvanic technic to coat it. The only solution is to use vacuum coating, This parts are often small and to put them on a holder is requiring a lot of time. Final cost is then not compatible with the utilization of the part. Using TiN coating is an efficient solution. TiN as the gold color required for this kind of application. This adjunction of carbon to the layer to decrease the yellow component of the color and to increase the red one which is more convenient for decoration purpose. TiN was obtain by reactive sputtering of titanium in an atmosphere of argon and nitrogen. A bias of 50 to 100 volts is necessary to obtain a reflective layer. If not the surface energy (temperature) is to low and the layer is dull and even grey. TiCN is obtain in an atmosphere of argon, nitrogen, and a small amount of acetylene or methane. The bias is also mandatory. The present system is not equipped with a mass spectrometer or an optical spectrometer to control the nitrogen partial pressure. For this reason it is necessary to always work with the barrel fully loaded, or the wall of the barrel will be coated with TiN. The layer will trap nitrogen and nitrogen will outgas of the wall during the next coating. In this case outgasing will be added to the flux of nitrogen and the resultant stoechiometry of the layer will vary from run to run.

Metal other than titanium can be coated with TiN or TiCN. The coating will begin with a titanium layer sputter with a high bias on parts and a low deposition rate to create a good interface. Then the coating will be converted to TiN with a high power on the cathode and a low bias.

- **Stainless steel on metal parts** : Standard steel screws can be coated with stainless steel or pure chromium material for corrosive applications. The advantage of this technique is the low cost of stainless steel. The first experiments are presently under development. The first tests show already a better corrosion resistance than screw made of stainless steel, the main reason is that sputtering is increasing the homogeneity of the material. Further investigation is necessary and will be done like doping the layer for friction application .

Conclusion : An industrial system was developed to coat or treat parts in a barrel. Only a small area of the extended capabilities of the equipment is tested at this time. A large potential of applications is currently under investigation for corrosion resistant layers on small parts like automotive screws.

REFERENCES

1- patent # DD246011
2- patent # J61194177 and FR2530671
3- Teer D.G., IPAT proceeding, 303-306, 1991
4- patent # WO93/19217
5- Sproul W. , SVC proceeding, 1994
6- Liebel G., activation du plasma, SURFACE, 11-13, 1991
7- Kaplan S., Plastic Engineering, Vol 44 N° 5, 1988
8- Mason P., SVC proceeding, 192-197, 1994

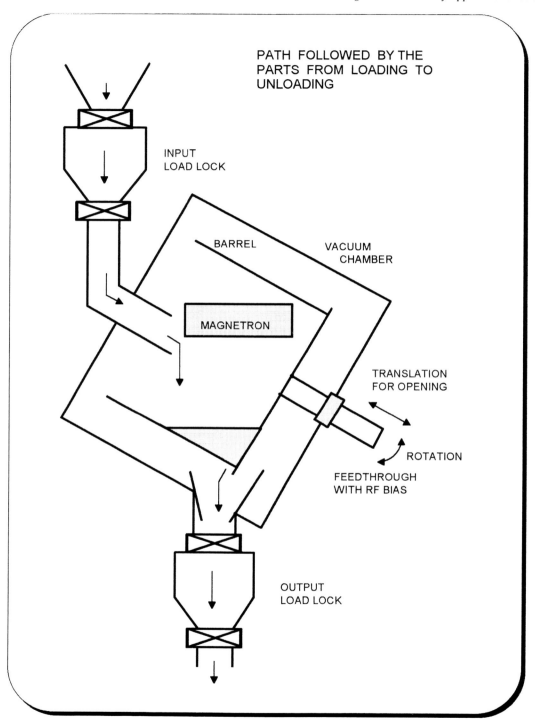

Figure 1

Electron Beam Curing: The Future of Coated Steel Production?

BYRON WILLIAMS

*British Steel Strip Products, Welsh Technology Centre,
Port Talbot, West Glamorgan SA13 2NG.*

Abstract.
The production of prepainted steels for building, automotive, and domestic appliance end products is a very large scale manufacturing process. The application of electron beam (EB) curing to this process is considered. EB processors are described, and the potential benefits of utilising the technology critically assessed. Case studies of processes utilising EB curing are considered, and show these processes to have lower energy requirements, and to be substantially more environmentally friendly than existing processes.

Synopsis.
La production d'aciers prépeints destinés aux produits finis des domaines de la construction, de l'automobile et des appareils ménagers est un procédé de fabrication à très grande échelle. L'étude qui suit porte sur l'application de rayons à électrons en traitement à ce procédé. Les processeurs de rayons y sont décrits et les avantages potentiels de l'utilisation de cette technologie évalués. Des études de cas de procédés utilisant le traitement par le rayons à électrons sont considérés et montrent que ces procédés demandent moins d'energie et sont considérablement plus respectueux de l'environment que les procédés actuels.

Introduction.
Prepainted steel has been produced commercially since the early 1960's, and since that time the market has shown a remarkable rate of growth. Figures published by the ECCA show that 2.376 million tonnes of prepainted steel was sold in Europe during 1991; a 746% increase compared with the figures for 1970. Recent market reports show no slowdown in demand, and suggest that the market will continue to grow as more varied applications are found. The material is used in applications as diverse as domestic appliances and prefabricated building materials. In addition to the decorative function, the paint coating also provides a protective barrier, which prevents corrosion of the steel substrate.

Though the nature of the product has become increasingly sophisticated, in terms of the metallic coating, surface treatments and paint technology, the production technology used in its manufacture has not changed significantly since its earliest inception, and a modern coil coating line would not look unfamiliar to personnel involved with the earliest production process. A schematic diagram of British Steel's No.2 organic coating line at Shotton Works is shown in Fig. 1. Conventional coil coatings consist of a polymeric resin and other additives dispersed in a solvent. In the curing process, the solvent evaporates leaving behind a coating film. The evaporated solvent is burned in the paint oven, supplementing the gas supply. The solvents, however, are expensive and give off small quantities of volatile organic compounds (VOC's), in addition to CO_2, which is produced in the combustion process.

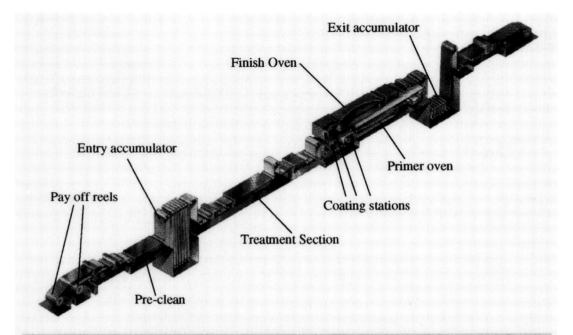

Figure 1. No.2 Organic coating line at Shotton Works. The line is one of the most modern in Europe and can run at a maximum speed of 100m.min^{-1}. The line measures over 250m from end to end. Coatings are cured using two large gas fired ovens.

Radiation curing processes offer a cheaper and more environmentally sound option for coated steel production. In anticipation of future legislation requiring reduced VOC and CO_2 emissions, British Steel have been developing electron beam (EB) curing as a processing technique for coil coating. Whilst this technology is not new, it is only during the last 10-15 years that low energy, self shielded EB processors have become widely available. With the advent of these devices, the technique has become a very attractive option for the processing of large quantities of material.

2. Electron Beam Processors.

Though a number of different types of processor are available, each will consist of essentially two components. Firstly, an electron source with a facility for producing high kinetic energy electrons and secondly, a processing zone, with a facility for moving the material to be cured through the unit. The Broadbeam® processor can be considered as typical. The processor is shown in schematic in Fig. 2. Electrons are produced from a modular array of short ca 60cm long heated cathodes, housed in a vacuum, and aligned in the direction of the moving product. Electrons are drawn off the cathodes by a relatively small voltage applied to the common extraction grid mounted between the filament-grid modules and the output window. The grid current also maintains the beam current. Electrons are accelerated into the process zone by the electric field which rises from the large negative potential on the housing terminal. This voltage can be varied between 100 and 250 kV. The electron beam produced from each cathode is essentially gaussian in shape and beams from each cathode overlap to produce a very intense and uniform beam. Delivering a dose of 1Mrad[1], the Broadbeam processor can operate at a line speed of 1500 m.min^{-1}. If higher doses are required, then the line speed must be reduced accordingly, e.g. for a dose of 3 Mrads, the maximum line speed falls to 500 m.min^{-1}.

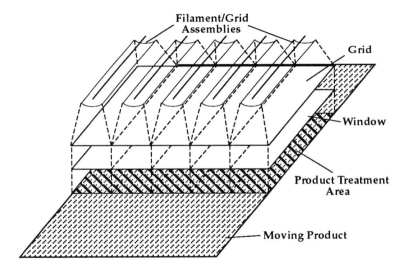

Figure 2. Schematic of Broadbeam type EB processor. The overlapping electron beams from each of the cathodes means that a very intense and uniform beam can be produced.

The high kinetic energy electrons pass into the processing zone via a thin titanium window. During normal operation the processing zone contains a fast moving product; therefore, the maintenance of a vacuum in this region is not feasible. Oxygen is removed from the zone by flushing with high purity nitrogen. This has the effect of preventing the formation of ozone which has attacks the titanium foil, and further, oxygen can act as a free radical scavenger which would inhibit the curing reactions of the coating material.

As the high speed electrons pass through the processing zone and into the denser phase of the coating, the number of interactions between the kinetic electrons and bound electrons increases.

The corollary of this process, is that the depth to which the kinetic electrons can penetrate is dependant on both the accelerating voltage and the density of the curable medium. For example, for a coating of density = 1g.cm^{-3}, and an accelerating voltage of 170 kV, electrons can reach a depth of 260 μm before they are slowed to the speed of thermal electrons. However, once a beam has lost 50% of its energy it is believed to be ineffective at initiating cure. Therefore, for the above specifications, cure can only be effected at depths up to 150 μm. Figure 3. shows the dependence of beam intensity as a function of depth for a series of accelerating voltages, and a coating density of 1g.cm^{-3}.

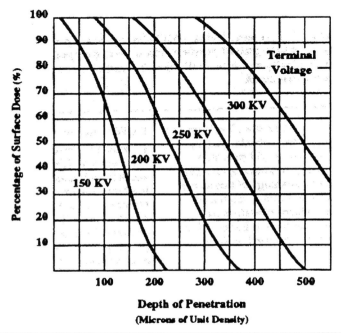

Figure 3. Depth/dose curves for electron beams with kinetic energies typically used in self shielded processors, penetrating a material of unit density (1g.cm^{-3}).

On impacting on the coating material, the electrons initiate cure by catalysing polymerisation reactions. The type of polymerisation reaction initiated by the fast electrons will depend, of on the chemical nature of the resin. The cure reactions are discussed at some length in a number of excellent publications[2], and will not be reiterated here. As a generalisation; however, the majority of commercial systems are based on either acrylates, which undergo free radical polymerisations or epoxides, which cure via cationic polymerisation. Acrylates have some inherent advantages over epoxide based systems in that they do not require an initiator to promote curing. Also epoxide systems will often require moderate heating (post baking) to improve adhesion and flexibility.

The Advantages of Electron Beam Processing.

In order to assess the potential benefits of EB processing in comparison with thermal drying, we will consider two typical case studies. Before moving on to these; however, it is as well to consider the putative improvements promised by electron beam processing, and to critically evaluate the validity of these. The following is a list of the advantages promised by EB processing, which are commonly quoted by resin and equipment manufacturers.

(a) Cleaner Technology - A 100% conversion of resin to coating material can be achieved. i.e. the process is waste free.
(b) Faster Throughput - Line speeds in excess of 500 m.min^{-1} are commonplace.
(c) Reduction in Line Space - A unit will typically occupy <10% of the space of a thermal oven.
(d) Low Temperature Process - In general, products do not require pre or post heating, and the cure process occurs at ambient temperature.
(e) Reduced Processing Time - Instant cure allows immediate shipping, handling or forming.
(f) Novel Coating Systems - Allow improved systems to be developed.
(g) Reduced Energy Consumption - 20-50 times less energy than thermal drying.

The majority of claims made regarding the process benefits withstand close inspection; however, the latter two claims warrant further discussion. Concerning the claims made for novel chemistries, resin manufacturers will often quote statements such as "radiation curable formulations use novel chemical systems; therefore, the potential exists to produce coatings with improved performance !" Unfortunately, the specifics of how to produce coatings with improved performance, or even a performance comparable to that of conventional coil coatings, are as yet a little thin on the ground. As far as steel, and zinc coated steel are concerned, achieving the requisite combination of adhesion, flexibility, and corrosion resistance on these substrates is non-trivial. Furthermore, the coil coating systems currently available are the product of over 20 years of intensive research and development; therefore, to produce an EB curable coil coating which gives comparable performance with existing systems will require considerable further effort.

The latter point, of reduced energy consumption, is a rather throwaway statement which obfuscates a number of important issues. A comparison between EB curing and thermal drying, merely in terms of the energy requirements for each process, cannot be made. Firstly, space costs and maintenance need to be considered, as well as the initial outlay required to purchase the equipment. Furthermore, the costs of high purity nitrogen need to be compared against the cost of fuel gas, and the cost of the coating materials themselves is an aditional factor which should be included in the equation. To address this question, we will consider two case studies where an evaluation of these factors has been made.

1. Production of Painted Wheel Hubs.

This work was carried out by Volkswagen and Polymer Physik[3] during the 1980's. The process steps in the production are outlined below;

(a) Steel wheels degreased and phosphated -
(b) Cathodic dip painting -
(c) Baking for 14 minutes at 180°C.
(d) Cooling on chain conveyor.
(e) Automatic spray, EB curable cover coat. 25-28 g.m^{-2}.
(f) Electron beam curing, 3 seconds under nitrogen.

Type of Cost	Thermal Drying (%)	Electron Beam Curing (%)
Investment Costs	46.9	51.5
Space Costs	10.5	2.0
Wages	7.4	7.4
Servicing	1.6	1.9
Energy, Inert gas	16.8	8.1
Paint Costs	16.8	23.8
Total	100	94.7

Table 1. Comparison of production costs for electron beam curing vs thermal cured coating on wheel hubs. (Figures from reference 3)

A relative cost comparison, as a percentage, is given in Table 1. This clearly shows that when all the factors contributing to the overall running costs are considered, the energy reduction figures can be seen in better perspective. The fact that an overall cost saving can be made; however, needs to be stressed. As this indicates that although the raw material and investment costs are higher, significant cost savings can be made in other areas which negate these.

Table 2. shows a comparison of the annual energy requirements for the production of 2x10^{6}m^2 of material produced by the two processes. A further consideration is the daily solvent emission con-

Energy Requirements	Thermal Drying	Electron Beam Curing
Heating Oil	320 Tonnes	-
Natural/Inert Gas	-	94 Tonnes
Electric for Machines	500 MWh	192 MWh
Electric for Beam	-	128 MWh
Electric for inert gas	-	12 MWh

Table 2. Annual energy requirements for the production of two million square meters of material using thermal drying and EB curing. Painted steel wheel production.

verted for the same production. For thermal drying the paint requirement is 1600kg of material, of which 770 kg is solvent. This compares with a paint requirement of 1000 kg and 20 kg of solvent for the EB process. The solvent is added to the formulation as a viscosity reducing agent, but, its ultimate fate is not elucidated. In a conventional process, the solvent would be burned in the curing oven, but in the absence of a combustion source, as in the EB process, it is not clear if the solvent enters the atmosphere, or is incorporated into the coating. It should be stressed that low viscosity formulations containing reactive diluents, i.e. systems where a 100% conversion to solids occurs, can be readily formulated. The paint/solvent figures imply that the loss of 150,000 kg of solvent is prevented by using the EB process.

2. Paper Coatings.

The second study we will consider is a comparison of costs for polyethylene extrusion coating and EB curing of special papers[4]. The production costs for both materials - in Deutsch Marks - are outlined in Table 3. Coating the paper with polyethylene or the EB curable material increases the chemical resistance. Using EB processing, as against extrusion, and utilising a coating with the same characteristics as polyethylene, the production speed can be increased to 300m.min^{-1}, which is a 250% increase in speed compared with the extrusion process. In comparison to polyethylene, the surface roughness of the EB curable coating is decreased by an order of magnitude, and furthermore, papers coated with the EB curable material can be recycled.

	Extrusion of PE	Coating and EB Curing
Amortisation Costs	170 DM.h^{-1}	284 DM.h^{-1}
Productivity	14,400 m^2.h^{-1}	36,000 m^2.h^{-1}
Electric	46 DM.h^{-1}	31 DM.h^{-1}
Nitrogen		48 DM.h^{-1}
Utility Costs	46 DM.h^{-1}	79 DM.h^{-1}
Coating Costs PE; 30g.m^{-2} 3.2 DM.kg^{-1}	1,382 DM.h^{-1}	
EB Coating 10g.m^{-2} 7.0 DM.kg^{-1}		2,520 DM.h^{-1}
Total Costs	1,598 DM.h^{-1}	2,883 DM.h^{-1}
Total Costs per m^2	0.11 DM.m^{-2}	0.08 DM.m^{-2}

Table 3. Cost comparison of production processes for extruded polyethylene and electron beam processed speciality papers.

Clearly, for both the processes considered, there are advantages in using EB processing technology in place of more traditional processing equipment. This, in terms of both reduced energy consumption and a more environmentally friendly process. Though other case studies have been published[4] which show that EB processing can be more expensive than thermal curing, there is no instance in which the process has not been shown to produce a significant reduction in energy consumption, or to reduce waste.

Currently, companies are not penalised for burning fossil fuels; however, this situation may change with the introduction of levies aimed at reducing CO_2 emissions. Given the increasing influence of Green pressure groups within Europe, the likelyhood of the former becoming law seems high. In this case, or in the case of an increase in the price of fossil fuels, there would be an additional incentive to switch to EB based processes.

As a major supplier of organic coated steels, British Steel is very much aware of the implications of these developments, and has implemented development programs both through European Coal and Steel Community sponsored research programs, and via joint development work with coil coating suppliers. The Welsh Technology Centre is admirably well equipped to sponsor these developments, having a production scale Broadbeam EB curing unit incorporated into a pilot scale coil coating facility.

Conclusion.
Electron beam curing offers a low energy, more environmentally friendly production process than current systems. The EB process is inherently faster than conventional processing methods, and has considerably lower space requirements. Though EB processing has seen widespread adoption by industries such as furniture and packaging manufacturers, and on a limited scale for coil coatings by Nippon Steel, its introduction on a large scale by coil coating manufacturers is limited by the quality of the available coating systems. British Steel is actively involved in the development of electron beam curable coil coating systems via joint research projects with paint producers, and *in house* developments. This program will maintain British Steel's market position as a producer of high quality prepainted coil well into the next century.

Acknowledgements.
The author would like to thank Dr B.J. Hewitt, Technical Director - British Steel Strip Products, and Mr E.F. Walker, Manager, Technical Coordination, Welsh Technology Centre, for permission to present this paper.

References.
1. The rad is a unit of absorbed radiation, and is equal to 0.01 joules per kilogram of material.
2. See for example; Howard, J.; *Polymers Paint and Colour Journal*, 1992, 182, 247.
3. Häring, E.; *Metallöberflache*, 1988, 42, 7.
4. Mehnert, R.; RadTech Europe 94' Conference Proceedings. pp 828-837.

Changeover from an electrodeposition technique based on low current density to the utilization of high current densities for copper coatings.

ALBERT VOLVERT   BUNDY. S. A.   Liège, BELGIUM.

With one short theoretic reminder, we show the way that was used by Bundy S A. We explain the reasons which lead that factory to use electrical high current density for continuous copper and nickel electro-coating on steel strips. Before 1989, we had 15000 A capacity to copper strike with 18 m² of electro-plating area, 60000 A to copper acid electro-plating on 53 m² of area and 20000 A for nickel Watts on 22 m² of area. The major disadvantage was the impossibility to use strike and nickel Watts together. During 1989, we have increased nickel coating capacity to 60000 A with high current density (HCD). In 1995, we have increased copper strike capacity to 20000 A (LCD) with soluble anodes and 76000 A for copper acid electro-plating (LCD) but with insoluble anodes. To 1996, we should run in copper acid plating line with high current density and 90000 A capacity.

Par un bref rappel simplifié de la théorie, nous montrons la démarche réalisée par la société Bundy S A. Nous expliquons les raisons qui ont conduit cette entreprise à utiliser les hautes densités de courant électrique pour revêtir en continu, de cuivre et de nickel, des feuillards d'acier. Avant 1989, nous avions une capacité de 15000 A en précuivrage pour une surface d'électrolyse de 18 m², de 60000 A en cuivrage acide pour une surface d'électrolyse de 53 m² et de 20000 A en nickelage Watts pour une surface d'électrolyse de 22 m². En 1989, nous avons porté la capacité du nickelage à 60000 A (HCD). En 1995, nous avons porté la capacité de précuivrage à 20000 A (LCD) avec des anodes solubles et 76000 A en cuivrage acide (LCD) avec des anodes insolubles. En 1996, le cuivrage acide sera effectué à haute de densité de courant avec 90000 A.

Depuis 1966, nous effectuons en continu du revêtement électrolytique de cuivre sur des feuillards d'acier, et depuis une dizaine d'années nous déposons en continu du nickel.

Les cellules de cuivrage et de nickelage sont intégrées au sein d'une même ligne de fabrication. Les feuillards d'acier revêtus sont utilisés, dans la fabrication de tubes brasés pour les circuits de freinage dans l'automobile, dans la fabrication de tubes soudés pour l'électroménager et les circuits d'alimentation en carburant, dans la fabrication de piles et accumulateurs rechargeables.

ORIGINE DU CHANGEMENT.

En 1989, suite à la demande accrue du marché, nous nous sommes trouvé dans l'obligation d'accroître notre capacité de production. Notre unité de production avait deux inconvénients majeurs, le manque de place et l'impossibilité d'utiliser le précuivrage et le nickelage en même temps. Nous faisions de l'électrodéposition sous de faibles densités de courant électrique (LCD) avec des anodes solubles, nous devions choisir entre différentes techniques d'électrolyse avant d'effectuer nos investissements.

DEMARCHES EFFECTUEES EN 1989.

Nous avions décidé d'accroître la capacité et la productivité du nickelage. Le manque de place nous obligeait de construire de petites cellules, celles-ci nous imposaient une haute densité de courant (HCD). Etait-ce compatible avec la qualité du revêtement demandée par nos clients ? Nous avons entrepris, avec l'aide de l'Université Libre de Bruxelles dans le service du Professeur R WUINAND et le laboratoire de recherche de Cockerill-Sambre, la mise au point d'électrolytes et une étude complète portant sur les phénomènes suivants :

Passivation anodique. Si le procédé à anodes insolubles peut-être utilisé avec des faibles ou des hautes densités de courant, il n'en va pas de même avec les anodes solubles car sous l'effet des hautes énergies, elles se passivent et deviennent insolubles *(augmentation des surtensions de décristallisation et de dissolution)*. Bien sûr certains produits organiques peuvent diminuer les risques de passivation. Les « Fig.1 et 2 », montrent les différences entre anodes solubles et insolubles.

Cristallisation: De nombreux chercheurs ont donné une explication aux mécanismes de l'électrodéposition et de la cristallisation notamment le professeur R WUINAND de l'ULB. Celui-ci a classé l'ensemble des structures de croissance dans un tableau à double entrée « Fig.3 ».

Vitesse de l'électrolyte:	loi de VIELSTICH	$Re = v.d/\nu$
Couche limite de diffusion.	Equ transfert matière	$J_{limNi++} = 2 F. C^{o}_{Ni++}.(D_{Ni++}/\delta)$
	loi de VIELSTICH	$\delta = Re^{-0.9}. Pr^{-1/3}$

Rendement électrique: loi de FARADAY, loi d'OHM
Adhérence:

Ces tests furent réalisés sur ligne pilote et en laboratoire, nous nous sommes assurés de la qualité des dépôts en les comparant avec les revêtements classiques au moyen de techniques d'investigation perfectionnées (MEB, diffraction X, AUGER, etc ...).

On aurait pu croire en 1989, que l'utilisation de la haute densité était un choix défavorable, l'énergie utilisée étant pénalisante (tension plus élevée, chute Ohmique plus importante dans la bande d'acier, perte par effet Joule, besoin énergétique plus conséquent pour faire circuler l'électrolyte à grande vitesse). Malgré ces divers inconvénients, nous avons construit notre première ligne HCD et avons pu vérifier que la HCD nous offrait de nombreux avantages. Nous avons pu construire quatre cellules compactes, pouvant supporter des densités de courant de 230 $A.dm^{-2}$ pour une capacité de 60000 A, avec une quantité réduite de l'électrolyte (10 m^3). L'entretien de ces cellules est fortement réduit. Nous avons pu nous affranchir de plusieurs paramètres de l'électrolyse grâce aux hautes énergies apportées et du niveau d'impureté dans le bain. Nous avons obtenu, une meilleure régularité d'épaisseur du revêtement, des dépôts moins fragiles et moins poreux, un rendement de courant plus élevé.

Grâce à ces cellules HCD, nous avons, avec nos partenaires ULB et RDCS, étudié l'utilisation d'électrolytes pour le cuivre.

La nouvelle configuration de la ligne d'électrolyse nous a permis de développer de nouveaux produits à base de cuivre et de nickel, notamment un tube particulier pour les circuits de freinage. Elle nous a permis également d'entrevoir la possibilité de moderniser et d'améliorer la partie utilisée pour le cuivrage acide. Cette partie date de 1966, le procédé est à anodes solubles et il consomme environs 50 tonnes de billettes de cuivre par mois. La consommation de billettes est pour nous financièrement pénalisante et de plus nous devons supporter les variations de formats. Il était nécessaire pour nous de rénover.

DEMARCHES EFFECTUEES DEPUIS 1991.

Validation par nos clients de revêtements électrolytiques de cuivre HCD.

Comme nous voulions nous affranchir des contraintes du marché des billettes de cuivre, nous avons étudié le procédé à anodes insolubles en utilisant des anodes d'alliage plomb argent. Puisque dans ce cas nous devons remplacer le cuivre électrodéposé, nous avons étudié et construit un procédé de lixiviation du cuivre « Fig.4 », avec comme partenaires les sociétés Air Liquide, Praxair et Robin industries.

De 1991 à 1994, nous nous sommes consacrés aux différentes études fondamentales, nous avons mis au point un programme nous permettant d'effectuer l'optimisation électrique de la nouvelle ligne « Fig.5 », nous poursuivons celle-ci par l'utilisation conjointe d'un programme d'analyse par éléments finis QUICKFIELD^{TM} « Fig.6 ».

Pour satisfaire aux lois de VIELSTICH et de transfert de matière, nous avons commandé au Centre de Recherche de l'Institut Gramme une étude sur l'optimisation de l'injection de l'électrolyte (SIMULOG logiciel TIGRE), étude terminée en Aout 1995.

A partir de 1995, nous avons démarré la lixiviation du cuivre par oxydation et mise en solution par l'acide sulfurique, puissance utilisée 40 kW

Nous avons remplacé, dans les sections d'électrolyse en milieu acide, toutes les anodes de cuivre par des anodes en alliage PbAg et validé le système par des contrôles (analyses des bains, des dépôts). En Juin 95, nous avons construit dans la cuve du bain de WATTS une cellule expérimentale à l'échelle industrielle, grâce à elle nous avons pu corriger nos erreurs et valider la qualité du dépôt. Après cet ultime test nous avons récupéré la cuve du bain de WATTS et nous l'avons transformé en précuivrage alcalin, ce qui a augmenté notre capacité. Nous développons, pour accroître nos connaissances, une cellule de laboratoire capable de traiter des échantillons de 15 dcm^2 en haute densité de courant.

En Décembre 1995, la capacité en cuivrage est de 96000 A, mais toujours en basse densité de courant. Le remplacement définitif de la section de cuivrage acide par la haute densité de courant sera effectif en Aout 1996 « Fig.7 ».

L'ampérage total utilisé en électrodéposition sera passé de 80000 A en 1988
à 170000 A en 1996.

Electrolyse avec anodes solubles.

Dans ce procédé il y a consommation de l'anode durant le fonctionnement.

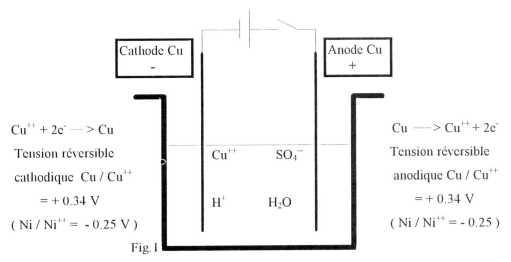

$Cu^{++} + 2e^- \longrightarrow Cu$

Tension réversible
cathodique Cu / Cu^{++}
$= + 0.34$ V
($Ni / Ni^{++} = - 0.25$ V)

Fig.1

$Cu \longrightarrow Cu^{++} + 2e^-$

Tension réversible
anodique Cu / Cu^{++}
$= + 0.34$ V
($Ni / Ni^{++} = - 0.25$)

La tension de la cellule à l'équilibre (sans courant) est égale à la différence entre les deux tensions réversibles c-à-d **nulle**.

Electrolyse avec anodes insolubles.

Dans ce procédé il y a appauvrissement de la concentration en métal de l'électrolyte.

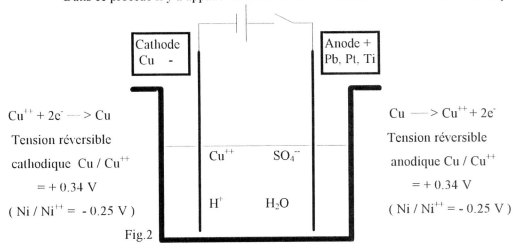

$Cu^{++} + 2e^- \longrightarrow Cu$

Tension réversible
cathodique Cu / Cu^{++}
$= + 0.34$ V
($Ni / Ni^{++} = - 0.25$ V)

Fig.2

$Cu \longrightarrow Cu^{++} + 2e^-$

Tension réversible
anodique Cu / Cu^{++}
$= + 0.34$ V
($Ni / Ni^{++} = - 0.25$ V)

Tension de la cellule à l'équilibre (sans courant) est égale à 1.23V - 0.34V = 0.89V pour le cuivre
à 1.23V-(-0.25V) = 1.48V pour le nickel

C'est la tension minimale qu'il faut fournir au système avant de produire du cuivre ou du nickel.
Nous constatons qu'il est plus avantageux de travailler avec des anodes solubles, mais de grande pureté.

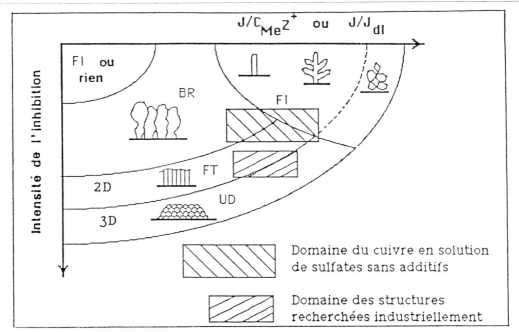

Fig.3 Domaines d'existence des différents types de dépôts electrolytiques en fonction du transport de masse et de l'intensité de l'inhibition.

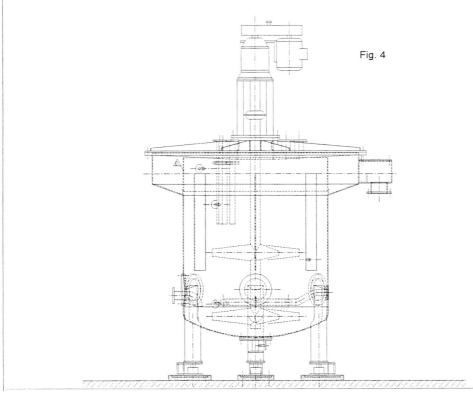

Modélisation de la cellule HDC cuivre

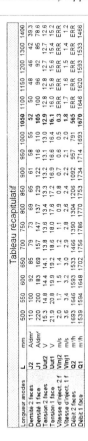

Données

Tôle			
l	682 mm		Largeur de tôle
e	0.35 mm		Epaisseur de tôle
Tt	80 °C		T° de tôle
Vt	1 m/s		Vitesse de tôle

Electrolyte			
Te	70 °C		T° électrolyte

Cellule			
I2	7500 A		Intens. par demi cellule (2 faces)
I1	7500 A		Intens. par demi cellule (1 face)
la	600 mm		Largeur anode
Lt	550 mm		Distance anode –C.R.
Le	20 mm		Distance interpolaire

Tension			
Vcath	-0.34 V		volt Cathode
Van	1.23 V		volt Anode
Vsurt	1.5 V		surtensions
Vbar	1.5 V		Chute de tension dans bus bars

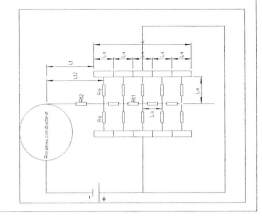

Tableau récapitulatif

Fig 5

Longueur anodes	L	mm	500	550	600	650	700	750	800	850	900	950	1000	1050	1100	1150	1200	1300	1400
Densité 2 faces	J2	A/dm²	110	100	92	85	79	73	69	65	61	58	55	52	50	48	46	42	39.3
Densité 1 face	J1	A/dm²	220	200	183	169	157	147	137	129	122	116	110	105	100	96	92	85	78.6
Tension 2 faces	Uut2	V	15.3	14.8	14.4	14.1	13.8	13.6	13.4	13.3	13.2	13.1	13.0	12.9	12.8	12.8	12.7	12.7	12.6
Tension 1 face	Uut1	V	21.9	20.8	19.9	19.2	18.6	18.0	17.6	17.2	16.9	16.6	16.4	16.1	16.0	15.8	15.6	15.4	15.2
Vitesse d'inject. 2 f	Vinj2	m/s	2.0	1.7	1.5	1.4	1.2	1.1	0.9	0.8	0.7	0.6	0.5	0.3	ERR	ERR	ERR	ERR	ERR
Vitesse d'inject. 1 f	Vinj1	m/s	3.6	3.4	3.2	3.0	2.9	2.8	2.6	2.4	2.2	2.1	2.0	1.8	1.7	1.6	1.5	1.4	1.2
Débit 2 faces	Q2	m³/h	1693	1646	1593	1533	1466	1390	1304	1206	1092	957	791	565	ERR	ERR	ERR	ERR	ERR
Débit 1 face	Q1	m³/h	1539	1594	1648	1702	1756	1786	1770	1753	1734	1714	1693	1670	1646	1620	1593	1533	1466

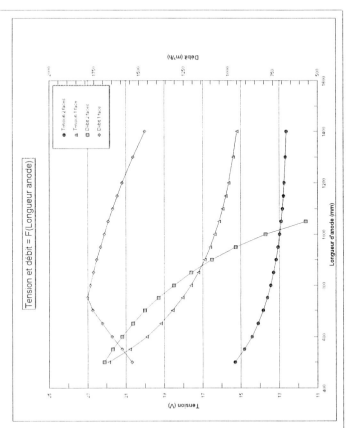

Tension et débit = F(Longueur anode)

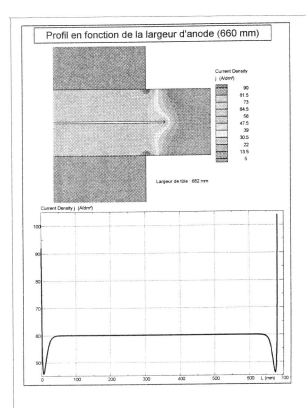

Fig.6 : Exemple d'analyse donnée par Quickfield™.

Fig.7 : Projet pour la nouvelle cellule à haute densité de courant.

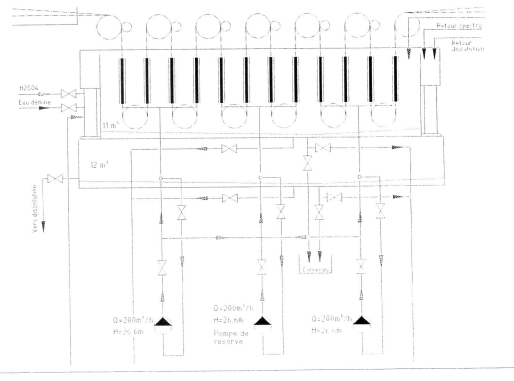

New Magnetodielectric Materials For Induction Heating Applications

V.S. NEMKOV, R.S. RUFFINI

Centre for Induction Technology, USA

ABSTRACT

Magnetodielectric materials (MDM) composed of soft magnetic powders and special binders become more and more popular in the induction heating industry due to unique combinations of properties and to the possibility of their variation to satisfy specifications of any particular application. They are used for magnetic flux control and concentration resulting in strong improvements in induction coil efficiency and in workpiece temperature pattern.

At the same time MDM allow for the shielding of stray magnetic fluxes preventing undesirable heating of components in close proximity to the coil. The report is devoted to the results of long-term study and production of MDM and their application to a variety of induction systems for heat treating, mass heating, brazing, curing and other technologies.

RESUMÉ

Les matériaux magnétodiélectriques (MDM), constitués de poudres magnétiques douces et de liants spéciaux, sont de plus en plus répandus dans les industries utilisant le chauffage par induction ; eux seuls assurent la combinaison de certaines propriétés et la possibilité de les faire varier pour les adapter aux exigences propres à n'importe quelle application. Utilisés pour le reglage et la concentration fes flux magnétiques, ils améliorent fortement le rendement des inducteurs et la répartition des températures dans la charge. Les MDM permettent en outre de limiter les fuites de flux et d'éviter l'échauffement parasite d'éléments proches de l'inducteur.

Cette communication est consacreé aux résultats d'une étude a long terme des MDM et de leurs applications à de nombreux systèmes à induction utilisés pour le chauffage en masse, le brasage, les traitements termiques, et d'autres applications.

INTRODUCTION

Magnetodielectric materials consist of magnetic particles and dielectric material which serves as a binder and electric insulator of the particles. Magnetic properties of a composite depend on the properties of the particles, their shape, packing and volume space factor. Mechanical and thermal characteristics depend mainly upon a type and percentage of binder. All the properties are critical to production technology.

MDM have been known for a long time but were used in special fields such as radio engineering and electronics because they are poor magnetics and poor dielectrics. But MDM have a precious combination of electromagnetic, thermal and mechanical properties variable in a wide range. Modern materials and technologies permit us to improve drastically the properties of MDM and to provide their effective use in power electrical engineering.[1] Induction heating is one of the fields where MDM find numerous applications.[2,3]

PROPERTIES OF MDM

Two forms of MDM are available on the market: solid and formable. Solid forms now cover almost all the market demands. They are produced by pressing of a magnetic powder and a binder with subsequent thermal treatment.

Low frequency solid materials for induction applications are made from atomized or electrolytic powder of iron alloys with 2-3 wt. % of binder. Carbonyl iron or ferrite powder are used for frequencies higher than 20-30 kHz with binder percentage up to 15% at radio frequencies.

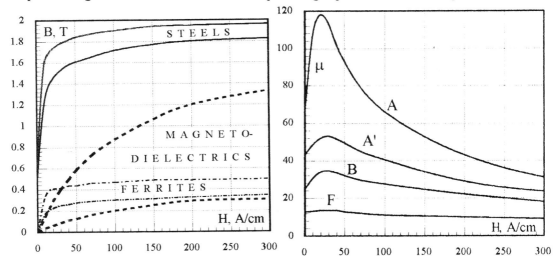

Figure 1 B-H curves for laminations, ferrites and MDM used in induction systems

Figure 2 Permeabilities of MDM materials of Fluxtrol type

The resulting B-H curves of MDM vary in a wide range,' Fig. 1'. Typical curves for steel laminations and ferrites used in induction systems are shown for comparison. MDM are quasi-linear materials without explicit saturation at high field intensities such as 300 A/cm. Only at field intensities of H>1000 A/cm go they demonstrate real saturation with B_s=1.6 T for the upper curve. Permeabilities of MDM are shown on Fig. 2. For low frequency material (Fluxtrol A) a curve of permeability has a well pronounced maximum which is absent on the curves for high frequency material (Fluxtrol F). The values of permeability in most conditions of induction heating (H=10-100 A/cm) are in the range 15-100.

Compared to ferrites MDM with iron powder base have a higher Curie point and flux densities in strong fields. Machinability, good mechanical properties and stability to thermal shocks are the other advantages of MDM. Information on power losses in MDM at high field intensities and frequencies is very scarce but it may be affirmed that they have comparable levels with laminations and ferrites in the fields of their proper application.

Laminations have much higher saturation induction and permeabilities but their use is limited typically by frequencies below 10 kHz, field densities B=0.5-0.8 T and relatively simple geometries. Very high permeability of laminations is not of a big importance in many induction applications. Low resistance to hot water and polymer quenchants set additional restrictions to their use.

APPLICATIONS

Magnetodielectric materials as well as laminations and ferrites may play different roles in induction heating systems:

- concentration of the magnetic field;
- improvement of the coil efficiency and power factor;
- heat pattern control due to the redistribution of the magnetic field and/or of the currents in the coil turns and in the workpiece;
- protection of the workpiece or heating machine components against unintended heating;
- elimination of the external magnetic fields in close proximity to the coil;
- improvement in power source matching to the coil and in efficiency of the supplying circuitry and its cost.

Often two or more functions are being fulfilled simultaneously by the same magnetic part and different terms are used for them: concentrators, diverters, guides, screens, cores, impeders.

A lot of publications are devoted to the influence of magnetic flux concentrators on the field and current distributions in different electromagnetic devices since the beginning of the century. But the overall results of their use in induction heating systems are not widely known and recognized. It may be explained by a complexity of the analysis which must deal not only with the electromagnetic problem and energy saving but with thermal, technological and mechanical problems also.

MAGNETIC FLUX CONCENTRATION AND EFFICIENCY IMPROVEMENT

Magnetic circuits of induction coils are usually non closed. Therefore a sufficient part of the coil current must be consumed to push the magnetic flux through the "back path" of the magnetic circuit.[4] For internal, pan-cake and hair-pin coils, as well as for short external cylindrical coils typical for heat treating, brazing and many other technologies this additional magnetizing current may be equal or even exceed the "working" current necessary for energy transfer into the workpiece.

One example of the concentrator application to a linear coil is shown on Fig. 3. The magnetic flux and the workpiece power produced by a bare coil are 1.67 and 2.5 times less than for a coil with a concentrator if the total currents of the coils are equal (Fig. 3E). The current of the bare coil must be increased 1.67 times to provide the same specific power on the workpiece surface under the coil. Then the voltages of both coils will be approximately the same while the power factor is 1.6 times less for the bare coil.

It might be expected that the current rise k times will result in the coil losses increasing $k^2 = 2.8$ times but it is not true especially for single-turn induction coils. At audio and radiofrequencies an additional magnetizing current flows on the side and back surfaces of the tubing (Fig. 3C). For this reason an effective resistance of the bare coil is 2.1 times less and the power losses only 1.3 times more (instead of 2.8 times) than for the coil with the concentrator. But the real energy and cost savings due to the flux concentration are usually many times higher.

The losses in the coil legs, in the copper of the matching transformer and in the tank circuitry for the bare coil are in fact k^2 times higher. As a result the power supply losses and rated power will increase also. An analysis of an energy balance for the above case shows that for the same *useful* power (equal to 100 kW) the total power savings are about 70 kW instead of relatively modest savings of 9 kW in coil losses.

Two thirds of these savings are due to a reduction of the coil current demand and one third - to the improvement of power distribution in the length of the workpiece. Really, only a central part of a

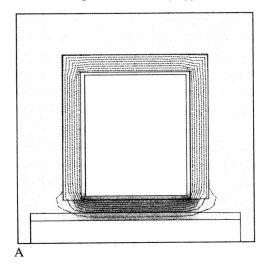

A

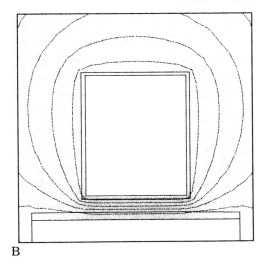

B

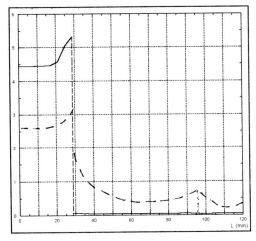

C

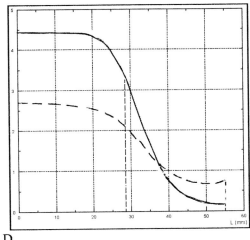

D

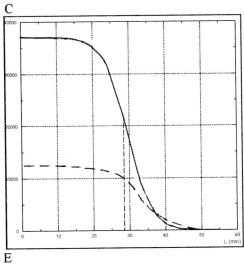

E

Figure 3 Effect of the concentrator on magnetic field, current and power distributions in a linear coil. Total coil current is constant. Dash lines on the graphs refer to bare coil.

A, B. Flux density lines and magnetic field intensity pattern.

C. Current distribution in a half of perimeter of the coil tubing.

D. Induced current distribution in the workpiece.

E. Energy distribution in the workpiece.

bell-shaped power distribution (in the length of 20 mm, Fig. 3E) is useful. The rest of the power is useless or even harmful because it may cause a back temper of adjacent workpiece elements previously hardened.[5] This excess power is equal to 25% and 40% for the coils with and without a concentrator. For odd shaped parts these losses may be even higher if a concentrator is not used. Moreover due to less current demand the apparent power of the matching transformer and the tuning tank capacitance decrease 1.8 times. The rated power of the chosen energy converter may be 1.4 times less if the coil has a concentrator. This results in less investment and in maintenance cost savings simultaneously.[3]

It is important that the effect of the concentrator does not change dramatically with the magnetic material permeability after a certain limit. If the permeability is higher than 30-50 for audio frequencies or 15-20 for most radiofrequency applications, the coil parameters do not change significantly. The higher values of permeability are of a big importance only in special induction systems mainly with a closed magnetic circuit. Calculations and practical tests show that hysteresis and eddy-current losses in MDM are small compared to energy savings and do not notably influence the energy balance.

POWER DENSITY AND HEAT PATTERN CONTROL

For case hardening a good heat pattern control and a reliability of the process are even more important than energy savings. Application of magnetic guides allows us to improve the power distribution control providing the required case pattern and metallurgical properties. On many parts of complicated geometries such as shafts with tight fillet areas a required case pattern may not be received without flux concentrators.

MDM have big advantages in these applications. Machinability or formability of MDM permit us to apply them successfully to the odd-shaped coils where it is difficult to use laminations or ferrites. Except for mechanical problems in manufacturing and low space factor, laminations can not work in three-dimensional fields because of their heating by the flux component perpendicular to sheets. For this reason in some applications such as heavily loaded linear coils for seam annealing of HF welded tubes, a combination of MDM and laminations can be used successfully. MDM are applied in the end zones where a 3D magnetic field is present.

Transverse Flux Heating is another process where MDM are used successfully for magnetic flux concentration and control.

Other advantages of MDM is that they may be used in narrow spaces between the workpiece and the coil for flux distribution and part heating control. Thin layers (2-4 mm) of MDM attached directly to the coil surface can work here at high field intensities due to an intimate thermal contact with the copper.

For some other applications mechanical properties of MDM may be of decisive importance. Magnetic guides can concentrate or control the magnetic field and be used simultaneously as constructive components of the coil design. The walls of the quenching or vacuum chamber, the plates rigidly supporting a coil assembly or even punches for pressure transfer to a stack of induction heated composite parts, may be made of MDM.

The use of the MDM is especially convenient in development and laboratory tests of the induction coils. Pieces of MDM may be applied, removed, cut, glued and reapplied again to reach the desired results without remanufacturing the coil.[2] Some coils for brazing or heat treating have parallel connected turns and it is difficult to predict and provide the correct current distribution between them. Small parts of MDM applied to winding legs allow you to smoothly control the current distribution and hence the heat pattern.

TECHNOLOGY OF MDM APPLICATION

Solid MDM are the prevailing form in induction applications due to their higher magnetic, thermal and mechanical properties. They are produced in net-shape forms according to customer

Figure 4 Quick disconnect induction coil for scan heat treating
Left- Bare coil and machined pieces of MDM (Fluxtrol B);
Right- Complete coil assembly.

specifications or in standard shapes and sizes. For Fluxtrol materials round shapes with outside diameters up to 280 mm and thickness up to 51 mm as well as plates up to 51x104x394 mm are available. They may be easily machined (saw cut, milled, turned, drilled and ground) using regular tools 'Fig. 4'.

In the case of single turn coils net-shape or machined MDM pieces may be applied to the coil tubing directly. Good thermal contact to water cooled copper is essential for heavy duty applications to remove power losses and absorbed radiation from MDM. Soldering of solid forms of MDM to the coil copper provides the best thermal contact. But this method is possible only for some types of MDM such as Fluxtrol A when temperature resistance and adhesion properties are sufficient for soft soldering. A thin layer of thermally conductive epoxy glue with alumina, silica or metal powder filling applied to the copper provides excellent bonding and superior thermal conduction.

In multi-turn coils the MDM pieces must be insulated from the coil copper in order to avoid electric and thermal breakdown. Typical breakdown tracks have a form of worm-holes with conductive walls. MDM materials, especially low frequency types, have electric resistivity insufficient for withstanding significant external electric voltage. One-two millimeters of silica-filled epoxy is usually sufficient for concentrator attaching to multi-turn coils for heat treatment. For high voltage coils better electrical insulation must be provided and a separate means of MDM cooling may be required. Properly applied MDM work for many years without degradation or notable changes of physical and magnetic properties withstanding elevated temperatures up to 300°C and the attack of hot water and polymer quenchants.

REFERENCES

1. V.A. TROITSKIY, A.I. ROLIK, A.I. YAKOVLEV: 'Magnetodielectrics in Electrical Engineering', Technika, Kiev, 1983.

2. R.S. RUFFINI: 'Production and Concentration of Magnetic Flux for Induction Heating Applications', *Industrial Heating*, November 1994.

3. R.S. RUFFINI: 'How Power Supply Efficiencies Are Improved by Use of Flux Concentrators', *Industrial Heating*, September 1995.

4. V. NEMKOV et Al.: 'The Use of End and Edge Effects for Induction Heater Design', *Proceedings of the XII-th Congress UIE*, Montreal, Canada, 1992.

5. V. RUDNEV, R. COOK, D. LOVELESS: 'Keeping your temper with flux concentrators', MAN, November 1995.

Thermal Stresses in the High Frequency Induction Pulse-Hardening of Non-Symmetrical Workpieces

BLINOV Yu., KACHANOV B., KOGAN B., FEDOROVA V.

Department of Electrotechnology and Converter Engineering, Electrotechnical University
Prof.Popov Str., 5; 197376 St.Petersburg, Russia. fax:+7-(812)-2342758; E-mail:
root@post.etu.spb.ru

Abstract. High frequency (HF) induction pulse-hardening technology is discussed. An advantage of the technology is heat removal without external cooling with velocities of quenching up to $10^4 \, °C*s^{-1}$. The rapid heating of workpieces especially with non-symmetrical geometry can lead to relatively high level of thermal stresses. The analysis of the stressed-deformation state of the workpiece and an approach for the unified simulation of electromagnetic field, transient temperature distributions, thermal stresses and pulse-power supply are described as well as typical experimental results are shown.

Résumé. La technologie de HF de trempe impulsion ce discuter. Le plus grand avantage est une derivation de chaud sans extérieur réfrigerant par une vitesse jusqu'a $10^4 \, °C*s^{-1}$. Particulierement la vitesse du chauffage de modèle du travaille, avec la geometrie asymetrique peut mener dans quelqu-uns de cas vers relativement â haut niveau de chaud tension. Ce decrire: analyse d'état tension-deformation du produit, approche pour en commune modeler le champ electro-magnetique, distribution des temperatures dans le regime transitoire, la tension de chauffage et des sources d'alimentation d'impulse. Or les resultats d'expérience se montrer.

INTRODUCTION

High frequency (HF) induction hardening technology (surface and volume) is very wide used in the manufacture of different production. Usually the power range is 10÷400 kW and frequency range is 10÷440 kHz in the technology. There are publications[1,2] devoted to the novel approach of the HF induction heating application with the specific power up to 20 kW*cm^{-2}. Theoretical understanding of the whole processes in the workpiece including inductor construction and gap should be reviewed.

The work was supported by State Committee of Russian Federation on High Education

The level of the specific power demands to limit the energy transferred into the workpiece that leads to the necessity of pulse modes of power supply use. The continuous energy transmission into the workpiece leads to melting of a surface of workpiece. The principle advantage of the induction pulse hardening process is the use of heat removal without external cooling means, with velocities of quenching up to 10^4 °C∗s^{-1}. Besides the high efficiency of the equipment and little power consumption provide higher economic parameters in comparison with the similar process in which the laser beam is used. Moreover in the laser hardening the thickness of hardened layer can not be greater then 0.5 mm because the thermal sources are located only on the surface of a workpiece. The increase of the hardened depth can be reach only with the part-melting of the workpiece surface. In the HF induction pulse hardening technology specific power equals to 1÷10 kW∗cm^{-2}. Heating times usually is below then 400 ms. Power range is 60÷100 kW, frequency range 66÷440 kHz. In the process there are not restrictions of heat speed of a workpiece surface because the thermal sources in the HF induction pulse hardening are placed directly into a workpiece on a current penetration depth. The thickness of hardened layer in the technology is defined by frequency and specific power and can reach 1.5 mm. The rapid heating of a workpiece with non-symmetrical geometry leads to local overheating and relatively high level of thermal stresses. The analysis of the stressed-deformation state of the workpiece is necessary for the application of the pulse-hardening process, since these stresses, added to preexisting external stresses, can give rise to the appearance of a microcracks. Despite on the above listed advantages the realisation of the technology requires the decision of a lot of problems for its wide introduction in an industry. There are: development of construction principles of the industrial equipment for the technology; development of typical design of inductors, ensuring operation at specific power 10 kW∗cm^{-2}; theoretical and experimental study of a physical processes in workpieces at high specific power.

EQUIPMENT NECESSARY FOR THE TECHNOLOGY

Realisation of the HF pulse-hardening technology demands to overview: the operation modes of the power supply, construction of inductors, matching conditions of power supply and inductor.

<u>Power supply.</u> For the technology two types of power supplies can be used. In the frequency and power range taking into consideration pulse mode of operation it's easy to realise the technology with the use of vacuum-tube generator. These modes can be received simply enough and do not require significant changes in the scheme of generator[3]. The influence on the grid circuit of generator is the base of pulse-periodic modes. The use of transistor generators demands to realise a pulse-periodic

mode. The first way is the use of frequency change control that can give a good results for high Q-factor of a load. The problem of overvoltage on the generator elements must be solved. The last problem does not permit to decrease the times on-off of generator and at hardening of a number of workpieces does not give the wishes results. Special units or control laws must be suggested for the solving of the delivered problem. The vacuum-tube generators were used for the experiments.

Construction of inductor. The increased reliability of operation at large specific powers is one of the main requirements for the inductor design. It can be achieved by the use of the inductor with increased surface of a heat removal. The decision of a delivered problem is complicated as the area of an active surface is as a rule essentially limited and not exceed 2÷4 cm^2 and inductor has one turn. It is caused by rather small power of supply sources and necessity to execute local heat of workpieces. At the development of inductors it's necessary to take into account high thermal conductivity of cooper, to increase the sizes of heat removal parts of inductor leaving an active surface of inductor without changes. An active surface of inductor should be made as bulky details and has a qualitative contact, which is provided at the use of profiles cooling tubes and soldering of high-temperature alloys. The size of tubes should be increased not less than at 20%, providing of a condition for reducing hydraulic resistance of a cooling system. Magnetic concentrators (yokes) is the important factor, influencing to a quality of thermal treatment and to an opportunity of concentration of energy in small volume. Yokes enable to move the current on an active surface of inductors, as well as enable to increase their inductance, that important for choice of matching conditions. At the technology power supplies with frequency not below 66 kHz are applied, that results to application of ferrite yoke. In case of pulse-periodic mode of operations use a great attention should be given to a temperature field of ferrite, which operates in strong fields, having rather bad cooling, that is connected with their low thermal conductivity at the absence of special systems of a heat removal. A unique way, enabling to ensure reliable conditions of operation, is the increase of working volume of a yoke. The technology is effective at operation with minimum gaps between an active surface of inductor and workpiece as thus intensity of a magnetic field is maximum, that in turn permits to ensure the heaviest depth of distribution of thermal sources, increasing a speed of heating. The specified conditions of operation result to necessity of special preparation of active surfaces of inductors and yoke, providing their high accuracy of manufacturing. The gap in the majority of cases makes 0,1 mm and there is necessity of putting of special dielectric covers having high mechanical properties. It is possible to relate a covers on the basis of mineral-ceramics or oxide of aluminum.

Matching conditions. The choice of matching conditions is the key problem under realisation of the HF hardening technology. Low voltage inductors with a low value of equivalent inductance applied at the technology demand to make some changes at the HF generator. The application of the vacuum-tube as well as transistor generator leads to the necessity of the use of matching HF transformer to decrease the value of voltage that has the positive fact because the value of equivalent inductances on the primary winding of the transformer is increased. The Q-factor of the resonance circuit included condensers and inductor is more bigger for the vacuum-tube generator then for transistor generator. In case of vacuum-tube generator it leads to simplicity of power matching conditions of oscillate circuit and vacuum-tube and to operation of generator in the permitted by standard of frequency range. (For frequency 440 kHz the tolerance of frequency according to the standard equals to ±2.5%) Moreover there are additional adjustable elements in oscillate circuit, for example controller of feedback, controller of power, that make more simple the power matching at the change of inductors and workpiece. Transistor generator with low Q-factor demands the exact way of generator parameters choice, but the change of a frequency during the heating will be bigger then for vacuum-tube generator. Besides the bigger change of frequency during the technological cycle for the transistor generator leads to the change of the power matching conditions that compel to review of the power of the generator. The negative aspects of the last problem is known i.e. the possible overheat of some parts of the non-symmetrical workpieces.

DESCRIPTION OF SIMULATION PROCEDURE

The simulation algorithm of hardening process can be divided on four basic parts:

electromagnetic and thermal problems. Induction heating process can be described by Maxwell's and Fourier's equations. Methods of the equations solving were developed. Electromagnetic characteristics of workpieces depend on temperature and electromagnetic field intensity. In the work for the electromagnetic and thermal problems solving the next assumptions were made. The particularity of the technology is the use of minimum gaps with a big specific power that leads to a high level of electromagnetic field intensity. The last fact makes right assumption about the necessity to describe ferromagnetic workpiece in the range under inductor with relative magnetic permeability equals to 1. Besides there is yoke that gives the form of current distribution in the workpiece as almost rectangle. The sizes of current conductivity range is defined by both the narrow of induced conductor and the depth of current penetration in metal under the pre-given frequency, specific resistivity and relative magnetic permeability. Thus the thermal sources distribution in the workpiece can be define without the electromagnetic problem solving only for the case of the HF induction pulse-hardening technology. 2D-model based on finite elements method is applied for the definition of temperature distribution in the workpiece. The calculations were carried out for the heating and cooling of the workpiece. The temperature distribution in the workpiece near the inductor is calculated only because heating time is little and heating is carried out for limited range. The choice of number and size of finite elements is defined by geometry of the workpiece and depth of current penetration in metal. In further the same elements are used for calculation of stress-deformation state;

HF generator simulation problem. The final aim of HF generator modeling is steady-state process, which gives load voltage and frequency. In the work the Cauchy Problem solving was used. The vacuum-tube or transistor model should be used. Vacuum-tube is a non-linear element for which anode $i_a = f_1(u_a, u_g)$ and grid $i_g = f_2(u_a, u_g)$ currents are known. The model of the tube is two dependent current sources the values of which are defined in accordance with the handbook characteristics. The similar model of MOSFET transistor should be used for simulation of solid-state generator. The characteristics of MOSFET transistor also is shown in handbook and taking into attention valve mode of transistor operation its characteristic can be defined as dependent current source that gives at simulation a good accuracy. Ordinary differential equations (ODE) system in the state variables is formed automatically. These basic principles were used in the program of HF generator simulation[3];

thermal stresses problem. Non equal plastic deformation and a non equal temperature in different points of workpiece are the main cause of internal stresses. The connection between elastic deformation and heat transfer is negligible under heating and it is not taken into account. The last one usually is described by six components which are the forces per square and are applied to perpendicular surfaces passing through investigated point. Additionally boundary conditions are written. Definition of stressed state in workpiece is possible under the solving of differential equation system of equilibrium. Also the deformation of elastic body is described. The last ones and boundary conditions give the system which is valid for calculation of whole distribution of stresses;

coupling simulation problem. As seen a simulation of this unified process is very complicate, because the different types of equations describe the coupling process. Simulation of induction heating process demands the use of partial differential equations. Simulation of stresses demands also the use of similar equations and simulation of power supply demands the use of ODE. The unification is the main problem of simulation because before the software for the separate simulation of these processes had been developed[4,5]. The iterative procedure has been created for simulation of the technology[4]. The optimum heating process can be simulated for the choice of the minimum strains in the workpiece.

DESCRIPTION OF SIMULATION RESULTS

On fig. 2 temperature fields, received at heating of boundary of a workpiece, submitted on a fig. 1 are

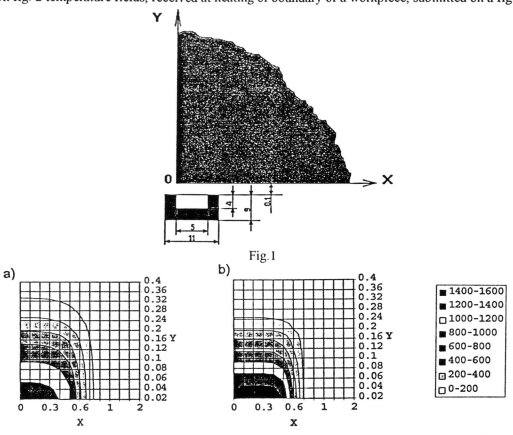

Fig.1

Fig.2 Temperature fields °C a) mode I - 4,5 kW*cm^{-2}; b) mode II - 7,5 kW*cm^{-2}

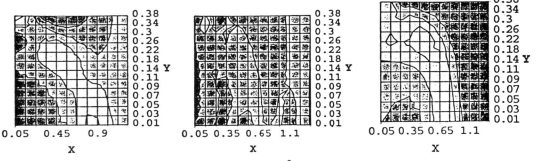

Fig.3 Thermal stresses (kg*cm^{-2}). Mode I. a) σ_x, b) σ_y, c) σ_z.

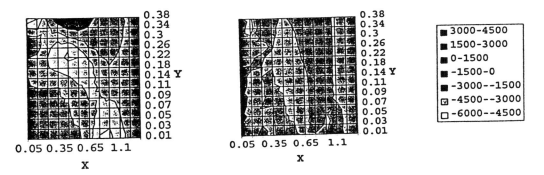

Fig.4 Thermal stresses (kg*cm^{-2}). Mode II. a) σ_x, b) σ_y

indicated. The temperature field is changed essentially and depends on a level of specific power and heating time. The increase of specific power up to 7,5 kW*cm^{-2} and reduction of pulse duration up to 150 ms (fig.4) provides on a' surface of a workpiece high temperatures, though the area of temperatures is appropriate to a range 600÷800°C. The field of thermal stresses is differed at comparison of first two variants (fig.4), though the general character has not undergone significant changes despite on differences in absolute values. Thus it's possible to make the conclusion about inexpediency of realization of technological process at a pulse duration less than 100 ms because in this case it is possible to melt a surface of metal. For maintenance of given of a hardened layer depth at absence of melting on a surface the specific powershould be decreased. The results are indicated on fig.3÷4. Availability of stresses placed in various areas of a workpiece, permits to conclude about an opportunity of destruction of workpiece on a border of section, that essentially at operation on high frequencies with large specific power.

DESCRIPTION OF EXPERIMENTAL RESULTS

In the Table 1 results of experiments are indicated for ferromagnetic steel U8 which were received on vacuum-tube generators. Mechanical properties of metal are not worse then 64 HRC for indicated samples in a zone of a hardened layer. The size of a hardened layer has been defined as result of metalgrafic researches on change of a structure of metal. The increase of pulse duration at constant specific power results to the increase of hardened layer depth up to finite values and then on a surface melting area is formed. The value of a hardened layer depth is decreased, that is explained by an increase of a zone of thermal influence and results to decrease of cooling speed. Reception of a thermal treatment zone by operation of power supply on frequency 440 kHz at large pulse duration results to formation of a metal splashing due to electrodynamic forces, influencing on melting metal.

CONCLUSION

Unified approach for simulation of electromagnetic field, thermal distribution, thermal stresses, power supply is suggested. The explanation of high microhardness was done under heating with high specific power. Experimental results confirm a validity of the model. The real technological application of pulse induction heating were explained.

The authors would like to thank our Italian colleagues Prof. Lupi S. and Dr. Dughiero F. (University of Padua) for fruitful and useful discussion of theoretical and experimental results.

Table 1
Values of hardened layer depth at pulse hardening of a plate boundary by thickness of 45 mm for steel U8 at specific power not less than 4,7 kW*cm^{-2}.

sample	time of processing, ms	Depth of hardened layer in mm	
		at frequency 66 kHz	at frequency 440 kHz
1	100	1.087	0.535
2	150	1.315	0.730
3	200	1.380	0.865
4	250	1.425	0.975
5	300	1.830	1.065
6	400	1.680	----

REFERENCES

1. M.Zukov, V.Schukin, V.Neronov etc.: 'High-Frequency Pulse Hardening of Steel'*Physics and chemistry of materials treatment*, 1994(6). 98-108. (In Russian).

2. A.Vasiliev, B.Kogan, B.Kachanov: 'Pulse Induction Heating'*Electrotech'92, Proceeding of XII-th Congress UIE, Montreal (Canada), 14-18 June*, 1992. 36-39.

3. Yu.Blinov., B.Kogan, A.Vasiliev etc.: 'Methods for Analysis and Control of Vacuum-Tube Oscillators for Electrothermy Applications', *Soviet electrical engineering*, 1987, **58**(8). 33-36.

4. Yu. Blinov, S.Lupi, F.Dughiero: 'Mutual Influence Between Load and Frequency Converter in the Induction Heating of Steel', *Proceedings of Int.Conf. Industrial Electronics, Control and Instrumentation (IECON'94), 5-9 September, Bologna (Italy)*, 1994. 679-683.

5. F.Dughiero, S.Lupi: 'The Control of Temperature Transients in the Induction Heating of Steel', *UIE-Seminar, Heat Transfer in Electroheat, Lodz (Poland), 22-25 Oct.* 1991. 61-67.

Effect of Temperature on Vibration and Magnetostriction in Mild Steel Heated by Induction

R T BAKER and T N OLIVER

Electric Power Research Group, Department of Mechanical and Electrical Engineering, Aston University, UK

ABSTRACT

When a ferromagnetic steel billet is heated by induction, a large increase in the amplitude of longitudinal vibration often occurs as a result of resonance. The temperature at which resonance occurs depends on many factors including billet length and heating power. Resonance was most frequently observed when the surface temperature of the billet reached the Curie point. It is well established that magnetostrictive vibrations occur in a ferromagnetic material subjected to an alternating magnetic field, but existing data suggests that linear magnetostriction diminishes towards the Curie point. Linear magnetostriction was measured in a sample of mild steel up to 800°C using a high temperature strain gauge. It was discovered that magnetostriction was responsible for resonance at temperatures below 600°C but not for temperatures near the Curie point.

Lorsqu'une billette d'acier ferromagnétique est chauffée par induction, on observe une augmentation importante de l'amplitude des vibrations longitudinales résultant souvent de l'effet de résonance. La température à laquelle l'effet de résonance se produit est liée à différents facteurs, dont la longueur de l'élément et la puissance de la chaleur. Lorsque la température de surface de la billette atteint le point de Curie on observe que le phénomène de résonance se produit plus fréquemment. Bien que l'on ait déterminé l'existence de vibrations magnétostrictives lorsque une matière ferromagnétique est en présence d'un champ magnétique alternatif, des données actuelles suggèrent que la magnétostriction linéaire décroit en approchant le point de Curie. On a mesuré la magnétostriction linéaire dans un échantillon d'acier demi-doux jusqu'à 800°C en utilisant une jauge de contrainte. On a découvert l'importance du rôle joué par la magnétostriction produisant l'effet de résonance pour des températures inférieures à 600°C mais non pour des températures proches du point de Curie.

INTRODUCTION

Induction heating engineers may be aware of an increase in vibration in the form of a change in sound as a ferromagnetic steel billet passes through the magnetic transformation or Curie point (760°C), but little is known about the phenomenon. Pierce showed that nickel-iron rods could be made to resonate at their natural frequencies in an alternating field at ambient temperature due to magnetostriction.[1] Pike and Moses report a decrease in magnetostriction with temperature rise for grain-oriented silicon-iron.[2] Similar trends are observed for iron as a function of temperature such as that by Honda and Shimizu.[3] Work concerning the generation and detection of ultrasound in hot steels using electromagnetic acoustic transducers (EMATs), such as that by Cole and Lee and Ahn, report increases in the amplitude of longitudinal ultrasonic pulses just before the Curie point.[4,5] The enhancement was reported to be due to an increase in magnetostriction or the concentration of the steady magnetic field in a thin ferromagnetic layer of the material. It is postulated in this work that this phenomenon and the increase in vibration in steel heated by induction are related.

LONGITUDINAL VIBRATION MEASUREMENTS IN MILD STEEL BARS

A 3.2 kHz, 60kW variable power Crossley induction heater was used to investigate vibration in mild steel bars of BS970-En3b.[6] The field strength at the centre of the coil was 70 kA m^{-1} (rms). Data was recorded and processed using the software package LabView. Various lengths of mild steel bar were heated at a range of powers from different starting temperatures, Figure 1. The longitudinal acceleration was measured at one or both faces of the bar using accelerometers. For bars shorter than the coil length, a laser vibrometer measured the velocity of the face of the bar.

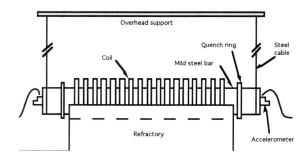

Figure 1: Apparatus to Measure Longitudinal Vibration.

Surface temperature was measured at the mid-point of the bars using sheathed thermocouples. The ends of the bar were water cooled to protect the accelerometers. The vibration signal was rectified to give the magnitude and plotted together with the heating power and surface temperature. Figure 2 shows the results for a 40 mm diameter, 600 mm long mild steel bar when heated at full power.

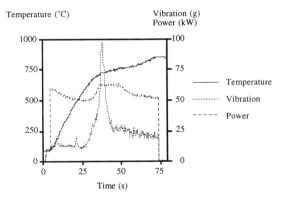

Figure 2: Temperature, Vibration and Power During Induction Heating of a 600 mm Long Mild Steel Bar.

The longitudinal acceleration was seen to increase rapidly to a maximum as the surface temperature approached the Curie point. The acceleration measured was typically in the region of 100g corresponding to a displacement of 0.3 µm. The power increased as the bar reached the Curie temperature due to increases in the skin depth and resistivity of the mild steel. Figure 3 shows the frequency components of acceleration recorded at 740°C when the acceleration was greatest.

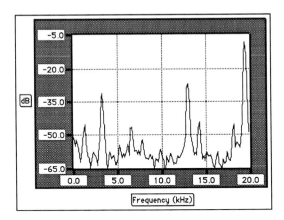

Figure 3: Frequency Components of Vibration at 740°C for a 600 mm Long Mild Steel Bar Heated at 60kW.

It was discovered that an increase in vibration in other lengths of mild steel did not always occur at temperatures near the Curie point, Table 1. Tests were all conducted at 60kW and 3.2 kHz. The longitudinal natural frequencies of the lengths of heated mild steel bars were measured as functions of temperature. It was found that the increase in vibration during heating was due to a longitudinal resonance of the bar being excited. The two faces of a bar were found to vibrate 180° out of phase at resonance, that is both moved inwards or outwards at the same time.

Length of Bar	Frequency of Main Component	Temperature (°C)
200 mm	None	-
400 mm	6.4 kHz	20
450 mm	6.4 kHz, 19.2 kHz	520, 760
500 mm	19.2 kHz, 12.8 kHz	540, 810
600 mm	12.8 kHz, 19.2 kHz	300, 740
1050 mm	12.8 kHz, 19.2 kHz	300, 740
1200 mm	6.4 kHz, 12.8 kHz	540, 730

Table 1: Vibration in Lengths of Mild Steel Bar.

Other factors affected the magnitude of resonance at the Curie point, such as initial bar temperature. The nearer the bar was to the Curie point prior to heating, the lower the vibration and the temperature at which resonance occurred, Figure 4.

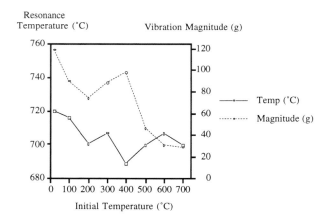

Figure 4: Resonance Temperature and Magnitude as a Function of Initial Temperature for a 600 mm Long Mild Steel Bar Heated at 60kW.

The power at which the bar was heated also affected the magnitude of vibration at the Curie point and the temperature at which resonance occurred, Figure 5.

Different diameter mild steel bar, but of the same length 600 mm, was heated at 60kW. The frequency components were recorded at resonance near the Curie point showing an increase in the magnitude of resonance with an increase in bar diameter, Table 2.

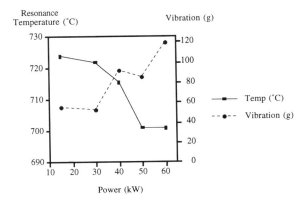

Figure 5: Resonance Temperature and Magnitude as a Function of Power for a 600 mm Long Mild Steel Bar.

Bar Diameter (mm)	Acceleration (g) at Frequency (kHz)				
	3.2	6.4	12.8	19.2	25.2
25	1.5	4.6	7.0	39.2	11.1
40	5.9	9.0	13.3	71.7	20.0
50	1.0	13.6	16.7	91.8	33.8

Table 2: Frequency Components at Resonance for Different Diameter Mild Steel Bars.

It was discovered that doubling the diameter of bar doubled the power input. The acceleration component therefore increased proportionally to power input. Materials such as austenitic stainless steel, copper, titanium and monel (70%Ni/30%Cu) were found not to resonate when heated electromagnetically. Since titanium, which has a phase transformation at 902°C did not show resonance, it is hypothesised that resonance is not caused by phase transformation alone.

DISCUSSION

Resonance occurs in a billet heated by induction, at 3.2 kHz in this case, when a natural frequency coincides with twice the heating frequency (6.4 kHz) or a subsequent harmonic (12.8 kHz, 19.2 kHz and 25.6 kHz). It appears that resonance occurs most frequently at the Curie point simply because the natural frequency of the bar changes most quickly with reference to surface temperature. Initial temperature is one of the main factors affecting the resonance at the Curie point leading to a reduction in the magnitude when the initial temperature is increased. The magnitude of resonance was also found to be directly proportional to power input. Magnetostriction was hypothesised to be the mechanism responsible for the longitudinal vibrations for a number of reasons; the longitudinal vibration was at twice the heating frequency or harmonics thereof, resonance only occurred in ferromagnetic steels and the modes most frequently excited were those with the faces of the heated bar vibrating out of phase. Magnetostrictive vibrations at the mid-point of the bar would produce all the above characteristics. It was unclear as to why vibrations should occur at or above the Curie point where magnetostriction was frequently reported to diminish.[7]

MAGNETOSTRICTION MEASUREMENT

Joule magnetostriction in mild steel was measured using a high temperature strain gauge similar to that used by Pike and Moses.[2] A Kyowa KHCS-10 gauge was welded onto a 200 mm length of mild steel (BS970-En3b) of diameter 50 mm. The bar was placed inside a magnetising coil which was located inside an electric furnace, Figure 6. The field generated was 3000 A m^{-1} (rms) at 50 Hz. Magnetostriction was measured at 100 Hz, 200 Hz and 300 Hz as functions of temperature.

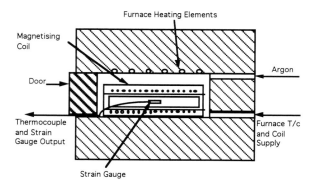

Figure 6. High Temperature Strain Gauge Apparatus.

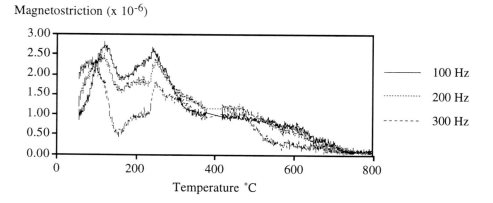

Figure 7. Magnetostriction at 3000 A m^{-1}

Figure 7 shows that increases in magnetostriction occur at 120°C and between 200°C and 300°C. The latter increase is almost a step change and may well be related to the Curie point of cementite (210°C). It is not clear as to why there is an increase in magnetostriction at 120°C. The most important observation is that magnetostriction diminishes towards the Curie point.

In order to compare the experimental results with theory the magnetisation process in the mild steel bar was considered. It was shown by Simmons and Thompson that for a single grain with a misorientation θ in the plane of the sheet the a.c. magnetostriction λ is approximately given by:[8]

$$\lambda = \frac{3}{\sqrt{2}} \lambda_{100} \frac{B}{M_s} \sin\theta \left(\sin^2\frac{\theta}{2} - \cos^2\theta\right) \qquad (1)$$

If it is assumed that the grains are perfectly aligned in the [0 1 1] direction giving $\theta = 90°$, then:

$$\lambda = \frac{3}{2\sqrt{2}} \lambda_{100} \frac{B}{M_s} \qquad (2)$$

The grains in mild steel are not perfectly aligned but randomly orientated. An average approximation was calculated from equation 1.

$$\lambda(av) = \frac{3}{\sqrt{2}} \lambda_{100} \frac{B}{M_s} \frac{1}{\pi} \int_0^{\pi} \sin\theta \left(\sin^2\frac{\theta}{2} - \cos^2\theta\right) \qquad (3)$$

$$\lambda(av) = \frac{1}{\pi\sqrt{2}} \lambda_{100} \frac{B}{M_s} \qquad (4)$$

It must be noted that equation 4 is an approximate equation assuming that the grains in the mild steel specimen are randomly and equally arranged. Using equation 4 the magnetostriction constant was calculated, Figure 8. It is shown to be in close agreement, considering the assumptions used, as that measured using single crystals of iron.[7]

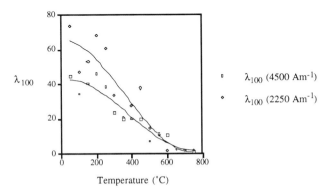

Figure 8. Magnetostriction Constant λ_{100} as a Function of Temperature For Mild Steel.

Much work has been investigated regarding the use of EMATs for on-line monitoring applications in hot steel up to 800°C.[5] An enhancement in the longitudinal ultrasonic signal is frequently reported to occur at the Curie point. Many of the hypotheses relating to EMATs were tested on mild steel bars heated by induction. None were found to account for resonance near the Curie point. The work in this paper may be significant to studies made of ultrasound amplitude in hot steel.

CONCLUSION

Longitudinal resonance in steel bars, when heated by induction, occurs when a natural frequency coincides with a multiple of twice the heating frequency. A number of factors affect the magnitude of resonance such as bar diameter or power and initial temperature. Joule magnetostriction has been measured in mild steel to 800°C. The mechanism responsible for resonance below about 600°C is concluded to be Joule magnetostriction. Near the Curie temperature Joule magnetostriction is unlikely to be the cause of resonance.

ACKNOWLEDGEMENT

This research is supported by the UK EPSRC and Midlands Electricity plc, Halesowen, West Midlands, UK. The authors are also grateful to Professor W.T. Norris, Mr J.M. Thomas and Dr. M. Booth for their advice and assistance in this work.

REFERENCES

1. G. PIERCE: 'Magnetostriction Oscillators', *Proc. Inst. Radio Engineers*, 1929, **17**. 42-88.
2. E.C. PIKE and A.J. MOSES: 'Effects of Temperature on a.c. Magnetostriction in Grain-Oriented Silicon-Iron', *Journal of Materials Science*, 1977, **12**. 187-191.
3. R. HONDA and S. SHIMIZU: 'Change of Length of Ferromagnetic Substances Under High and Low Temperatures by Magnetisation', *Phil. Mag. [6]*, **6**. 392-400.
4. P.T. COLE: 'The Generation and Reception of Ultrasonic Surface Waves in Mild Steel at High Temperatures', *Ultrasonics*, 1978, **16** (4). 151-155.
5. S.S. LEE and B.Y. AHN: 'EMAT Applications at High Temperature', *Non-Destructive Testing and Evaluation*, 1992, **7** (1). 253-261.
6. BS970: 'Bright Bars for General Engineering Purposes', *British Standards Institution*, 1991.
7. R.M. BOZORTH: 'Ferromagnetism', Van Nostrand, 1951.
8. G.H. SIMMONS and J.E. THOMPSON: *Proc. IEE*, 1971, **118**. 1302.

Effective micro-metallurgical induction heating process for surface macro-reinforcing of mechanised parts.

V. VOLOGDIN
FREAL Ltd Company, St. Petersburg, Russia
V. GANJUCHENCO
Electrotechnical University of St. Petersburg, Russia
S. MAMYKIN
SPLAV Ltd Company, Moscow, Russia

ABSTRACTS

Comparative analysis of technological parameters of different surface hardening methods are rewiewed. Mathematical model for heating process for hard-facing of steel parts is described and some technological applications of induction macro-reinforcing method are considered.

L'analise comparativ des parameters technologiques des metodes differentes du renforcement supersufiel des pieces metalliques, et aussi des characteristiques d'exploit des pieces renforcés est representé dans l'ouvrage. La discription du modele mathematique du processus est présenté. L'utilisation du modele permet d'effectuer l'optimisation du processus technologique et de la construction de l'installation du choffage par induction.

In the range of hardening processes and maintenance engineerings the methods based on electric technologies are of special value. The processes based on electric heating allow to variate in the wide range technological parameters such as temperature, zone of thermal influence, width and depth of applied hardening layer, surface raggedness. They also allow to use corrosion-, heat-, chemical-, wear - resistance coeval materials and easily yield for monitoring control and mecanisation.

Among the technologies used for hardening and repair some of them use alloyed powder materials to be applied to get fused layer on the surface of the part. Main specifications for them are provided in Table 1.

The analysis of these used for friction wear protection shows that most of them contain a number of unperfect features: high cost of equipment and consumed materials low productivity, non-compleate realisation of powder materials properties, high energy losses.

Table 1.

N	Maintanance enginering processes	Productivity, Kgh^{-1}	Hardeness of HRC	Hard-faced layer thickness, 10^{-3} m
1	Gas flamed facing	1.5	13 - 42	0.5 - 1.5
2	Gas flamed powder facing	2.5	40 - 54	0.1 - 0.5
3	Electrode arc facing	0.4 - 4.0	26 - 62	0.5 - 3.0
4	Hard-facing under flux layer	1.8 - 2.0	17 - 62	0.5 - 20
5	Electroslag hard facing	1.0 - 6.0	51 - 57	10 - 20
6	Induction hard-facing	0.5 - 20	40 - 57	0.1 - 2.5
7	Plasma hard-facing	2 - 12	30 - 65	0.1 - 2.0

Powder materials analysis demonstrates that leading part belongs to self-fluxing firmly alloyed Ni ,Co containing powders but the task is in the rumen for more inexpensive materials. The most advanced is application of metal ceramic and ceramic compositions on metal oxides base such as Al_2O_3 and TiO_2 as well as elements and carbides B and C, TiC, CrC and others.

But it is a very difficult task to apply such compositions to steel or cast iron parts surface by hard-facing process because of their poor sousing, low diathermic and electric conductivity.

The objective of future investigations is both in searching for new materials, compositions and also in searching for new low cost methods to decrease of thermal influence zone in details as well as to increase their hardness and wearproof properties.

Micro-metallurgical Induction heating method is proposed in this paper for surface macro-reinforcing of metal parts. The pattern of this process is given schematically in fig.1 as a crosssection of hard-facing zone. The required layer of powder materials mixture (3) is applied on the flat surface z=0 of massive part (1), ($z \to \infty$) to be hardened. Induction coil (2) moves along the part in the flat parallel to the surface of the part (Z=0) in the direction (-x) and creates zones of surface heating $q_1...q_N$ (N - number of the wire of coil). The distribution of sources of heat density is assigned to be known. The eddy current heating of powder material is to be neglected because of it's large resistance value. Heat is transmitted from the surface of part to the mixture of powder materials. The melting of powder batch begins in the point $x = x_{N+1} - b_{N+1}$, $T(x) = T_m$, $q_m = q_m(x(T))$. Melting comes to the end in the point $x = x_{N+1} + b_{N+1}$ and open mirrow of melted alloy appeares and heating losses specific power $q_1 = q_1(x(T))$ is taken into consideration.

Cristallisation begins when $x = x_{N+2} - b_{N+2}$ in point where surface temperature reaches $T(x) = T_m$, and comes to the end at the point $x = x_{N+2} + b_{N+2}$, where balance of hidden cristallisation heat, heat conducted to the depth of part and heat losses from the surface take place.

The following assumption have been taken into consideration:

- the frequency of current supplied to induction coil is high enough to consider $q_1...q_N$ to be surface type;

- thickness of applied alloys (h) is to be small enough to consider the surface of alloyed part to be flat;

- phisical properties of metal part, melted zone and alloyed surface are to be average and do depend on temperature (T).

Solution of the equation for temperature field $T=T(x,z)$ is obtained like a sum of interval of integrated function for lineary moving source of heat [1] :

$$T = 2(\pi\lambda)^{-1} \sum_{i=1}^{N} \int_{x_i-b_i}^{x_i+b_i} q_i(x') \exp\left[U(x-x')\cdot(2a)^{-1}\right] \times$$

$$\times K_0\left\{U\left[(x-x')^2 + z^2\right]^{0.5} \cdot (2a)^{-1}\right\} dx' +$$

$$+ 2(\pi\lambda)^{-1} \sum_{i=N+1}^{N+3} \int_{x_i-b_i}^{x_i+b_i} \alpha_i\left(T(x')(T-T_0)\right) \exp\left[U(x-x')\cdot(2a)^{-1}\right] \times$$

$$\times K_0\left\{U\left[(x-x')^2 + z^2\right]^{0.5} \cdot (2a)^{-1}\right\} dx' \quad (1)$$

where: $K_0(y)$ - modified Bessel function; x_i - coordinate of zone centers of induction heating ($i = 1...N$) and heat exchange ($i = N+1...$); $2b_i$ - width of i- strip of heat and heat exchange; $\alpha_i(T(x))$ - nonlinear coefficient of heat transmission characterising (at $i = N+1$) heat exchange between melted metal and melting batch (at $i = N + 2$) heat losses from melted metal and hard-faced layer to space around, λ - heat conductivity.

In cristallisation zone of melted metal ($x_{N+2} - b_{N+2} \leq x \leq x_{N+2} + b_{N+2}$) temperature is known on the surface $z = 0$:

$$T(x,0) = T_m \tag{2}$$

Temperature distribution on the surface $z = 0$ is of interest to obtain technological parametres and for process optimisation.

Temperature distribution on the surface of detail must satisfy equations (2) ... (4):

$$T(x,z) = 0 \qquad x, z \to \pm\infty \tag{3}$$

$$T(x,0) - 2a(\pi\lambda U)^{-1} \sum_{i=1}^{N} \int_{x_i - b_i}^{x_i + b_i} \exp(U) q_i(U) K_0(U) dU -$$

$$-2a(\pi\lambda U)^{-1} \sum_{i=N+1}^{N+3} \int_{x_i - b_i}^{x_i + b_i} \alpha_i(T)(T - T_0) \exp(U) K_0(U) dU = 0 \tag{4}$$

and additional conditions of conjunction (continuety of temperature field on unknown boundaries of heatexchange zone.

There is no precise solution of this problem to be known.

Temperature distribution for mentioned above case is obtained by numerical method of sequential approach. For the initial approach the temperature field from surface induction sources is taken.

Because of low heat conductivity of powder batch heat losses from it can be neglected. Heating of local portion of batch above surface of detail at temperature T' can be considered as a row of consequensial cycles of heating comming one after another to the surface of thermophisically thin layers of batch from starting temperature T_0 up to melting temperature of batch T_m during time $\Delta\tau$.

Temporal change of temperature $T'' = T''(\tau)$ of upper layer (near to detail) is described by following equation:

$$T'' = T' - (T' - T_0) \exp\left(-\tau \cdot \Delta\tau^{-1} \ln\left((T' - T_0)(T' - T_m)^{-1}\right)\right) \tag{5}$$

Average batch temperature of layer applied to the surface of detail is obtained from the following:

$$T''_{av} = \Delta\tau^{-1} \int_0^{\Delta\tau} T''(\tau) d\tau \tag{6}$$

According to the experimental data the heat exchange on the melting front has a character of a radiant exchange. Therefor the average quantity of specific local power transmitted from the detail to the powder batch is [2]

$$P_{av} = \delta \varepsilon T_m^4 \ln^{-1}\left[(\theta - \theta_0)\cdot(\theta - 1)^{-1}\right] \times$$
$$\times \left[(1 - \theta_0^4)\cdot 4^{-1} + \theta(1 - \theta_0^3)\cdot 3^{-1} + \theta^2(1 - \theta_0^2)\cdot 2^{-1} + \theta^3(1 - \theta_0)\right] \quad (7)$$

where $\theta = T' \cdot T_m^{-1}$, $\theta_0 = T_0' \cdot T_m^{-1}$, δ - Boltzmann's constant, ε - radiation coefficient, T_0 - ambient temperature.

The local powder batch melting rate is

$$U_m = P_{av}(\rho_b H)^{-1} \quad (8)$$

where ρ_b - batch specific density, $H = L + cT_m$, c - specific heat, L - specific melting heat.
The local coefficient of heat transmission thus

$$\alpha_{N+1} = P_{av}\left(T' - T_{av}''\right)^{-1} \quad (9)$$

This relationship has been used in problems (2)...(4) solution.

The example of quasi-stationary temperature field T=T(x,0) on the detail surface in the hard-facing melting process is represented on the fig.2.
Induction melting coil represents a binary conductor horisontal forming two heating zones with the identical widths $2b_1 = 2b_2 = 10^{-2}$ m and with the equal specific power density $q_1 = q_2 = $ Const. The distance between the centres of these zones is $\Delta x = x_1 - x_2 = $ = $3 \cdot 10^{-2}$ m. The powder batch melting point is $T_m = 1350\,°C$.

The numerical calculations cycle made by the described method shows that the mathematical model describes correctly the main phenomena in the hard-facing induction high-friequency melting process.

Particularly the model account for the threshold characyers of the hard-facing process. This process character is universal for phases transfer creations in the body treated electotermic process.

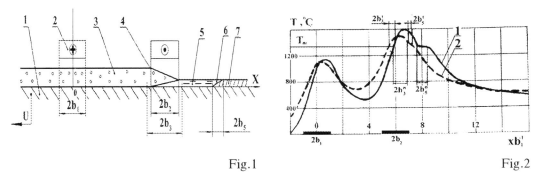

Fig.1　　　　　　　　　　　　　　　　　　Fig.2

Fig.1 Induction hard-fasing process scheme: 1 - part treated; 2 - induction coil; 3 - initial powder batch layer; 4 - powder melting front; 5 - melted zone; 6 - solidification front; 7 - layer applied.

Fig.2 Surface temperature distribution diagram: material of part steel 20XG2SL; $q_1 = q_2 = q$; curve 1: $U = 21 \cdot 10^{-4}\,mc^{-1}$, $q = 1.06 \cdot 10^7\,Wm^{-2}$;

curve 2: $U = 84 \cdot 10^{-4}\,mc^{-1}$, $q = 2.00 \cdot 10^7\,Wm^{-2}$.

Mathematical model of heating process allowes to design of induction system and to develope optimum technological regimes of induction heating for surface reinforcing process.

Micro-metallurgical induction heating process has found wide application for hardening of parts and friction assemblies of railroad transport.

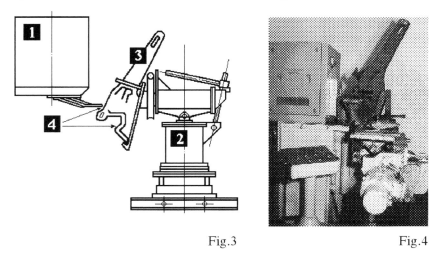

Fig.3　　　　　　　　　　　　　　Fig.4

Fig.3, Fig. 4. Process of induction automatic coupling hard-facing:
1 - HF power supply circuit; 2 - manipulator; 3 - coupling unit; 4 - surfaces to be hard-facing.

In the middle of 1980 Engineering Centre of Ministry of transport "Alloy" pioneered a solution of the problem of restoration and hardening of rolling stock parts first for the lock and then for the unit (Fig.3) of the automatic coupling [3]. The high frequency equipment - universal power supply generator has been developed by V.P.Vologdin institute specialists (VNIITVCh). Induction coil which is performing the heating by continuons-sequential method, removed heating circuit, which allows to perform hard-facing of large parts including the surface difficult for access (Fig.4) have been developed by "FREAL Ltd" company [4].

Micro-metallurgical induction heating process proved to be most effective and has several advantages for economy, production and environment.

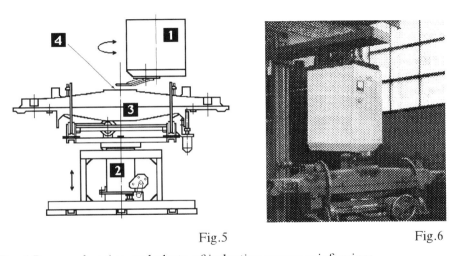

Fig.5   Fig.6

Fig. 5,Fig. 6.Process drawing and photo of induction macro-reinforcing:
1- HF power supply circuit;2 - manipulator;3 - owerspring beam;4 - macro-reinforcing area.

Economic advantages. The longevity of hardened parts increases 5 - 10 times. Maintenance costs decrease, with savings of hard-facing materials and power consumption. Labour costs are cut due to a reduction in repair time and prophilactic measures.

Production advantages. There is an increase in productivity in comparison with the argon are welding method and other methods. There is no need for a finishing operation after induction heating. There is an improvement in working conditions; a lower level of noise; the absence of ultra-violet radiation and lower vibration.

 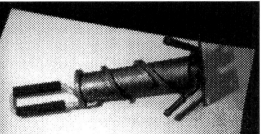

Fig.7 Fig.8

Fig. 7. Micro-metallurgical induction heating process for surface macro-reinforcing of steel parts.

Fig. 8. Induction coil.

Induction coils designed for performing heating by continuons-sequential method. Active zone of coil is provided by flux field concentrator. This results to melting process improvement and reduces from 20% to 50% of energy losses.

REFERENCES

1. Y. S. CARSLAW, J. C. JAEGER. *Conduction of heat in solid*. Oxford, Cearendon Press, 1959.
2. V. M. GANJUCHENCO, E. I. HAIKINA, Calcul of oxide material melting induction heating devices, *Trans.Electrotechnic.Inst.*,v.255,1979,Leningrad,21-25.
3. J. A. ZAICHENKO, *Technological complexes for restoration and hardening parts by induction metallurgical method*. Railroad Transport. Series "Locomotives and locomotive household". Maintenace of Locomotives. Moscow,1991.Issue 2, 1-26.
4. V. V. VOLOGDIN. New Advanced Technology and Equipment for Hard - facing Automatic Coupling.*J. Railway Technology International*,1995.

The use of pulsed radiant flux with variable time structure for laser technologies.

V. M. GANJUCHENKO, S. G. VOLOGDINA
Electrotechnical University of St. Petersburg, Russia
F. KOSTRUBIEC
Technical University of Lodz, Poland

ABSTRACTS

The conception of the physical phenomena control and optimisation during the laser treatment of materials using the pulsed laser beams of controled complex time characteristics is developed in the paper on the example of deep fusion laser welding technology. In this technology the prevention of the negative influence mutual of the fusion, vaporisation and flux-material interaction is possible by means of separation of these physical phenomena in the time. The time characteristics of the laser beams for the such "optimal" deep fusion laser welding process have been obtained by the joint solution of the correspondings heat conductions, evaporation and hydrodynamics problems. The energetics parameters of this welding have been analysed and compared with the electron beam and vacuum laser beam welding parameters. An algorithm of welding conditions and laser beam parameters determination is discussed.

Le conception du réglage et de l'optimisaton des phénomenon physiques lors du treatment des matériaux par laser en utilisant le rayonnement du laser a pulsation avec les caractéristiques temporels reqlés est developpé dans l'article sur l'exemple de la technologie du soudage a pénétration profonde par laser. Dans cet technologie le prévention de l'influence négatif mutuel de la fusion, de l'evaporation des matériaux et de l'action dy rayonnement du laser sur les matériaux est possible par voie de la répartition des ces phénomemon physiques dans le temps. Les caractéristiques temporels du rayon du laser pour tel procédé "optimale" du soudage á pénétration profonde par laser sont calculés a l'aide de la resolution compatible des problemes de la transmission de la chaleur, de l'evaporation et hydrodynamique. Les parameters énergétique du soudage "optimum" sont analysés et comparés avec les caractéristiques correspondants des soudage a faisceau d'electron et par laser á bide. L'algorithme de la determination des parametres temporels du rayonnement du laser et des conditions du soudage est discuté.

1. INTRODUCTION

The idea of laser treatment of materials using the cycle of two succesive pulses of different width ($\tau_1 > \tau_2$) and different power density ($q_1 < q_2$) appeared in mid-eighties [1] for laser drilling. The starting point for the development of this conception in the field of the laser beam deep fusion welding is the energy threshold character of phase transitions "solid-liquid" and "liquid-gaseous phase" occuring in the flux-metal interaction zone. The control of the energy absorbed on the interaction zone allow to regulate the speed and the intensivity of the phase transitions indicated and of the phisical phenomena accompaning these transitions (gaseous and hydrodynamics, appearence of the laser torch, etc.).

2. DEEP FUSION LASER WELDING: PROCESS WITH THE CONTROL OPTIMAL OF THE PHENOMENA PHYSIC.

Let's enumerate briefment the factors responsible of the character and efficiency of the deep fusion laser welding [2,3,4] (fig. 1).

- absorption of laser irradiation on the vapor-gaseous hollow walls;
- laser beam canalisation along the hollow;
- stability hydrodynamic of molten zone around the hollow;
- metal destruction on the front part of the hollow and its stability;
- metal destructed (liquid and vapor) transfer on the back part of the hollow;
- plasma torch formation near the hollow outfall; laser beam defocusation by the plasma and beam generation conditions alterations.

The optimal controled welding process organisation thus must guarantee: minimisation of the loss energetic in the time of metal destruction on the hollow front part and of its transfer, plasma torch formation minimisation and molten zone stability hydrodynamic.

1. The last condition demand the creation of the metal vapor pressure in the vapor-gaseous hollow (at least enought oftenly)

2. On the other side the metal vapor dynamic pressure is the mechanism effective for the metal destructed (molten) transfer from the front cavity part to the back one [3].

3. The loss energetic in the metal destruction process phase may be minimized on condition that tne solid-state metal destruction take place only by means of the metal melting without its vaporisation [4] in order to exclude the interaction "laser irradiation-vapor" in and near the vapor-gaseous hollow.

The evident contradiction between demands 1 and 2, from one side and 3 from other side may be decided by the utilisation in the laser beam deep fusion welding process of the pulsed laser beam with the complex-time power structure.

Every pulse include three zones ensuring consecutive optimal control of the physical processes in the laser beam hollow on the different stages:

a) metal melting on the front hollow wall part,

b) melting layer transfer on the bask one by means of the melt part evaporation and using the vapor pression,

c) pause for the vapor-gas evacuation from the hollow for its "cleaning" before the cicle repetition begining.

3. METHODOLOGY OF PULSE PARAMETERS OPTIMISATION.

The calculation of the laser beam parameters optimal control in time is based on the determination of the laser beam intensity and energy necessary for the different welding stages.

1. Stage one, during the time interval $\Delta\tau_1$, in which the process of the front wall hollow melting takes place. The parameters of this stage will be analysed using the heat equation with appropriately selected boundary and initial conditions.

2. Stage two, including the time interval $\Delta\tau_2$, in which the transport of the liquid metal from the front part to the back part of the hollow takes place, until the front part is completely "dried". To analyse this stages, the method of superficial balance of energy shall be used.

3. Stage three, comprising the time interval $\Delta\tau_3$, is the hollow "cleaning" stage, in which the beam intensity is equal to 0.

For the first and third stages, the problem of parameters determining of the radiant flux resolves itself into solving the following heat boundary problems [7]:

$$\text{div}[\lambda(T)\text{grad}T] = c\rho \frac{\partial T}{\partial \tau} \qquad (1)$$

$$\lambda(T)\frac{\partial T}{\partial n}\bigg|_{S_k} = q_0 A(T|_{S_k}) \qquad \begin{cases} \tau \leq \tau'(n) + \tau_c(n-1) \\ \tau > \tau_c(n-1) \end{cases} \quad (2)$$

$$T\big|_{S_k} = T_{max} \leq T_v \qquad \begin{cases} \tau \leq \Delta\tau_1 + \tau_c(n-1) \\ \tau > \tau'(n) + \tau_c(n-1) \end{cases} \quad (3)$$

$$\lambda(T)\frac{\partial T}{\partial n}\bigg|_{S'_k} = 0 \qquad \begin{cases} \tau \leq n\tau_c \\ \tau > \Delta\tau_1 + \tau_c(n-1) \end{cases} \quad (4)$$

$$T(x, y, z, 0) = 0 \qquad (5)$$

$$T(x, y, z, t) = 0 \qquad \begin{cases} x, y \to \pm\infty \\ z \to \infty \end{cases} \qquad (6)$$

$$\lambda(T)|\text{grad}T|\Big|_{T>T_m} - \lambda(T)|\text{grad}T|\Big|_{T<T_m} = L\rho \frac{\partial T}{\partial \tau}|\text{grad}T|^{-1} \qquad T = T_m \qquad (7)$$

There: $\lambda = \lambda(T)$ - thermal conductivity of the metal welded; $c = c(T)$ - specific heat; $\rho = \rho(T)$ - specific gravity; q_0 - irradiation beam intensivity maximal; τ_c - cycle interval of pulse laser irradiation; L - latent melting heat; n - normal to surface; T_v, T_m - vaporisation and melting temperatures; k=1,2,3... - cycle numeration; $A = A(T|_{S_k})$ - absorbtion coefficient; S_k - laser cavity surface heated at the cycle beginning; S_k' - cavity surface at the cycle end; T_{max} - temperature maximal admitted on the surface S_k.

The equestion (1) with boundary, Stephan's and initial conditions (2)...(7) describe the non-stationary temperature field around the cavity on condition that the most effective heating of the hollow front part is carried out under condition that the cyclically repeated irradiation beam has the intensivity maximal limited $q_{max} = q_0$ and the temperature maximale admitted on the surface S_k is $T_{max} < T_v$.

The resolution T=T(x,y,z,t) is utilised for the time variation determination of the laser beam power density at the final part of the first welding stage and compet "melting" pulse part in the interval $\Delta\tau_1$ is:

$$q(t) = q_0 \qquad T\Big|_{S_k} < T_{max} \qquad (8)$$

$$q(t) = -\lambda \frac{\partial T}{\partial n}\Big|_{S_k} A^{-1}(T_{max}) \qquad T\Big|_{S_k} = T_{max} \qquad (9)$$

The problem (1) for the cavity environs (fig.1) is simplified usially by its reduction to a two-dimentional problem. In ours case for the short pulse cycle ($\tau_c << Ra^{0,5}$, where R - cavity radius, a - thermal diffusivity ^{3}) it's possible the problem reduction to the one-dimensional equations in the cylindrical system of coordinates (fig.1). Such a procedure has been used in the examples described below.

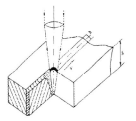 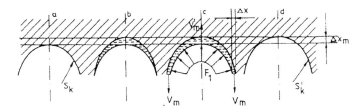

Fig.1 Cross-section of a vapor -gaseous hollow in deep- fusion laser welding.

Fig.2 Stages of the welding cycle

As has been endicated above, the transport of the molted metal from the front cavity wall to its rear part takes place in the small time interval $\Delta\tau_2$ and result from a momentary increase in the pulse power density. The temperature of the surfase being heated reaches a higher value than that of the temperature of boiling of the metal, and the metal vapours pressure in the cavity increases rapidment. The liquid metal is "squeezed out" in the direction of the rear cavity wall through two fissures of the thickness Δx each (fig.2). The presence of the surface tension gradient on the cavity wall melt is an additional factor supporting the transport of liquid. The work of the vapours of metals, which is done during the liquid metal transport from the front to to the back holloy wall, is equal to the sum of the kinetic energy of the moving liquid and energy losses to overcome viscosity forces.

Making simplifying assumptions that the surface S_k is flat the thickness of the melted layer is uniform on the whole surface and equal to Δx, the superficial energy balance concerning the evaporation of the unitary surface is equal to [5,6,7];

$$W_2 = p \cdot \Delta x + L_v V_v \rho_m \Delta \tau_2 \qquad (10)$$

in which the vapour pressure

$$p = 0{,}5 \cdot V_m \cdot \Delta x^{-1} (\rho V_m \Delta x + \eta) \qquad (11)$$

where: L_v - latent heat of vaporization; V_v - speed of motion of evaporated front; ρ_m - specific gravity of the melted metal; V_m - speed of motion of the transfered metal; η - viscosity of liquid metal.

Under these conditions, the relationship between the rate V'_m of the decrease in the thickness of the melted layer and the rate V_m of the tansport of liquid, is expresed by the dependence

$$V_m = V'_m \cdot 2b_m \cdot \Delta x^{-1} \qquad (12)$$

in which b_m is the width of the interaction zone on the from cavity wall (fig.2).

The final formula for the unitary energy (specific energy) of the liquid metal mass transport from the front to the back cavity wall assumes the form[6,7]:

$$W'_2 = W_2 \cdot [A(T)]^{-1} = [A(T)]^{-1} [b_m(2b_m \cdot \rho \cdot V'_m + \eta) \cdot \Delta x^{-1} + \rho \cdot L \cdot V_0 \cdot \Delta \tau_2] \qquad (13)$$

The average rate of welding for such a process is equal to

$$V_w = \Delta x \cdot \tau_c^{-1} \qquad (14)$$

where: $\tau_c = \Delta \tau_1 + \Delta \tau_2 + \Delta \tau_3$ - total duration of the welding cycle; $\Delta \tau_3$ - time of pause between successive pulses nessesary for remjving erosion products and preventing from uncontrolled formation of plasma during the melting of metal in the next cycle.

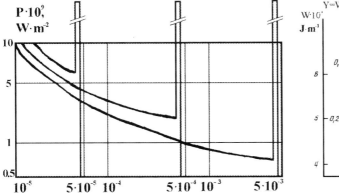

Fig.3 Time vaporation of the beam intensivity (calculed).

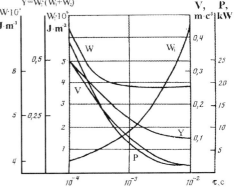

Fig.4 Energetic parameters of deep-fusion laser welding (calculed).

The time characteristics of the laser beam intensivity are represented on fig.3 for $\tau_c = 5 \cdot 10^{-5}$, $5 \cdot 10^{-4}$ and $5 \cdot 10^{-3}$ c. for the welding of X18H9T steel. Other parameters energetic, including the energy specifique by onem of the weld W, welding rate V, W_1, $Y = W_2 \cdot (W_1 + W'_2)^{-1}$, where W_1 - energy density of the "melting" stage pulse, W'_2 - energy density of the second pulse stage, are represented on fig.4. The following assumption have been made for the calculations: temperature dependences of steel parameters $\lambda(T)$, $\rho(T)$ and $c(T)$ have been taken into account; the values $T_{max} = 3000\,°C$, A=0.8, b=$5 \cdot 10^{-4}$ m, $q_0 = 10^{10}$ $W \cdot m^{-2}$ have been assumed; filling factor

$$Q = \tau_c \cdot (\Delta\tau_1 + \Delta\tau_2)^{-1} = 1,1.$$

The comparison of the metods a) proposed in this article, b) the electron beam welding[8] and c) the vacuum laser welding[9] is represented in Table 1. It's carried out by the comparison of the beam power P, welding rate V and welding capacity N calculated

$$P = b_m h (W_1 + W_2')(A \cdot \tau_c)^{-1} \qquad (15)$$

$$N = A \tau_c b_m^{-1} (W_1 + W_2')^{-1} \qquad (16)$$

where h - welding depth, with the parameters experimental corresponded.

Table 1. Welding parameters calculated and experimental[8,9].

variant	a	b	c
P, kW	1.8	2.0	11.0
V, m·c^{-1}	0.03	0.03	0.05
N, m·kW^{-1}	0.00266	0.00155	0.002

Therefore, the discussed laser welding with approriately selected pulse parameters is characterised by greater energetic efficiency than traditional welding.

REFERENCES:

1. R.V.ARUTIUNIAN, L.A.BOLSHOV and V.V.VITIUKOV: 'Reseach calculed and theoretic of pulse-periodic regimes of beam-material interactions', *Preprint of IAE-4023/16M*, Kurchatov's Institut of Atomic Energy, 1984, 23.
2. S.I.ANISIMOV, J.A.IMAS, G.S.ROMANOV and J.V.CHUDAKOV: 'The powerful radiation action on the metals', Science, Moscow, 1970.
3. A.A.VEDENOV and G.G.GLADUSH: 'Physal phenomena in the materials laser treatment', Energoatomizdat, Moscow, 1985, 208.
4. M.I.TRIBELSKIY: 'On the liquid surface form in laser fusion of medium', *Quantum electronics*, 1978, 4, 804-812.
5. V.A.BATANOV, F.V.BUNKIN and A.M.PROHOROV: 'Metal target evaporation by powerful optic radiation', *Journal of experimental and theoretic Physic*, 1972, 63, 586-608.
6. P.G.KLEMENS:'Heat Balanse and Flow Conditions for Electron Beam and Laser Welding', *J. of Appl. Physics*, 1976, 47, 2165-2174.
7. V.M.GANJUCHENKO: ' On the laser irradiation parameters in deep-fusion welding', *Physic and Chemistry of the Material Treatment*, 1987, 2, 40-46.

8. E.BEYER, G.YTRZIGER, C.HOLLT ets.:'Comparation of Laser and Electron Beam Welding', Lecture of 2-nd International Conference "Beam Technology", Essen,1985, 25-28.

9. N.ABE, J.ARATA and T.ODA: 'High Power CO_2-Laser Welding ',Lecture of 2-nd International Conference "Beam Technology", Essen,1985, 17-20.

Resources Saving Technologies of Thermal Treatment with High-Frequency Induction Heating in Infrastructure of Railway Transport

F.V.BEZMENOV, V.V.VOLOGDIN, K.P.FILIPPOV, V.I.CHERVINSKY

VNIITVCH, St. L.Tolstoy 7, St.Petersburg, 197376, Russia

ABSTRACT

For countries with extensive railway network and its high loading, of interest are the developed in Russia processes ensuring the increase of serviceability of rails, wheel pairs, parts of the rolling stock.

The paper presents studies, technology and the construction of the commercial installation for thermal treatment of welds in rails.

Butt welding allows to get over from rails with the length of 12,5 m and 25 m to rails of 800 m long which are track laid without bolted joints.

The paper also presents the technology of thermal treatment and technical parameters of the installation for thermal treatment of roll surface of wheel pairs.

RÉSUMÉ

Les pays disposant d'un large réseau ferroviaire à tonnage élevé peuve avoir de l'intérêt pour des techniques conçues en Russie qui assure l'augmentation de l'aptitude fonctionnelle des rails, des paires de roues et des éléments du matériel roulant. La rapport est consacré aux études théoriques, à la technologie et à la structure de l'installation industrielle de traitement thermique des soudures sur rails. La soudage bout à bout permet de passer des rails longs de 12,5 et de 25 m aux barres longues de 800 m, lesquelles sont mises en place sans boulonnage.

On décrit egalement les techniques de traitement thermique et les paramètres techniques de l'installation pour le traitement thermique de la face de roulement des paires de roues du convoi.

Rail lengths can be increased from the normal 12.5 or 25m to 800 m and laid without bolted joints by butt welding the sections. The benefits include reduced metal requirements and lower cost track maintenance.

The welding of high carbon content (up to 0.83%) unhardened steel produces large grain structure and reduction of strength in the weld region. Stress relieving and increase of strength can be achieved by normalization of the region of the weld.

The welding of hardened rails causes reduced strength due annealing in the weld zone. The effect can be corrected by repeated hardening of the weld region.

High frequency induction heating of the weld region produces more effective and better controlled heat treatment compared to conventional flame or contact heating methods.

The use of this method achieves the required temperature distribution in the weld heating zone and high accuracy in heating, to obtain the necessary heat drop along the rail cross-section and also between the rail head, rail web and rail base.

The required temperature distribution in the rail head and base was achieved using different constructions of slot inductors. By selecting the position of turns relative to the rail contour, and also by altering the magnetic circuit of the inductor we achieved necessary distribution of specific power along the rail cross-section. In its final form the inductor producing combined longitudinal and transverse electromagnetic field was made in the form of two symmetrical halves adjoining the rail from the left and right sides and fed by one source via two independent matching transformers.

Metallographic studies of rail samples, measurement of metal hardness and conductance of natural bend tests of thermally treated welded rails under static, impact and cyclic loads established that thermal treatment of welds by using induction heating ensured their mechanical properties in contrast to other methods of thermal treatment.

MECHANICAL PROPERTIES OF WELDED RAILS WITH BULK SURFACE HARDENING

Characteristics	Without thermal treatment	With thermal treatment
1. Hardness, HB (at width of 40 mm into both directions from the weld)	250	350-360
2. Ductility at static bend: - sag, mm after removing the load	(below the requirements by 40%)	(corresponds to requirements)
3. Brittle strength, kGm.10 at the temperature:		
- 0 °C	2,3	6,6
- minus 20 °C	2,1	4,8
- minus 40 °C	1,5	3,7
- minus 60 °C	1,0	3,2
4. Fatigue strength (tension in the head). Endurance limit, kg/mm2	31	49
Load at the endurance limit, 10/3 kg	45	70

In hardening with repeated induction heating of welds the recovery of sorbite small grain structure in hardening is ensured in the welded places along the total zone of thermal treatment.

To carry out thermal treatment of a weld in rails under industrial conditions the experimental sample has been developed and then serial samples of induction installations.

The installation contains a common frame on which are assembled thyristor power supply of 2500 W, frequency 2400 Hz, a bank of capacitors, control board, cabinet for preparation of cooling water-air mixture, two matching transformers with connected to them halves of inductors with sprayers, roller table for feeding a rail with a system of rigid fixation of a rail part with a weld relatively to the inductor, pneumocylinders, feeding transformers with inductors and sprayers into a working position and a mechanism for moving screens to protect the rail base against cooling air-water mixture to get on it after the end of heating. The whole cycle of thermal treatment consisting of rail feeding, its fixation, placing of inductors, turning on and off heating, supply and removal of cooling and removal of inductors is automatic.

Normalization regime of the weld for a rail of type P65 is:
- Maximal heating temperature   -   850C
- Heating time   -   4-4,5 min
- Cooling time   -   1-1,5 min
- Power supply   -

Fig. 1.   External view of the installation inductors.

INSTALLATIONS FOR THERMAL TREATMENT OF ROLL SURFACE OF WHEEL PAIRS

While using the railway wheels their wear occurs both along the perimeter and locally; cavities, rolls, chipping wear hardening and partial hardening of the roll surface are produced. To give a wheel geometrical shape necessary for its normal operation the roll surface of a wheel is machined on special lathes after its separation from a rolling stock. Normally, the depth of a hard layer is 1,502 mm per side, therefore machining is made for a depth up to 3-5 mm. It should be noted that 1 mm thickness of a tire ensures about 30000 km of run.

Thermal treatment of a roll surface of wheel pairs permits to reduce hardness of the surface layer and made cutting to the depth of 1,502 mm preserving with this the hardness necessary for normal use of railway wheels. The service life of wheels increases, durability of a cutting tool increases several times in increased cutting regimes.

When developing the installation for thermal treatment of the roll surface of wheel pairs a heating regime has been worked out, construction parameters of the inductor were defined, batch-type mean-frequency equipment was selected.

Metallographic studies and tests of wheel pairs confirmed the correctness of selecting the heating regimes.

The installation includes:
- power supply - thyristor converter SChG 3 - 100/10 (power at mean frequency 100 kW, current frequency 10 kHz;
- mean frequency current transformers;
- capacitors;
- inductors;
- mechanism for installing and rotating q wheel pair;
- mechanism for suspending and adjusting the inductor position;

The suspension mechanism permits automatically to place the inductor on the heated surface and maintain the established gap between the inductor and the worn-out roll surface of a wheel.

Electric circuit of the installation permits to work both in manual and automatic regimes.

TECHNICAL SPECIFICATIONS OF THE INSTALLATION

1. Power of mean frequency, kW	100
2. Frequency, kHz	10
3. Capacity, pcs/h, not more than	8
4. Wheel dia along the roll circle, mm	
- maximal dia	950
- minimal dia	850
5. Pressure of cooling water, kGs/sq cm	3
6. Water consumption together with the power supply, cu.m/h	4,5

Upon agreement with the Customer it is possible to thermally treat wheels of other sizes.

Position of the equipment is shown approximately and could be changed after the agreement with the Customer. The distance from the frequency converter to the heating units should not be over 10 m.

The basic version is the installation IT14-100/10 in which one source feeds two heating units of 50 kW power each and the roll surfaces of two wheels are heated simultaneously. The variant of unipost installation is possible, in this case the wheel pair should be turned for treatment of the second wheel.

If it is necessary to increase the capacity it is possible to connect each heating unit to a separate power supply.

Approximate cost of the installation of type IT14-100/10 designed for simultaneous heating of the roll surface of two wheels, having the power of 100 kW is 46 000 USD.

INDUCTION INSTALLATION IT15-100/10
(for annealing of wheel pair tires)

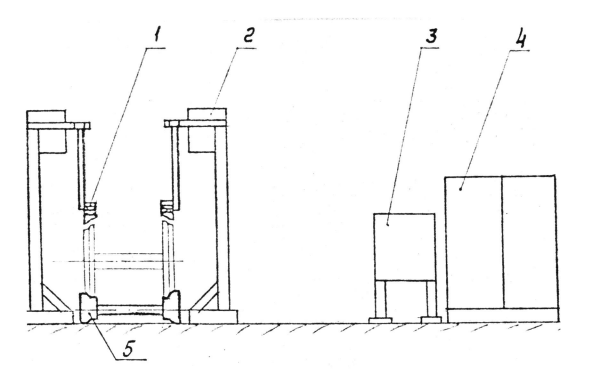

1. Inductor
2. A bank of capacitors
3. Control board
4. Oscillator SChG3 - 100/10
5. Rotator

MIV 78

INDUCTORS FOR DISMANTLING AXLE BOX INNER RINGS IN RAILTRUCKS

1. *PURPOSE*

They are designed for heating the bearing inner rings when removing them from the axle box axis of railtrucks.

Two types of equipment and technological processes with the inductor for frequency 50 Hz and 60 kHz were developed.

2. *TECHNICAL SPECIFICATIONS*

of the equipment for frequency 50 Hz

Dia of the bearing outer ring	154 mm
Voltage supply (should be indicated when ordering)	220/380 V
Current at 220 V	180 A
at 380 V	105 A
Working frequency	50 Hz
Heating temperature	180-200 °C
Heating time	30-40 sec
Inductor weight (depending on the design)	10-13 kg

3. *DESIGN*

The inductor is a multi-layer winding made of aluminium wire APSD 4-6,3 mm fixed between two glass/fabric-based laminated rings. Winding has thermal and additional electric insulation inside and is held with quides with which the inductor rests on the bearing ring.

4. *ADVANTAGES*

Small weight;

high reliability;

possibility to remove the first, second and labyrinth packing rings;

high electric safety ensured by the design made in a non-electroconducting body.

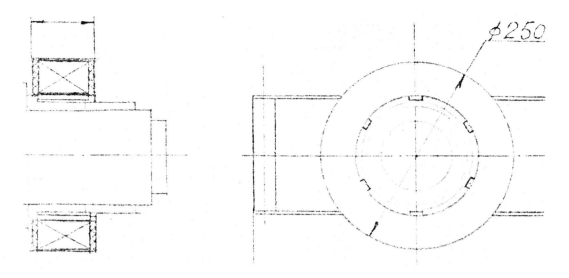

Inductor for frequency 50 Hz

5. DISADVANTAGES

- Large volume of heated metal resulting in the necessity to wash rings before their removal;
- Noise and vibration when heated.

6. PRICE

Cost of the inductor is 2200 USD.

TECHNICAL SPECIFICATIONS

of the equipment for frequency 60 kHz

Dia of bearing outer ring	154 mm
Voltage supply	380 V three-phase
Converting power up to	15 kHz
Working frequency	50-60 kHz
Heating temperature	180-200 °C
Heating time	15-20 sec
Inductor weight	4-5 kg

DESIGN

The installation includes:
- power supply
- high frequency matching transformer
- inductor
- set of connecting cables

Power supply - transistor high-frequency oscillator is installed stationary and is connected by a high-frequency cable of 15 m long with a unit of high-frequency transformer.

High-frequency transformer is placed on a hand truck which can move to the wheel pair.

Inductor is connected to the transformer by a cable of 2 m long and is put on a heated ring.

Toggle switch of heating and signal lamp indicating switch-on state of the inductor are fixed to the inductor.

ADVANTAGES

Small inductor weight

Absence of noise, vibration

High reliability of equipment

Possibility to remove the first, second and labyrinth packing rings

Cost of the installation is 12000 USD

Etude et Optimisation de l'Installation de Coupage des Métaux par la Torche à Plasma d'Air

VICTOR POGORA, ION PROTSOUC, TUDOR STANTCHOU

Université Technique de Moldova (U.T.M.),168, bd. Stefan cel Mare, Chisinau, 277012,République de Moldova

ABSTRACT / RESUME

The increase in cutting efficiency of metals and alloys with the help of plasma requires new technologies and constructive elaborations of plasma generators. These aspects were set off in the installation of metals cutting with an air plasma generator with copper electrodes elaborated by the Technical University of Moldova (T.U.M.).

The bids made by the authors had as a goal the optimisation of energetic, gasodynamic, erosive and technological parameters of the plasma generator. As a result the working activity of the electrode increased to 25 - 30 hours but the cutting velocity of the metals increased 1.4 - 1.5 times according to the plasma generators with thermochemical electrodes (Hf, Zr). The maximal thickness of the steels cut with this equipment is of 200 mm ensuring at the same time a high cutting quality.

La majoration de l'efficacité de coupage des métaux et des alliages à l'aide de plasma d'arc nécessite des nouvelles technologies et des élaborations constructives des torches à plasma plus avantageuses. Ces aspects ont été reflétés dans l'installation de coupage des métaux avec la torche à plasma d'air avec l'électrode creuse en cuivre élaborée à l'Université Technique de Moldova (U.T.M.) dont les résultats d'études sont donnés dans le présent ouvrage.

Les études effectuées par les auteurs ont eu comme but l'optimisation des paramètres énergétiques, gazodynamiques, d'usure et technologiques de la torche à plasma. Comme résultat de ces optimisations nous avons obtenu la majoration de la durée de travail de l'électrode creuse de la torche à plasma jusqu'à 25...30 heures et la vitesse de coupage des métaux a augmenté de 1.4...1.5 fois par rapport à la torche à plasma avec les électrodes thermochimique (Hf, Zr). L'épaisseur maximale des aciers coupés par cette installation est de 200 mm en assurant en même temps une qualité superbe de la coupure.

1. INTRODUCTION

L'un des domaines de l'utilisation de l'énergie électrique est le traitement thermique des métaux et des alliages à l'aide de plasma d'arc électrique et, en particulier, leur coupage. Mais l'utilisation de ce processus nécessite de résoudre une série de problèmes lies à:

- l'utilisation du gaz plasmogène qui doit assurer une efficacité accrue de coupage, une durée maximale de fonctionnement des électrodes et en même temps être bon marché, ininflammable et antidéflagrant;
- l'utilisation des matériaux accessibles et bons marchés pour la chambre de décharge de la torche;
- l'augmentation de la productivité et de l'épaisseur de coupage.

Le gaz le plus avantageux de ces points de vue est l'air. Néanmoins, l'air comme gaz plasmogène nécessite l'utilisation pour la chambre de décharge des électrodes spéciales avec une pièce d'insertion de Zr ou Hf[1-3], elles sont chères et se caractérisent par un nombre limité de commutations et sont prévues pour les courants jusqu'à 350A.

Compte tenu des désavantages de ces torches à plasma à l'U.T.M. a été élaborée une torche à plasma d'air avec l'électrode creuse en cuivre pour le coupage des métaux[4-5].

2. PRINCIPE CONSTRUCTIF ET FONCTIONNEL DE LA TORCHE

Le schéma de principe de cette torche est donné dans la fig.1. La tension de la source de courant continu 8 est appliquée à l'électrode creuse 1, la buse 6 et la pièce à couper 7. L'air comprimé est apporté dans la chambre de décharge par l'intermédiaire du générateur de tourbillon 5 qui en même temps joue le rôle d'isolateur entre les électrodes. Lors de l'application d'une impulsion de haute fréquence et de haute tension un arc électrique pilot 4 de 30...50A s'amorcera entre l'électrode creuse 1 et la buse 6. Sous l'influence de l'écoulement tourbillonnaire de l'air l'arc 4 va s'allonger jusqu'à la pièce 7. A la suite un arc électrique de travail 3 va s'initier entre l'électrode creuse 1 et la pièce 7. Par contre, l'arc pilot sera déconnecté par le contact K.

La partie radiale *ab* et la tache d'appui de l'arc électrique tournent et se déplacent à l'intérieur de la cavité de l'électrode creuse à une certaine

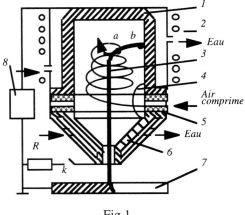

Fig.1

distance de son entrée. Cette distance est determinée par l'équilibre des forces gazodynamiques de l'écoulement tourbillonnaire et électrodynamiques qui apparaissent lors de l'intéraction entre la partie radiale et les champs magnétiques propre de l'arc électrique et du solénoïde 2 parcouru par le courant de travail.

3. CARACTERISTIQUES ENERGETIQUES DE LA TORCHE

Un des paramètres qui caractérise une torche à plasma est l'efficacité de transformation de l'énergie électrique en celle thermique qui est déterminée par les processus de l'interaction de l'arc électrique avec la paroi de la chambre de décharge. L'information détaillée concernant le transfert de la chaleur dans la chambre de décharge est d'une importance majeure car les températures moyennes de masse du courant de plasma dans la couche frontière peuvent dépasser la température de fusion des électrodes. Le paramètre qui caractérise l'efficacité thermodynamique de la torche à plasma est le rendement thermique déterminé d'après l'équation

$$\eta = \frac{G_a \bullet \Delta h}{U \bullet I} \bullet 100\% = \left[\frac{U \bullet I - (\Delta P_e + \Delta P_b)}{U \bullet I} \right] \bullet 100\%, \qquad (1)$$

où G_a est le débit massique de l'air, $g \bullet s^{-1}$; Δh - l'accroissement d'enthalpie de l'air lors du passage à travers la chambre de décharge, $J \bullet g^{-1}$; U-la tension dans l'arc, V; I-le courant d'arc électrique, A; $\Delta P_e, \Delta P_b$ - sont les pertes de chaleur respectivement dans l'électrode creuse et dans la buse, kW.

Les pertes de chaleur dans l'électrode et la buse sont calculées conformément à l'expression

$$\Delta P = C_e \bullet G_e \bullet (T_2 - T_1) \bullet 10^{-3}, \text{ kW}, \qquad (2)$$

où C_e est la chaleur spécifique de l'eau, $J \bullet (g \bullet K)^{-1}$; G_e - le débit massique de l'eau, $g \bullet s^{-1}$; T_1 et T_2 - la température de l'eau respectivement à l'entrée et à la sortie de la voie de refroidissement, K.

Les pertes de chaleur dans l'électrode creuse et dans la buse de la torche à plasma sont dûes au transfert de la chaleur par convection, radiation et la conductibilité thermique dans la tache d'appui. Conformément aux calculs analytiques[5] les pertes de la chaleur dûes au transfert par la convection dans cette torche constituent jusqu'à 85% et celles dûes au transfert par la conductibilité thermique ne dépassent pas 7% de toutes les pertes.

Malgré les valeurs insignifiantes des pertes de la chaleur par l'intermédiaire de la conductibilité thermique dans la tache d'appui notamment avec ce mécanisme est directement lié le processus d'érosion de l'électrode. Pour ce but à l'U.T.M. des études expérimentales ont été effectuées afin de définir la dépendance de la densité de courant et du flux thermique dans la tache d'appui de la polarité de l'électrode, du débit d'air et du diamètre du canal de la buse. Cet étude a été effectuée en utilisant une électrode sectionnée[6]. L'électrode sectionnée tourne et au dessus d'elle la torche à plasma avec l'arc électrique se déplacent parallèlement à l'axe de rotation de telle façon pour que la deuxième tache d'appui de l'arc électrique traverse la ligne de séparation des sections. Les paramètres enregistrés sont les températures de l'eau à l'entrée et à la sortie ainsi que le courant pour chaque section. L'étude a été effectuée pour le courant I=100A, le débit d'air (1.08...1.42) g•s^{-1}, et le diamètre du canal de la buse (2.7...4.0) mm, pour les deux polarités de l'électrode sectionnée.

La distribution radiale de la densité de courant et du flux thermique dans les taches d'appui(fig.2) a été obtenue lors de la dérivation graphique des courbes de courant et de flux thermique dans une de sections de l'électrode sectionnée. Pour la résolution numérique de l'équation intégrale d'Abel nous avons utilisé la méthode donnée dans[7].

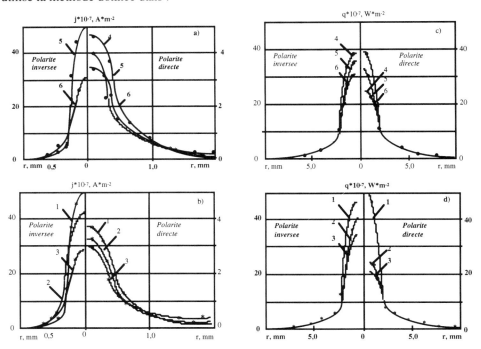

Fig.2. Distribution radiale des densités de courant (a,b) et du flux thermique (c,d) dans les taches d'appui de l'arc à plasma pour: le diamètre de la buse, mm: 1-2,7; 2-3,4; 3-4,0; le dédit d'air, g*s^{-1}: 4-1,08; 5-1,24; 6-1,42.

Pour la polarité négative de l'électrode sectionnée la densité de courant au centre de la tache cathodique est de 10 fois plus grande (fig.2) et le diamètre de la tache de courant est de presque 2 fois plus petit que pour la polarité positive. Lors de l'augmentation du diamètre du canal de la buse les densités du courant et du flux thermique au centre de la tache d'appui pour les deux polarités de l'électrode sectionnée diminuent (fig.2). La valeur et le caractère de distribution radiale de la densité du flux thermique dans la tache d'appui pour les deux polarités ne diffèrent pas beaucoup. Néanmoins, la composante du flux thermique due au transfert de la chaleur par les charges électriques n'a pas une influence importante.

Pour établir la dépendance du rendement thermique des paramètres constructifs de la chambre de décharge et du régime de travail de la torche nous avons effectué une série d'essais en utilisant la méthode mathématique de planification des essais. L'équation de régression du rendement thermique de la torche plasma avec l'électrode creuse s'écrit:

$$\eta_t = [80.7 - 0.6 \bullet G_a - 0.3 \bullet d_b - 0.6 \bullet (I \bullet w) - 0.6 \bullet l_e - 0.8 \bullet G_e^2 + 0.8 \bullet d_b^2 - 0.8 \bullet (I \bullet w)^2 -$$
$$- 0.7 \bullet l_e^2 - 0.3 \bullet G_a \bullet (I \bullet w) - 1.1 \bullet d_b \bullet (I \bullet w) + 0.4 \bullet d_b \bullet l_e + 0.3 \bullet (I \bullet w) \bullet l_e] \ \%, \qquad (3)$$

où d_b est le diamètre de la buse; l_e - la profondeur de la cavité de l'électrode creuse; w - le nombre de spires du solénoïde; I - le courant d'arc; G_a - le débit d'air; G_e - le débit d'eau. Tous les paramètres sont donnés en unités relatives.

4. CARACTERISTIQUES EROSIVES DE L'ELECTRODE CREUSE

Dans les conditions les plus difficiles du point de vue thermique se trouve l'électrode creuse et sa durée de fonctionnement est directement liée avec l'intensité d'érosion. Un des paramètres qui caractérisent l'intensité d'érosion est l'usure spécifique calculée par l'expression:

$$\gamma = (M_1 - M_2)/I \bullet t, \quad g \bullet C^{-1}, \qquad (4)$$

où M_1 et M_2 est la masse de l'électrode respectivement avant et après l'essai, g; t - la durée d'essai, s.

L'usure spécifique de l'électrode creuse est fonction d'une série de facteurs de la géométrie de la chambre de décharge et du régime de fonctionnement. Pour déterminer les valeurs optimales de ces paramètres qui assurent la valeur minimale de l'usure spécifique et, par conséquent, la durée maximale de fonctionnement de l'électrode creuse nous avons utilisé la méthode mathématique de planification des essais. L'étude a été effectuée en 5 étapes et chaque fois nous avons sélectionné les paramètres ayant le poids d'influence le plus important sur l'usure spécifique. A l'étape finale les essais ont été effectués pour l'intensité de courant I=300A, le diamètre de la cavité de l'électrode creuse d_e=13 mm et l'épaisseur de la paroi de cette électrode Δ=5 mm. Le modèle de régression de l'usure spécifique de l'électrode creuse pour cette étape est décrit par l'équation

$$\gamma = [1.05 + 0.72 \bullet l_e + 0.14 \bullet d_b + 0.38 \bullet G_a - 0.16 \bullet (I \bullet w) + 0.27 \bullet l_e^2 + 1.08 \bullet d_b^2 + 0.58 \bullet G_a^2 +$$
$$+ 0.5 \bullet (I \bullet w)^2 - 0.31 \bullet l_e \bullet d_b - 0.31 \bullet l_e \bullet G_a - 0.82 \bullet G_a \bullet (I \bullet w)] \bullet 10^{-6}, g \bullet C^{-1}, \quad (5)$$

où l_e, d_b, G_a, w sont les facteurs en unités relatives de même que dans l'équation (3).

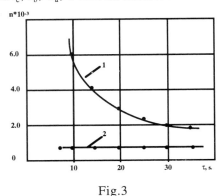
Fig.3

La commande gazodynamique et électrodynamique de la décharge en arc a permis d'augmenter considérablement la durée de service de l'électrode creuse par rapport à celle en Hf (fig.3). La fig.3 compare la variation du nombre d'enclenchement de la torche jusqu'au remplacement de l'électrode en fonction de la durée de commutation pour les électrodes: creuse (courbe 1) et en Hf (courbe 2).

5. CARACTERISTIQUES GAZODYNAMIQUES DE LA TORCHE

Un autre paramètre de la torche à plasma qui détermine l'intensité du processus d'échauffement, de fusion et d'évacuation du métal fondu dehors de la cavité de coupure est la chute totale du flux de plasma. La chute totale du flux de plasma a été déterminée par la méthode expérimentale à l'aide d'un capteur capacitif de pression avec le tube de Pitot qui traversait le flux de plasma.

La distribution radiale de la chute totale du flux de plasma aux différentes distances de la tranche de la buse l_{bm} pour les intensités de courant respectivement 200A et 300A est donnée dans la fig.4. La stabilisation dans l'espace de l'arc électrique s'effectue par la buse dont le diamètre est de 4 mm et l'écoulement tourbillonnaire de l'air ayant le débit constant et égal à 1.5 g•s^{-1}.

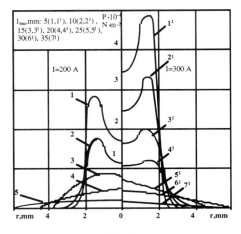
Fig.4.

D'après les courbes on peut mentionner la présence d'une zone avec une pression plus détendue dans la partie centrale du flux de plasma. Cet effet est dû aux températures plus élevées dans cette zone et à l'écoulement tourbillonnaire de l'air dans la chambre de décharge. Lors de l'éloignement de la buse la distribution radiale de la chute totale du flux de plasma se modifie et à la distance de (20...25) mm la distribution radiale devient parabolique (fig.4, courbes 4 et 5). Cette modification est le résultat de la dilatation thermique de l'arc électrique à la sortie de la buse et du processus de mélange turbulent de plasma avec l'air ambiant.

6. CARACTERISTIQUES TECHNOLOGIQUES DE LA TORCHE

Se basant sur les résultats d'étude, nous avons élaboré une gamme de torches à plasma munies d'une électrode creuse en cuivre qui sont utilisées dans des entreprises des différentes branches de l'industrie des pays de la C.E.I. Les torches sont destinées au coupage plasma automatique des aciers au carbone et inoxydables, des métaux non-ferreux et leurs alliages. L'épaisseur maximale de l'acier à couper est de 200 mm. Les torches font le découpage des tôles, le coupage des tubes et d'un produit rond, le coupage des profiles compliqués, le biseautage sous angle jusqu'à 45°. Elles se caractérisent par une grande puissance électrique et par conséquent, par une vitesse de coupage élevée (fig.5). L'orientation intensive du flux de plasma s'écoulant à travers la buse assure une qualité superbe de la coupure, une fente étroite formée lors du coupage, les surfaces latérales de coupure parallèles, nettes et lisses.

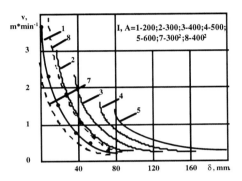

Fig.5. Vitesse de coupage de l'acier inoxydable

L'optimisation du processus de coupage des métaux a été effectuée par la méthode mathématique de planification des essais. La vitesse de coupage de l'acier inoxydable par la torche à plasma avec les paramètres optimisés est décrite par l'équation de régression

$$V=(16.67+0.52 \cdot l_e+1.35 \cdot d_b+0.97 \cdot l_b+0.43 \cdot I+0.48 \cdot G_a-1.72 \cdot l_{b-m}-5.67 \cdot \delta) \cdot 10^{-3}, m \cdot s^{-1}, \quad (6)$$

où l_{b-m} est la distance de la buse jusqu'à la surface du métal; δ - l'épaisseur du métal à couper.

Pour l'alimentation de la torche il est plus rationnel d'utiliser le redresseur avec la tension de marche à vide de 500V et la caractéristique externe brusquement décroissante.

7. SCHEMA - BLOC DE L'INSTALLATION DE COUPAGE PLASMA

Le schema-bloc de l'installation de coupage des métaux est donné par la fig.6. La tension de la source 1, l'air comprimé et l'eau de refroidissement sont apportés à la torche 4 depuis le bloc de commande 2. L'allumage de l'arc électrique se fait à l'aide d'un oscillateur 3 à haute fréquence. La torche est déplacée par la machine de coupage automatique 5 d'après un programme conformément à la forme de la pièce à couper 6. Pour assurer la protection du personnel ainsi que de l'environnement un dispositif de protection 7 est prévu.

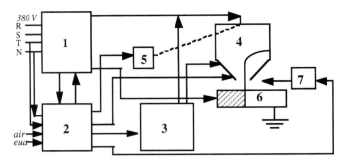

Fig.6. Schema-bloc de l'installation de coupage des métaux.

CONCLUSION

La torche à plasma d'air avec l'électrode creuse en cuivre pour le coupage des métaux est plus avantageuse que l'électrode en Zr et Hf car la durée de fonctionnement de l'électrode creuse, la vitesse de coupage et l'épaisseur des métaux à couper sont plus élevées. Les études expérimentales ont permis d'optimiser la géométrie de la chambre de décharge et les paramètres du régime de travail pour lesquels la durée de fonctionnement de l'électrode creuse, le rendement et la vitesse de coupage sont maximales.

BIBLIOGRAPHIE

1. E. M. ESSIBIAN: 'L'appareillage pour le coupage plasma', éd. Tehnica, Kiev, 1971.

2. D. G. BIHOVSKII: 'Le coupage plasma', éd. Machinostroenie, Leningrad, 1972.

3. K. V. VASSILIEV: 'Le coupage à plasma d'arc', éd. Machinostroenie, Moscou, 1974.

4. I. I. KISSELIOV, I. A. PROTSOUC, T. P. STANTCHOU, I. Z. TERZI: 'L'étude et l'optimisation de coupage plasma des métaux', éd. Stiinta, Kichinev, 1981.

5. T. P. STANTCHOU: 'L'élaboration et l'étude de la torche à plasma d'air avec l'électrode creuse pour le coupage des métaux', éd. Tehnica, Moscou, 1979.

6. S. P. POLIAKOV, P. F. BOULANII: 'Le capteur IQ pour la mesure de la densité du flux thermique et du courant dans la tache de l'arc électrique', Physique et Chimie de traitement des matériaux, 1980, 3, 137-139.

7. K. BOCKASTEN: 'Transformation of Observed Radiances into Radial Distribution of the Emission of a Plasma', JOSA, 1961, 51 (9), 943.

Index 1

Arc furnaces, electrical efficiency in	EE	Heat treatment of strip	MIII
Arc furnace electrodes	RE	Induction heating processes	RE, MIII
Arc furnace supply	MI	Induction heating of steel tubes	MII
Co-generation	EE	Induction heating for strip	MIII
Cold crucible melting	MII	Induction heating of wire	MII
Combined heat and power	EE	Induction motors	EE
Complimentary profile induction heating	MIII	Ladle drying	MII
Computer aided learning	RE	Laser technologies	MIV
DC arc furnaces	MI	Load forecasting	EE
Deformation fields in induction heating	MIII	Load management	EE
Demand side management (DSM)	EE	Mathematical modelling	RE
Direct resistance heating	MIII	In-line induction heating	MIII
Drying process modelling	RE	Levitation melting	MII
Education	RE	Metal processing	MI
Electric arc modelling	RE	Microwave heating	RE
Electric arc steel making	MI	Multilayer billet heating for rolling	MIII
Electric motors	EE	Neural networks	EE, RE
Electroforming	MII	Power quality	MI
Electromagnetic compatibility	MI	Plasma cutting	MIV
Electromagnetic pump	MII	Plasma,- transferred arc	RE
Electron beam curing	MIV	Pulse hardening	MIV
Electrodeposition	MIV	Quality, electrical supply	MI
Electrotechnology handbook	RE	Resistance heating	RE, MIII
Energy auditing	EE	Semi-conductor production	RE
Energy control	MI	Software	RE
Energy efficiency	EE	Strip heating	MIII
Energy management	EE	Steel disc heating	MIII
Energy optimisation and savings	MI	Surface hardening by induction	MIV
Energy savings in arc furnaces	MI	Surface hardening by lasers	MIV
Energy savings in industry	EE	Surface macro-reinforcing	MIV
Expert systems	RE	Surface treatment by plasma	MIV
Flicker	MI	Teaching Fellowship schemes	RE
Flicker level	MI	Thixotropic processing	MII
Flicker meter	MI	Total energy management	EE
Foundry practice	MII	Transverse flux induction strip heating	MIII
Fuel substitution	EE	Tundish heating	MI
Heat pumps	EE	Vacuum coating	MIV
Heat treament	MII		

Index

Index 3

UIE XIII Congress on Electricity Applications 1996

Alimentation et controlée des fours à arc	MI
Chauffage d'acier liquide par plasma d'arc	MI
Chauffage de bandes	MIII
Chauffage de disques d'acier	MIII
Chauffage de répartiteurs	MII
Chauffage par induction	MIII
Chauffage par résistance	MIII
Co-génération	EE
Découpage des métaux par plasma	MIV
Disciplines de recherche en Université	RE
Economies d'energie	EE
Electroformage	MII
Enseignmement assisté par ordinateur	RE
Etude thermique du chauffage d'acier liquid	MI
Flickermètre	MI
Fonderie	MII
Fours à arc	EE, MI
Fusion par creuset froid	MII
Gestion de la charge	EE
Gestion globale de l'énergie et MDE	EE
Induction et tréfilage	MII
Manuel d'electrotechnologie	RE
Metallurgie et traitement thermique	MII
Micro-ondes	RE
Modélisation des arcs	MI
Modélisation mathématique	RE
Moteurs électrique	EE
Moules de rotomoulage	MII
Optimisation énergétique	MI
Pompes à chaleur	EE
Plasma-froids microondes	MIV
Plasma d'arc tranféré	MI
Production combinée chaleur/force	EE
Production d'acier par fourà arc électrique	MI
Programmes de postes universitaires subventionné	RE
Promotion de la performance énergetique	EE
Réchauffage avant laminage	MIII
Research topics in Universities	RE
Réseau virtuels neuronaux	RE
Réseau neuronaux et prévision de la charge	EE
Revêtement sous vide	MIV
Séchage de poches	MII
Torche à plasma d'air	MIV
Traitement par bombardement électronique	MIV
Traitement superficiel	MIV
Traitement thixotrope	MII
Trempe en mode pulsé	MIV
Trempe superficielle par induction, par laser	MIV